Current Topics
in Survey Sampling

ORGANIZING COMMITTEE

DANIEL KREWSKI

Chairman
Ottawa Chapter of the American Statistical
Association
Ottawa, Ontario

DAVID L. BAYLESS

Survey Research Methods Section
of the American Statistical Association
Washington, D. C.

I. P. FELLEGI

Statistics Canada
Ottawa, Ontario

J. N. K. RAO

Department of Mathematics and Statistics
Carleton University
Ottawa, Ontario

Proceedings of the International Symposium
on Survey Sampling Held in Ottawa, Canada,
May 7-9, 1980

Current Topics in Survey Sampling

Edited by

D. KREWSKI
Health Protection Branch
Health and Welfare Canada
Ottawa, Ontario, Canada

R. PLATEK
Census and Household Surveys
Statistics Canada
Ottawa, Ontario, Canada

J. N. K. RAO
Department of Mathematics and Statistics
Carleton University
Ottawa, Ontario, Canada

ACADEMIC PRESS 1981

A Subsidiary of Harcourt Brace Jovanovich, Publishers
New York London
Paris San Diego San Francisco São Paulo Sydney Tokyo Toronto

ACADEMIC PRESS, INC.
111 Fifth Avenue, New York, New York 10003

United Kingdom Edition published by
ACADEMIC PRESS, INC. (LONDON) LTD.
24/28 Oval Road, London NW1 7DX

Library of Congress Cataloging in Publication Data
Main entry under title:

Current topics in survey sampling.

   Papers and abstracts of papers presented at the
International Symposium On Survey Sampling, Carleton
University, Ottawa, May 7-9, 1980, sponsored by the
Survey Research Methods Section and the Ottawa Chapter
of the American Statistical Association.
   Includes index.
   1. Social surveys--Methodology--Congresses.
2. Sampling (Statistics)--Congresses. I. Krewski, D.
II. Platek, Richard. III. Rao, J. N. K., Date.
IV. International Symposium on Survey Sampling (1980:
Carleton University) V. American Statistical Association.
Survey Research Methods Section. VI. American Statistical
Association. Ottawa Chapter.
HN29.C79   001.4'33   81-17544
ISBN 0-12-426280-5   AACR2

PRINTED IN THE UNITED STATES OF AMERICA

81 82 83 84   9 8 7 6 5 4 3 2 1

Dedicated to the Memory
of
WILLIAM GEMMEL COCHRAN
1909-1980

CONTENTS

IV SUPERPOPULATION MODELS

V VARIANCE ESTIMATION

VI   IMPUTATION TECHNIQUES

CONTRIBUTORS

Numbers in parentheses indicate the pages on which the authors' contributions begin.

Barbara A. Bailar (169), *Statistical Standards and Methodology, U.S. Bureau of the Census, Room 2031-3, Washington, D.C. 20233*

David L. Bayless (87), *Research Triangle Institute, Box 12194, Research Triangle Park, North Carolina 27709*

Arijit Chaudhuri (317), *Computer Science Unit, Indian Statistical Institute, 203 B. T. Road, Calcutta, India 70035*

James R. Chromy (329), *Research Triangle Institute, Box 12194, Research Triangle Park, North Carolina 27709*

Tore E. Dalenius (17), *Department of Applied Mathematics, Brown University, Providence, Rhode Island*

I. P. Fellegi (47), *Statistics Canada, B-8, 5th Floor, Jean Talon Bldg., Tunney's Pasture, Ottawa, Ontario, K1A OT6, Canada*

Barry L. Ford (413), *Statistical Reporting Service, U. S. Department of Agriculture, Washington, D. C. 20250*

Wayne A. Fuller (199), *Department of Statistics, Iowa State University, 221 Snedcor Hall, Ames, Iowa*

H. O. Hartley (9, 31), *Department of Mathematics, Duke University, Durham, North Carolina 27706*

Irene Hess (137), *Institute for Social Research, University of Michigan, Ann Arbor, Michigan 48106*

Christopher J. Hill (437), *Statistics Canada, C-7, 4th Floor, Jean Talon Bldg., Tunney's Pasture, Ottawa, Ontario, K1A OT6, Canada*

Tak-Kee Hui (227), *Department of Mathematics and Statistics, University of Toronto, Toronto, Ontario, Canada*

Cary T. Isaki (199), *Department of Statistics, Iowa State University, 221 Snedcor Hall, Ames, Iowa*

Graham Kalton (455), *Institute for Social Research, University of Michigan, Ann Arbor, Michigan 48106*

Daniel Kasprzyk (455), *Division of Survey Development, U. S. Department of Health, Education, and Welfare, 330 Independence Ave., S. W., Washington, D. C. 20201*

Douglas G. Kleweno (413), *U. S. Department of Agriculture, 0133 South Bldg. 14th & Independence S. W., Washington, D. C. 20250*

Nash J. Monsour (367), *6301 Shopton Crescent, Camp Springs, Maryland 20031*

R. Platek (105), *Statistics Canada, C-8, 6th Floor, Jean Talon Bldg., Tunney's Pasture, Ottawa, Ontario, K1A OT6, Canada*

Henry A. Puderer (437), *Statistics Canada, A-7, 4th Floor, Jean Talon Bldg., Tunney's Pasture, Ottawa, Ontario, K1A OT6, Canada*

J. N. K. Rao (247), *Department of Mathematics, Carleton University, Ottawa, Ontario K1S 5B6, Canada*

Poduri S. R. S. Rao (3, 305), *Department of Statistics MSB 508, University of Rochester, Rochester, New York 14627*

Robert Santos (455), *1531 McIntyre, Ann Arbor, Michigan 48105*

Carl Erik Särndal (227), *Department of Mathematics and Statistics, Université de Montréal, Montréal, Québec H3C 3J7, Canada*

A. J. Scott (247), *Department of Mathematics and Statistics, University of Auckland, Private Bag, Auckland, New Zealand*

A. R. Sen (349), *Environment Canada, Place Vincent Massey, Hull, Québec K1A 0E7, Canada*

Gary M. Shapiro (169), *Statistical Methods Division, U. S. Bureau of the Census, Suitland, Maryland 20233*

M. P. Singh (105), *Statistics Canada, C-7, 6th Floor, Jean Talon Bldg., Tunney's Pasture, Ottawa, Ontario, K1A OT6, Canada*

T. M. F. Smith (267), *Department of Mathematics, University of Southampton, Highfield Southampton, United Kingdom S09 5NH*

Robert D. Tortora (413), *Statistical Research Division, ESCS, U. S. Department of Agriculture, Washington, D. C. 20250*

Kirk M. Wolter (367), *U. S. Bureau of the Census, Room 3554, FOB #3, Washington, D. C. 20233*

PREFACE

In July 1978, the Survey Research Method Section of the American Statistical Association (ASA) sent a questionnaire to all its chapters asking them to indicate whether they would be interested in cosponsoring a regional meeting on the subject of survey sampling. The initial response by the Ottawa Chapter's President for 1978–1979, Phil Cohen, conveyed the enthusiastic support from the Ottawa Chapter for such a meeting to be held here. Around this time, the Statistics/Probability Group at Carleton University decided to hold an International Symposium on Statistics and Related Topics, which originally included a session on survey sampling. This decision provided further impetus to our holding an International Symposium on Survey Sampling in Ottawa. One of us (D. Krewski, President of the Ottawa Chapter for 1979–1980) subsequently presented Al Finkner of the Survey Research Methods Section of the ASA, with specific proposals concerning timing, program content, and publication of the proceedings. The decision to hold the symposium was made shortly thereafter. The symposium was subsequently held at Carleton University in Ottawa on May 7–9, 1980 in conjunction with the Symposium on Statistics and Related Topics. Several prominent survey statisticians from Canada and the United States and from as far as the United Kingdom, India, and New Zealand presented both invited and contributed papers. More than three hundred statisticians and scientists participated in the two symposia.

The present volume contains all the invited papers as well as abstracts of the contributed papers presented at the symposium on survey sampling. (A separate volume containing the invited papers given at the Symposium on Statistics and Related Topics will be issued by North Holland.) All papers were reviewed by one of the editors as well as by an independent referee and revised accordingly. The discussion following the invited presentations was recorded and edited for inclusion in this volume.

The topics convered here include nonsampling errors, current survey research activity, superpopulation models, variance estimation, and imputation techniques. While not all of the current problems in survey sampling could be discussed at a three-day symposium such as this, we hope that the topics included in this volume will be of interest to survey statisticians as well as to scientists in other disciplines involved in the application of survey sampling techniques.

A month before the symposium, the statistical profession was saddened to learn of the death of Professor William G. Cochran. Taking up a suggestion initially made by Dr. W. G. Madow, this symposium volume is dedicated to his memory. As part of this dedication, we invited two of Professor Cochran's close colleagues, Professors P. S. R. S. Rao and H. O. Hartley, to speak in his honor. Their papers appear in Part I of this volume. While preparing this volume for publication, we were deeply saddened to learn that Professor Hartley also passed away. The statistical profession has lost two of its truly distinguished statisticians. Their contributions to statistics, and to sampling in particular, are monumental. We are grateful to Professor Hartley for having presented two important papers, one dealing with response errors appearing in Part II of this volume and the other on hazard rate models in carcinogenic testing appearing in the volume on Statistics and Related Topics, which is dedicated to his memory. His very active participation was an important contributing factor in the success of the two symposia.

Part II of this volume contains three papers on nonsampling errors. T. Dalenius makes a case for the creation of an unified discipline of survey research with a view to serve as the basis for total survey design. H. O. Hartley presents general results on estimating variance components due to nonsampling and sampling errors, and proposes a "swapping algorithm" for interviewer assignment to minimize the effect of nonsampling errors. I. P. Fellegi's paper tackles the difficult question as to whether census counts should be adjusted for underenumeration when determining federal transfer payments to the provinces.

Research activities currently underway at four major survey orgranizations in North America are described in Part III. D. L. Bayless discusses the organization of the survey research centers at the Research Triangle Institute and reviews the costs and results of a number of survey research projects conducted there over the past two decades. R. Platek and M. P. Singh of Statistics Canada provide a comprehensive discussion of the various sources of error in surveys and discuss the cost effectiveness of different error control programs. Survey designs used in sociological research conducted by the Survey Research Center of the University of Michigan are reviewed by I. Hess. An overview of the research efforts underway at the U. S. Bureau of the Census, including small area estimation and the redesign of recurring household surveys, are outlined in the paper by B. A. Bailar and G. M. Shapiro.

The use of superpopulation models in survey design and inference is the theme of Part IV. This is an area currently of great interest, with the four papers here tackling a variety of problems. W. A. Fuller and C. T. Isaki develop optimal survey designs under appropriate superpopulation models and focus on estimators of the population mean that are design consistent, but not necessarily design unbiased. C. E. Särndal and T. Hui present Monte Carol results on a method of estimation of parameters of models for the response (or nonresponse) mechanism. A. J. Scott and J. N. K. Rao study the effects of clustering and stratification on standard chi-square tests for homogeneity and independence in two-way tables, employing actual data as well as some simple models for clustering. Finally, the paper by T. M. F. Smith reviews possible approaches to regression analysis with complex survey data and then compares the properties of these methods in a simulation study.

In Part V, a number of different problems in variance estimation are tackled. P. S. R. S. Rao discusses the conventional and weighted least squares estimators of the mean square error of the ratio estimator, as well as a modified version of the jackknife variance estimator. A. Chaudhuri reviews the work on non-negative unbiased variance estimators and proposes some new ones. Probability minimum replacement (PMR) sampling schemes are discussed by J. R. Chromy, along with a sequential procedure for their implementation. Alternative variance estimators for use with PMR sampling schemes are also outlined. A. R. Sen reviews the design of surveys of migratory game birds in North America and describes a method for obtaining improved estimates of error. Finally, K. M. Wolter and N. Monsour address the problem of variance estimation for seasonally adjusted time series. Results are given under two concepts of variability, the first treating the observed time series as fixed and the second assuming the series to be a realization of an underlying stochastic process.

Imputation techniques are discussed in Part VI. B. L. Ford, D. G. Kleweno, and R. D. Tortora compare six procedures for imputing for missing items using simulation studies based on actual data from an agricultural survey. C. J. Hill and H. A. Puderer provide a general discussion of the data adjustment procedures for missing and inconsistent data, which will be applied in the 1981 Canadian Census of Population and Housing. In the last paper, G. Kalton, D. Kasprzyk, and R. Santos describe the nonresponse problems encountered in the U. S. Survey of Income and Program Participation (SIPP) and discuss the development of nonresponse adjustment and imputation strategies for SIPP.

In preparing this volume, we received a great deal of assistance from many people. David Bayless and Ivan Fellegi served as members of the program committee. Michael Hidiroglou organized the session on contributed papers and helped out in many other ways as well. The local arrangements committee consisted of Karol Krotki (chairman), Mike Bankier, Jack Graham, and Andy McCleod. We are particularly indebted to Karol for his excellent work in this regard. We are also grateful to E. Saleh, chairman of the organizing committee for the International Symposium on Statistics and Related Topics, for his cooperation in working on matters of mutual interest. Graham Kalton provided considerable assistance in organizing the session on imputation techniques and in transcribing the discussion following that session. W. G. Madow presented an informal but stimulating address on "Statistical Models and Statistics" on the evening of May 7. We are grateful to D. F. Bray, D. Dodds, M. Hidiroglou, C. Patrick, and W. G. Madow for serving as session chairmen. Finally, we thank Judy Brennan and John Kovar for their help in organizing the material presented here, and Gill Murray for her excellent job in preparing the typescript.

D. Krewski
R. Platek
J. N. K. Rao

ORGANIZING COMMITTEE

DANIEL KREWSKI
Chairman
Ottawa Chapter of the American Statistical
Association
Ottawa, Ontario

DAVID L. BAYLESS
Survey Research Methods Section
of the American Statistical Association
Washington, D. C.

I. P. FELLEGI
Statistics Canada
Ottawa, Ontario

J. N. K. RAO
Department of Mathematics and Statistics
Carleton University
Ottawa, Ontario

I WILLIAM G. COCHRAN:
IN MEMORIAM

PROFESSOR WILLIAM GEMMEL COCHRAN: PIONEER IN STATISTICS, OUTSTANDING SCIENTIST AND A NOBLE HUMAN BEING

Poduri S.R.S. Rao

University of Rochester

Professor Cochran undoubtedly made substantial contributions to several basic areas of statistics. His first article, "The Distribution of Quadratic Forms in a Normal System", was published in 1934. In forty years of professional life, more than one hundred of his research articles of the highest quality were published in prestigious professional journals. The world will continue to benefit extensively from his research papers as well as his five books. His contributions to statistical theory, sampling techniques, design of experiments and statistical analysis will have everlasting effects on the proper use of statistics for the welfare of mankind.

Every equation and sentence that this distinguished scientist wrote was intended to be useful as a theoretical advancement or application to a practical problem. As one of his former students put it, "He was a genius" when it came to suggesting a simple and well-thought-of statistical procedure. As the readers of his work do not fail to notice, a useful solution should also appeal to common sense.

Born on July 15, 1909 in Rutherglen, Scotland, he passed away on March 29, 1980 at the Cape Cod Hospital in Hyannis. Educated at Glasgow and Cambridge, he worked with Fisher and Yates at Rothamsted. Thanks to him, outstanding programs in statistics

were developed at Iowa State, North Carolina, John Hopkins and
Harvard Universities.

"I am extremely grateful to Professor William G. Cochran
for the training, guidance and support I received from him throughout my academic career at Harvard University. As my principal
advisor, his constant encouragement, suggestions and criticisms
have been invaluable in preparing this thesis." This is only one
illustration of the expressions of deep gratitude towards him from
his students. Mrs. Cleo Youtz of Harvard recently wrote to me:
"As you well know, many students were close to Professor Cochran,
in addition to those who wrote their theses with him." Several
distinguished statisticians of today are either his former
students or received guidance from him.

He generously gave his time to his students for teaching,
suggesting research topics, discussing the results and recommending
improvements. He patiently made the necessary corrections to
every word and every equation that they wrote. When he politely
pointed out a mistake, if the student said that he was ashamed
that he did not notice the mistake himself, Professor Cochran
would quickly comfort him. If the student was not getting any
results in spite of his hard work, Professor Cochran would say,
"Give it a rest." Even if the student did only a reasonable job,
but had tried his best, he was quick to encourage him with "You
did a good job." If he realized that the student had tried hard,
he would say or write, "After all your hard work, you must feel a
sense of relief."

Professor Cochran did not look for 'diamonds in a coal mine.'
He compassionately gave his genteel hand to everyone who approached
him - especially the stumbling ones. He helped them to stand on
their feet and provided them with intellectual nourishment and
moral support. For all of this guidance and help that he offered,
he did not expect anything in return. About a year ago, I told
him, "You have done so much for so many people ...", but he would

not accept the compliment by saying, "Oh, these words you say...".

The students at Harvard Statistics Department were aware that Professor Mosteller consulted Bill Cochran whenever an important decision was to be made. Both of them, with Professors Dempster and Pratt, spent hours in planning the individual academic programs for the students and in looking after their welfare. National origins were no barriers for this personal attention. While learning with all the needed discipline continued in the Department, the students eagerly awaited the Christmas and New Year's parties as well as picnics arranged by the wives of the Faculty. The kind enquiries and the packets of cookies, pies and cakes to take home from Betty Cochran and the rest of them remain as the most affectionate of gestures, especially to a student ten thousand miles from home.

Professor Cochran loved life and loved everyone close to him. His friends and colleagues, and even his acquaintances, felt very close to this truly nice man. I have witnessed him to be very polite and considerate to women and young people and affectionate to children. He always encouraged them to enjoy life. Even after he had retired, he kept in touch with his colleagues, students and friends. In spite of his failing health during the last few years, he continued to be prompt in his correspondence with his colleagues, students and friends and offered his help in every possible manner. He usually started his letters with "Nice to hear from you" or "Nice to hear of your work." He then made enquiries concerning one's family before coming to the major part of the correspondence, and completed the letter with "Best wishes." This genuine interest from him was very much appreciated by his students and friends.

It may not be an exaggeration to say that perhaps it was his urge to help mankind that motivated most of his work - the design and analyses of experiments for higher agricultural yields and improved medical treatments, clinical trials for determining

correct doses and sampling procedures which helped the civilian
population get their ration cards or helped remote villagers with
health facilities.

Several honors in the form of degrees, memberships and
fellowships were bestowed on this world-renowned scientist. It
was only natural that he was entrusted with the statistical
analysis for the Surgeon General's Committee on Smoking and Health.
He was made the Chairman of the Advisory Committee of the U.S.
Bureau of the Census even after his retirement. Several statisticians from all over the world attended his retirement dinner, and
afterwards he wrote to them, "It was very pleasant to sit surrounded by friends and bask in the reflected glory of students."
Any amount of tribute paid to this true scientist would merely be
an inadequate expression of gratitude towards him.

It is sometimes said that a person looks bigger when he is
no more. Professor Cochran looked big when he was alive and
continues to look even bigger now. We were fortunate enough to
have known this noble human being. His presence in the classroom
or at a meeting or a gathering made all the difference. We miss
him around us, miss his advice, miss his affectionate looks and
kind words, and we miss his gentle smile. We also miss the
magnanimous aura that surrounded this marvelous personality.
Professor Cochran will continue to be with us in our minds and in
our hearts.

Awards and honours received by Professor Cochran include
the following:
Honorary A.M., Harvard University, 1957
Honorary L.L.D., Glasgow University, 1970
Honorary L.L.D., The Johns Hopkins University, 1975
Honorary Fellow, Royal Statistical Society, 1959
Guggenheim Fellowship, 1964-65
S.S. Wilks Memorial Medal, American Statistical Association, 1967
Member, American Academy of Arts and Sciences, 1971

Member, National Academy of Sciences, 1974

'Outstanding Statistician' Award, Chicago Chapter, American
    Statistical Association, 1974

Honorary Life Member, Biometric Society, 1976

Fellow, American Statistical Association

Fellow, Institute of Mathematical Statistics

Fellow, American Association for the Advancement of Sciences

President, American Statistical Association, 1953

President, Institute of Mathematical Statistics, 1946

President, Biometric Society, 1954 and 1955

President, International Statistical Institute, 1967-71

Vice-President, American Association for the Advancement of
    Science, 1966

Editor, Journal of the American Statistical Association, 1945-50

Emeritus Member, American Epidemiological Society

Emeritus Professor, Harvard University

IN MEMORY OF WILLIAM G. COCHRAN

H. O. Hartley

Duke University

I consider it an honor and privilege to speak in memory of
Bill Cochran, an outstanding statistician of world renown. So
distinguished and incisive, so diverse and far reaching are his
contributions to statistics that it is clearly impossible to even
recite them all on this short occasion. They will undoubtedly be
documented in official obituaries to be published prominently in
our journals.

Today, I would like to take the liberty to reminisce on a
more personal note about Bill Cochran as a colleague and friend.
In doing this I hope you will forgive me if I bring in my own
memories with some nostalgia.

I had the considerable fortune to cross the path of Bill's
industrious career on many occasions and quite early in his and
my lives. We both were research students at the School of Agriculture at Cambridge England working under the late John Wishart,
although Bill preceded me by some two to three years. He had left
his mark and the agronomists at the school (notably Hunter, the
barley expert and Sanders, the wheat expert) were full of praise
for Bill's cooperative spirit. Of course Bill had by then started
his career as a statistician by joining the famous team of R.A.
Fisher and Frank Yates at Rothamstead Experimental Station. He
was the junior member of this team and so characteristic of his
modesty is the following story against him which he told me many
years later.

9

Shortly after his arrival at Rothamstead, he had been given
the task of analysing a crop fertilizer experiment. The results
appeared to be very quaint. No main effects were significant,
only one interaction and that barely so. He wrote a very detailed
and careful report and submitted it to Frank Yates. It was
returned to him within a day. His report was struck out by red
pencil and there was just a two word comment by Yates: "crop
failed". Bill added somewhat wistfully, "had the results been
recorded in 'cwt per acre' I could have perhaps recognized this,
but a low 'pounds per plot' is not so obvious to someone new at
Rothamstead".

Bill's good humor and cooperative attitude soon made him
the most popular statistical consultant of the team. I had first-
hand experience of this when talking to agricultural scientists
at the Harper Adams Agricultural College where I had a temporary
job in 1936-37. Bill provided much council in many of the cooperative experiments in which Harper Adams was involved.

Of the many contributions that Bill made during this period,
mention should be made of the important paper "Problems Arising
in the Analysis of a Series of Similar Experiments" presented in
1937 to the Royal Statistical Society and published in what was
then called the Supplement to the Journal. This is a fitting
summary of the considerable experience he had gained at Rothamstead.

In 1938, he left England for an appointment at Iowa State,
first as a visiting professor and then as a permanent full
professor. By a fluke of chance I made exactly the same move in
1953. Although this was 15 years later (with World War II in
between) I found that all the older faculty, who remembered Bill
so well, were still full of praise for Bill's achievements,
particularly Snedecor, Homeyer and Jebe. Apparently, in his first
quarter, he gave two courses attended by all the Ames faculty,
one course on survey sampling and the other on design of experiments. Both courses were based on Bill's own lecture notes which
were later converted to the outstanding Wiley texts "Sampling

Techniques" and "Experimental Designs". The latter, of course,
was jointly authored with the late Gertrude Cox. "Sampling
Techniques" reveals the signatures of its Iowa origin. The Jeff-
erson County data on corn and farm acreages illustrate the use of
ratio estimators in stratified sampling. The farm acreage
distributions used in the exercises clearly peak at 80 to 160
acres, the two most popular Iowa farm sizes during that period.

    Bill's "sampling Techniques" is my favorite text. Some
say it is too mathematical. I do not think so, certainly not if
the book is used as it should be. There are clear and concise
statements of concepts and the mathematical theorems, if they are
just given as results, provide concise statements on the properties
of the concepts. The proofs are there in print for completeness
for those who need them. The outstanding feature of the text is
its concise narrative, sometimes enriched by good humored comments.
Let us read from the introduction to Chapter 4 on sample size
requirements.

    "A hypothetical example brings out the steps
involved in reaching a solution. An anthropologist
is preparing to study the inhabitants of some island.
Among other things, he wishes to estimate the percentage of inhabitants belonging to blood group O.
Cooperation has been secured so that it is feasible
to take a simple random sample. How large should
the sample be?

    This question cannot be discussed without first
receiving an answer to another question. How accurately
does the anthropologist wish to know the percentage of
people with blood group O? In reply he states that he
will be content if the percentage is correct within
$\pm$ 5% in the sense that, if the sample shows 43% to
have blood group O, the percentage for the whole
island is sure to lie between 38 and 48.

    To avoid misunderstanding, it may be advisable
to point out to the anthropologist that we cannot
absolutely guarantee accuracy within 5% except by
measuring everyone. However large n is taken, there
is a chance of a very unlucky sample that is in error
by more than the desired 5%. The anthropologist replies
coldly that he is aware of this, that he is willing to

take a 1 in 20 chance of getting an unlucky sample,
and that all he asks for is the value of n instead
of a lecture on statistics."

This admittedly hypothetical example puts the problem of
required sample size in a nutshell. Do you recognize the masterly
strategy of the experienced consulting statistician?

"Sampling Techniques" uses another major source of records:
Census data. This reflects Bill's association with Morris Hansen's
Panel of Statistical Consultants of which he was the chairman.
Early members were the late Fred Stephan, Bill Madow and myself.
Later, Ivan Fellegi joined the group. The late William Hurwitz
was the most prominent member of the Bureau Staff who regularly
joined the Panel meetings.

Bill had a remarkable talent to make people feel relaxed
in an informal and fruitful discussion. He had a way of giving
the answer to a problem by asking questions. Sometimes, after a
detailed presentation of a problem by a member of the Bureau
staff, a voice with a slow Scottish accent would say "That was a
very interesting presentation, but tell me why did you do this?...
Why did you do that or the other?" Often the answers to these
questions would point out a way to tackle the original problem.
I believe (it was before my time) the well known composite
estimators of the Bureau's CPS evolved in this way. I do remember
distinctly however one lunch time at the Bureau when I was holding
forth about unequal probability sampling without replacement and
Bill leaned back in his chair and asked "HOH, why don't you try
this?" 'This' was the key idea behind what is now known as the
Rao-Hartley-Cochran estimator. Well, did Jon Rao and I work hard
on Bill to make him agree to be a co-author after we had perfected
the mathematics of Bill's idea!

But let me return to the historical trail. Bill spent
about seven years at Iowa State and for a year or so he joined a
special group of distinguished statisticians at Princeton in a
World War II activity. In 1946, he was invited by Gertrude Cox

to join her Institute at N.C. State where he stayed for three years laying the foundations for "Experimental Designs", the leading applied text in this area.

He then became the chairman of the Biostatistics Department at Johns Hopkins. I remember visiting him in the summer of '49 breaking my train journey from Princetown to Raleigh. I really had the red carpet treatment and enjoyed the wonderful hospitality of Betty and Bill. The visit to the downtown campus of the Biostatistics Department was most impressive. This was the department that Bill had created, a perfect team of applied and theoretical statisticians. It was truly a proof of Bill's talent to cope with faculty having diverse interests and his skill in dealing with fiscal and administrative problems. The research output of this group under Bill's leadership was most impressive.

From here he went to Fred Mosteller's department at Harvard as a research professor with a strong link to the Harvard Medical School. I was very fortunate that Bill invited me to spend the spring semester of '61 at Harvard. During this time, I got to know Bill's and Betty's treasured cabin at Cape Cod. I had a most stimulating time. As might be expected, Bill's interests had turned more to biomedical problems including analytical studies of survey data, the superimposition of experimental analysis on non-experimental studies (including research on smoking and lung cancer), the handling of disturbing variables in operational studies as well as many others. He was very generous in his praise of the two-way stratification technique which I had developed with Bryant and Jessen. He had in fact applied this procedure to a two-way classification of hospitals in a special follow up study.

Some of these topics Bill covered in a most interesting lecture series delivered at Texas A & M in the spring of '66. I was very fortunate to persuade him to be one of the distinguished visiting lecturers sponsored by a National Science Foundation grant. Bill's sparkling performance as a lecturer and his witty

and good humored dialogue as a visitor at our parties will long
be remembered at Texas A & M.

Of course, Harvard was his last port and this brief narrative undoubtedly is only the tip of the iceberg visible to me
of the vast volume of Bill's activities there. Let me express
the hope that other friends will give a more competent account of
Bill at Harvard.

And now he is no longer with us. The loss to statistics is
immense, indeed, irreparable. We have lost one of our leading
scholars of world renown, a great gentleman and most of all a
genuine friend.

II NONSAMPLING ERRORS

THE SURVEY STATISTICIAN'S RESPONSIBILITY
FOR BOTH SAMPLING AND MEASUREMENT ERRORS

Tore E. Dalenius

University of Stockholm and Brown University

The author fears that survey research is on the verge of becoming divided into content-wise independent subdisciplines and consequently he blows the whistle. Thus, he formulates a goal for the survey research community, viz. the creation of a unified discipline of survey research, to serve as the basis for total survey design. Moreover, he outlines a two-phase program for the endeavors necessary if this goal is to be reached.

1.   INTRODUCTION

"... survey research has become, in some disciplines at least, the most widely used tool of empirical investigation"

-- Glock (1967).

"While perfection of sampling for all surveys must remain one objective, it alone will probably mean no more than an improvement of two or three percentage points, at the most. Errors of question interpretation, on the other hand, run the gamut"

-- Connelly (1945).

1.1 Two Vague Terms

Surveys, and hence *survey research,* are used in the statistical literature in a rather loose way. I will use *survey*

in accordance with the following quotation from Kendall and
Buckland (1960):

> "An examination of an aggregate of units, usually human
> beings or economic or social institutions. Strictly
> speaking, perhaps, 'survey' should relate to the whole
> population under consideration but it is often used to
> denote a sample survey, i.e. an examination of a sample
> in order to draw conclusions about the whole."

Furthermore, I will use 'survey research' to denote the
technology of the survey approach to scientific inquiry. This
technology is indeed rich in special methods. *Survey sampling*
and *survey measurement* are two major classes of methods; two
other classes of methods comprise tools for 'quality control' and
'evaluation', respectively.

1.2   Three Premises

My paper is predicated on the following three premises.

First Premise

In their scholarly work, survey statisticians tend to
focus on the *sampling* issues,[1] and they do so even in cases in
which there is an important measurement issue.[2]

Second Premise

As a consequence of what has just been said, a misleading
image of today's survey statistician as being one-sidedly
concerned with the *sampling* issues has been generated.

---

[1] In some instances, the titles of textbooks and theses reflect
this emphasis: Cochran's *Sampling Techniques* and Dalenius'
Sampling in Sweden are two examples. It is also of interest
to note that, while measurement issues are discussed in several
textbooks, they are relegated to the very end; moreover, only
a small fraction of the pages are taken up in discussing these
issues.

[2] It is interesting to note that at the 1977 symposium on
"Survey Sampling and Measurement", as reported in Namboodiri
(1978), sampling issues dominated the program.

I admittedly lack good statistics to support this premise. There are, however, several supporting indicators of interest. I will briefly discuss some of these; it is inevitable that I have chosen a format of sweeping statements.

a. The *Statistical Theory and Method Abstract,* published for the International Statistical Institute (and hence also for its section the International Association of Survey Statisticians), uses a section for classification of papers which identifies 12 classes of subjects, one and only one of which is to take into account papers on survey research. That class is designated *Sampling Design* and is further divided into 10 subclasses as follows:

0. General papers
1. Simple random; stratified; multi-stage
2. Sampling with unequal probability
3. Multi-phase sampling, double sampling
4. Natural (human, animal and biological) populations
5. Non-sampling problems
6. Censored, systematic and quota sampling
7. Nature and number of units; cost and efficiency
8. Acceptance inspection
9. Process control

While this division into subclasses has considerable scope with respect to *sampling* aspects, it appears to be insufficient with respect to *measurement* and other aspects; this critical viewpoint is in fact borne out by an examination of the issues published in the last few years.[3]

b. Courses in survey research taught in statistical departments focus on the *sampling* aspects; measurement aspects (such as design of questionnaires and interviewing) receive very little attention, if any. When survey research is taught outside of statistics departments, the situation is typically reversed.

c. Scanning a sizeable number of professional journals and

---

[3] The editor of this publication has informed me (in April 1980) that he is planning to revise the whole scheme of classification.

comparing them with respect to the space currently devoted to
sampling vs. measurement aspects reveals a significant difference: statistical journals devote much less space to measurement aspects than do other journals.

d. A perusal of *Survey Research* - the quarterly published by the
Survey Research Laboratory at the University of Illinois and
reflecting primarily the North American situation - is thought-
provoking, as shown by the following findings:
- academic survey research organizations are typically
  located outside of statistics departments;
- few directors of such organizations are members of the
  American Statistical Association;
- there appear, in addition, now and then, some 'juicy bits'
  in the job openings section; a recent note (Vol. 11, 1, 1979)
  concerning an opening at the study director level reads as
  follows: "Candidates must have strong academic backgrounds
  in survey research methodology and/or statistics/sampling".

Third Premise

Survey research - already organizationally divided in the
university setting[4] - may become divided into content-wise
(subject-wise) independent subdisciplines.

This third premise is clearly speculative in nature: it
simply identifies one possible future scenario.

1.3 Purpose of this Paper

In my view, it is in the interest of the whole survey
research community that survey research *not* become divided into
independent subdisciplines. What is needed is the amalgamation
of the relevant specialties (sampling, measurement, etc.) into a

---

[4] A similar observation applies in some instances to the government agency setting; the National Central Bureau of Statistics
in Sweden is an example in kind.

unified survey research discipline to serve as the basis for total survey design. Statistics - the technology of the scientific method - and only statistics can provide the theoretical and methodological linkage for this amalgamation.[5]

Consequently, I want to alert my fellow survey statisticians to the necessity of exercising a leadership which will pave the way for a unified discipline of survey research.

2. THE POINT OF DEPARTURE

2.1 Survey Sampling - Past and Present

The 1934 paper by Neyman on "the representative method" was a major force in stimulating research and development in survey sampling.

In the period up to around 1950, these activities took place largely in a non-academic setting, viz, in government agencies with programs for large-scale surveys, such as the U.S. Bureau of the Census and the Indian Statistical Institute; the research was typically part of ongoing surveys. This state of affairs has been succinctly described as follows: "...when survey sampling was in its infancy, most theorists were practitioners and most practitioners were theorists" (Namboodiri, 1978, p.xvii). To begin with, the focus was on the pros and cons of measurable sampling design ("probability sampling"). In the 1940s, great effort was devoted to the development of efficient sampling techniques for use in nationwide surveys. An important example in kind of special interest in this context (see section 3.4 below) concerns designs for multistage sampling, whereby primary sampling units are selected with probabilities proportional to some measure

---

[5] It may be worth remembering the role that measurement errors have played in the development of statistics, for example in the work of Gauss!

of size, and subsampling is carried out in a way which yields a
constant overall sampling fraction. At about the same time,
various problems associated with measurement errors[6] were addressed. The research in this period was summarized in a sequence of
textbooks which appeared around 1950.

In the 1950s, increased research attention was given to
measurement errors; this work resulted around 1960 in the development of *survey models* which provide a formal framework for survey
design aiming at striking a reasoned balance between sampling *and*
measurement errors, and hence a formal framework for total survey
design.

As pointed out above, the early evolution of survey sampling
took place mainly in government agencies. While a very sizeable
volume of research has been carried out during the period discussed here - the volume may very well be larger than before - the
picture has, nonetheless, changed in a significant way; thus, in
the last 20 years, we have witnessed a strong growth in university-
based research.

Special mention should be made here of research concerning
the foundations of survey sampling. It is perhaps not surprising
that much of this research is both highly theoretical and has a
rather narrow scope; especially, considerations having to do with
applications are rarely treated.

Mention should also be made here of the proliferation - at
least in the United States - of academic survey research organizations. While carrying out sample surveys may be their prime
charge, they are also actively involved in various research
activities.

---

[6] The terms 'measurement error' and 'nonsampling error' were
typically interchangeable.

2.2 Survey Measurement - Past and Present

As mentioned in section 2.1, survey statisticians have indeed not neglected measurement aspects in their methodological endeavors; my criticism concerns the fact that they have hidden their light under a bushel.

Much valuable work in the realm of survey measurement has been carried out by people affiliated with other disciplines than statistics, and especially by people in the social and behavioral sciences. In view of the expressed purpose of this paper - to alert my fellow survey statisticians to the necessity of exercising a leadership - I will be satisfied here to name some outstanding contributors, such as Hyman, Lazarsfeld and Likert.

2.3 Towards a Unified Discipline of Survey Research

The methodological developments reviewed in sections 2.1 and 2.2 may be viewed as important steps towards the goal: a unified discipline of survey research.

But we still have a long way to go. A unified discipline of survey research will not come about as the result of passivity or wishful thinking in the survey community. It calls for hard work, and it will be time-consuming and demand considerable resources. Hence a purposeful and detailed program for this work is necessary.

In developing such a program, account should be taken of the fact that some additional steps may be taken right now (first-phase steps, for short), while other steps (second-phase steps) may call for considerable planning. In other words, the program may be designed to reflect the distinction just made.

3.   THE FIRST-PHASE STEPS

3.1   First-Phase Steps - An Overview

Many issues which we traditionally look at solely from a *sampling* point of view may in addition be discussed from a *measurement* point of view. When dealing with these issues in the professional journals, in textbooks, and in courses, we may take advantage of this state of affairs by looking at them from both points of view. Doing so should serve to bring it into the limelight that survey statisticians naturally are concerned about both sampling and measurement errors.

In sections 3.2 - 3.5, I will give some examples of issues which naturally lend themselves to such an approach.

3.2   Total vs. Sample Survey

Textbooks on sampling theory typically use the following approach. An assumption is made that - associated with a population of N elements - there is a vector of unique values of some characteristic:

$$Y_1, \ Y_2, \ \ldots, \ Y_i, \ \ldots, \ Y_N$$

with mean per element $\bar{Y}$, the parameter to estimate. A sample of n elements and hence n Y-values provide the estimate $\bar{y}$. Thus, by virtue of the assumption stated above, a total survey is a special case of a sample survey, viz. the case where $n = N$, and therefore $\bar{y} = \bar{Y}$.

This approach - which may be referred to as the traditional approach - appears in a cursory analysis to be realistic in some applications, such as when a sample of census questionnaires is selected to prepare precensus estimates. On closer analysis, however, the basic assumption may prove far from realistic: Y-values may not be available for all units in the population, and some of the Y-values available may be changed in the course of the editing operation.

In other applications, the traditional approach may be far from realistic and even grossly misleading: there may at no stage exist a vector of unique Y-values. Hence, using the traditional approach may serve to downplay the fact that developing the survey design typically calls for developing a measurement design *and* a sampling design, and also an organizational framework, which will make it possible to have an implementation which is faithful to the survey design.

Thus, the traditional approach should - whenever possible - be replaced by a realistic approach, i.e. an approach which reflects in a comprehensive way both the sampling and measurement considerations.

In order to prevent misunderstanding, I hasten to add that assuming the existence of a vector of unique values may be perfectly realistic by reference to the sampling units at some level of a hierarchy of such units, while at the same time it may be perfectly unrealistic by reference to the elements.

3.3 Specific Sampling Techniques

The argument advanced in section 3.2 in favor of the realistic approach carries over naturally into discussions of specific sampling techniques.

Consider, for example, simple random sampling in a case where the data is subject to measurement errors. The presentation of the theory for this technique should include proper interpretation of, for example, such a result as "$E\bar{y} = \bar{Y}$" by making clear the operational measning of $\bar{Y}$ (say, in terms of "equal complete coverage"). And by the same token, it should be explained that the usual estimate of the variance of $\bar{y}$ comprises a contribution from the measurement process.

3.4 PPS-Sampling

In some contexts, it may prove intractable to develop a
theory which is consistent with the realistic approach advocated
in sections 3.2 and 3.3. This may explain why research concerning
the foundations of survey sampling typically postulates the existence of a vector of unique values of some characteristic, i.e.,
the traditional approach is used.

If using the traditional approach is judged necessary, it
is indeed desirable that the analysis be supplemented by a discussion of the measurement aspects whenever such a discussion is
crucial for an assessment of the design from a broader point of
view.

PPS-sampling is an example in kind. In the context of the
foundations, this technique is typically analyzed solely from a
sampling point of view; the design considered is typically one-
stage sampling of units. Such a setup is, of course, entirely
sufficient for an analysis of the purely foundation-oriented
issues (such as the choice of estimator). But it leaves out an
aspect of special relevance to applications.

Thus, the technique of PPS-sampling finds an important use
in multistage sampling designs, say for nationwide interview
surveys, where it is crucial to exercise firm control of the
fieldwork. The point will be illuminated by a simplified example
involving two-stage sampling. Thus, a sample of first-stage units
is selected, with probabilities of selection:

$$P_1, P_2, \ldots, P_i, \ldots, P_M$$

proportional to measures of the sizes of these units. In each
such unit thus selected, a sample of second-stage units (say
elements) is selected, with probabilities $P_{j|i}$ determined by
the constraint that

$$P_i \cdot P_{j|i} = \text{constant} .$$

The rationale of this scheme is that it provides a workload of (about) the same size in each selected first-stage unit. Such an outcome creates advantageous conditions for control of the fieldwork; it makes the scheme superior to some alternative schemes which have equivalent properties in terms of the sampling theory.

3.5 Sampling Designs

The point raised in sections 3.3 - 3.4 with respect to specific sampling techniques applies, in principle, to sampling *designs* used in surveys.

The design used by the U.S. Bureau of the Census for its Current Population Survey (up to the middle of the 1950s) serves well to illustrate this point. Thus, the design called for the selection of a small sample (m = 68) of large first-stage units and subsampling of households within these units. This setup was chosen to make it feasible, with respect to the budget, to hire supervisors to control the various field operations.

4. THE SECOND-PHASE STEPS

4.1 The Format of Section 4

The second-phase steps necessary to achieve the goal of creating a unified discipline of survey research will concern both content and organization of the forthcoming discipline. In sections 4.2 - 4.5, I will limit my discussion to *content* issues; in doing so, I will be content to list a few key steps and to comment on them briefly.

4.2 Generation of a Standard Terminology

Methodological development in the realm of survey research has long been and will continue to be a highly interdisciplinary endeavor. If it is to be successful, problems of communication

must be coped with. Especially, it is of decisive importance to
generate a standard terminology.

4.3 Developing New Curricula [7]

  I will point to three topics which should be part of any
curriculum.

a. *Survey Models*

  These models - also referred to as 'mixed error models' -
will play an instrumental role in the creation of a unified
discipline which encompasses both survey sampling and survey
measurement: they serve to *amalgamate* the theoretical components.

b. *Standards for Error Reporting*

  In discussing the accuracy of survey results, practitioners
too often are satisfied with what is easy to measure, viz. the
sampling error, leaving it to the user to account for other
sources of error (such as nonresponse).

c. *Total Survey Design*

  It is important to realize that the fact that there is not
(and is not likely to be) a formal theory for optimization does
not mean that total survey design must be dealt with as an art of
which some great survey statisticians are capable, but which
cannot be taught or learned. There do indeed exist paradigms for
the statistical engineering that total survey design in fact is,
and such paradigms should be part of any advanced curriculum of

---

[7] The following quotations from Glock (1967) may still hold true:
  i. "That survey research is not taught to any extent in many
      disciplines and is frequently mistaught in others is
      further evidence that survey research has not really
      penetrated deeply into the social science consciousness";
  ii. "The number of universities which provide systematic
      training in survey research can almost be counted on the
      fingers of one hand."
A survey of the teaching situation may be called for!

survey research.

4.5   Textbooks

If new curricula - incorporating the topics mentioned in sections 4.3 and 4.4 - are developed and tested and evaluated, we will no doubt in due time have available textbooks on survey research. They are badly needed.

ACKNOWLEDGEMENTS

The author wishes to thank Dr. Barbara A. Bailar, U.S. Bureau of the Census, and Mr. Thomas B. Jabine, U.S. Department of Energy, for stimulating discussions of the subject of this paper.

REFERENCES

Connelly, G.M. (1945). Now let's look at the real problem: validity. *Public Opinion Quarterly.* Spring 1945, 51-60.

Glock, C.Y. (ed.) (1967). *Survey Research in the Social Sciences.* New York: Russell Sage Foundation.

Kendall, M.G. and Buckland, W.R. (1960). *A Dictionary of Statistical Terms.* London: Oliver and Boyd.

Namboodiri, N.K. (ed.) (1978). *Survey Sampling and Measurement* New York; Academic Press.

ESTIMATION AND DESIGN FOR NON-SAMPLING ERRORS OF SURVEYS

H. O. Hartley

Duke University

A general methodology to estimate the total
variance of an estimator of population total is
provided. The variance formulae include all relevant
sampling and nonsampling variance components and
these are estimated directly from the survey data.
The optimization of the interviewer assignment is
also investigated.

1. INTRODUCTION

The importance of non-sampling or measurement errors has
long been recognized (for the numerous references see e.g., the
comprehensive papers by Hansen, Hurwitz and Bershad, 1961 and
Bailar and Dalenius, 1970). Briefly the various models suggested
for such errors assume that a survey record (recorded content
item) differs from its "true value" by a systematic bias and
additive error contributions associated with various sources of
errors such as interviewers and coders. The important feature of
these models is that the errors made by a specified error source
(say a particular interviewer) are usually 'correlated'. These
correlated errors contribute additive components to the total
mean square error of a survey estimate which do not decrease
inversely proportional to the overall sample size but only
inversely proportional to the number of interviewers, say. Consequently, the application of standard textbook formulas for the

estimation of the variances of survey estimates may lead to
serious underestimates of the real variability which should
incorporate the non-sampling errors.

Attempts have, therefore, been made to estimate the components due to non-sampling errors. The early work in this area has
concentrated on surveys specifically designed to incorporate
features facilitating the estimation of non-sampling components
such as reinterviews or interpenetrating samples or both, see,
e.g., the early studies reported in the pioneer papers by
Mahalanobis (1946), Sukhatme and Seth (1952). However later
examples in these papers and some of the more recent literature
(see e.g., Cochran, 1968; Fellegi, 1969; Nisselson and Bailar,
1976; Battese, Fuller and Hickman, 1976) have also treated surveys
in which such features are either lacking or limited, but these
results are restricted to simple surveys permitting the use of
analysis of variance techniques. Mention should also be made of
Hansen, Hurwitz, Marks and Mauldin (1951), Hanson and Marks (1958)
and Koch (1973).

In this paper we provide a general methodology applicable
to essentially any survey to estimate the total variance of
estimators of target parameters, such as the population total.
Our variance formulas include all relevant sampling and non-
sampling variance components and these are estimated directly
from the survey data and do not assume that estimates of non-
sampling errors for "similar" content items made in special
studies can be transferred to the current survey estimates. This
paper does not address non-sampling biasses and represents a
generalization of the work by Hartley and Rao (1978), Hartley and
Biemer (1979), in that it extends the approach to cover situations
of interviewer and coder assignments more commonly used in present
survey practice. The assumptions made and the conditions under
which variances can be estimated are summarized in subsequent
sections.

At the present time the technique has been applied to artificial data generated from special cases of models (1) and (2) below. It is not possible to apply it to survey data acquired in the past since

(a) the required information on interviewer and coder assignments is usually not available;

(b) even if (a) is available, interviewer and coder assignments have to satisfy certain estimability conditions (see below) which were not recognized in the past.

However, both the Bureau of the Census and the Research Triangle Institute have expressed an interest in trying the method in future survey operations.

2. MODEL ASSUMPTION FOR NON-SAMPLING ERRORS

This study is confined to non-sampling errors of quantitative content items. It is hoped to cover categorical items in subsequent studies. We adopt "additive error models" (also used in the more recent literature) in which the error made by (say) a particular interviewer are correlated through an additive error term. Confining ourselves to one particular content item, it is assumed that the true content item of the t-th respondent interviewed by interviewer i and coded by coder c has the following additive non-sampling errors:

$$\text{Interviewer error} = b_i + \delta b_t$$
$$\text{Coder error} \quad\;\; = c_c + \delta c_t \qquad\qquad (1)$$
$$\text{Respondent error} = \delta r_t$$

where

b_i = error variable contributed by i-th interviewer common to all units, t , interviewed by i-th interviewer,

c_c = error variable contributed by c-th coder common to all units, t , coded by c-th coder,

δb_t, δc_t, and δr_t = elementary interviewer, coder, and respondent errors respectively afflicting the content item of unit t (respondent t).

We assume that the b_i, c_c, δb_t, δc_t are random samples from infinite populations with zero mean and variances σ_b^2, σ_c^2 $\sigma_{\delta b}^2$ $\sigma_{\delta c}^2$ and that δr_t is a (nonobserved) error with zero mean sampled from the finite population of respondents by the survey design implemented.

Our method can easily be extended to cover other and/or additional sources of non-sampling errors.

3. THE TYPE OF SURVEY COVERED

The type of survey here covered is essentially a general stratified multistage survey with restrictions delineated in Section 4 below. To fix the ideas we first describe the concepts in terms of a three-stage survey and then outline the general multistage situation. Denote by η_{hpst} the true content item for tertiary t of secondary s of primary p in stratum h . Denote by y_{hpst} the corresponding recorded content item. Then we clearly have $y_{hpst} = \eta_{hpst}$ + error, and if we replace the units label t in (1) by the quadruple subscript hpst this equation can be written in the form

$$y_{hpst} = \mu_h + (\bar{\eta}_{hp..} - \mu_h) + (\bar{\eta}_{hps.} - \bar{\eta}_{hp..})$$
$$+ (\eta_{hpst} - \bar{\eta}_{hps.}) + \delta r_{hpst} \qquad (2)$$
$$+ \delta b_{hpst} + \delta c_{hpst} + b_i + c_c$$

where the μ_h, $\bar{\eta}_{hp..}$ and $\bar{\eta}_{hps.}$ are the population means of the true content item respectively for a stratum, primary and secondary. We now combine terms in the second and third lines of (2) and write

$$\varepsilon_{hpst} = (\eta_{hpst} - \bar{\eta}_{hps.}) + \delta r_{hpst}$$

and

$$e_{hpst} = \delta b_{hpst} + \delta c_{hpst} .$$

The characteristic feature of our approach (similar to that of Hartley and Rao, 1978) is that we recognize that for most survey designs the variance of the target parameter estimates only depends on the variances of the pooled terms ε_{hpst} and e_{hpst} so that such estimates can be computed *without* estimating the variances of the individual terms in (2). Moreover in the special case (also considered by Hartley and Rao, 1978) in which

(i)   the last stage (tertiary) units are drawn with equal probability,

(ii) the last stage (tertiary) populations are essentially infinite (finite population correction (fpc) negligible),

the pooled terms $\varepsilon_{hpst} + e_{hpst}$ are random samples from infinite populations with variances $\sigma_e^2(h,p,s)$ (say) and the variance of the target parameter estimates only depends on $\sigma_e^2(h,p,s)$. No assumptions need therefore be made about the independence of the five individual terms comprising $\varepsilon_{hpst} + e_{hpst}$ and randomizations (of say respondents) ensuring this independence are not required (note, however, the randomization of interviewers and coders ensuring the independence of the b_i and c_c from $\varepsilon_{hpst} + e_{hpst}$ discussed below).

The estimation of the variances of the estimates of the target parameters will depend on two design features namely

(a)   the survey design;

(b)   the design for the allocation of interviewers and coders to the units.

We have discussed the restrictions on (a) above. With regard to

(b) survey operations will differ with regard to practical feasi-
bilities of allocation plans. In this paper we handle the following restrictions on this allocation most frequently imposed:

> (A)   The complete work load of a psu is handled by one
>   selected interviewer and one selected coder.

Actually the above restriction makes it hardest to estimate
σ_b^2 and σ_c^2 and there are no difficulties in dealing with
situations in which the work load in some or all of the psu's
are shared by interviewers. For example a generalization to the
assignment specification, (B) below, is quite feasible:

> (B)   the complete work load of a secondary is handled by
>   one selected interviewer and one selected coder.

Situations where (A) is preferred are numerous, for example,
in an educational survey psu's may be schools, secondaries age
groups, and tertiaries students and it would in most situations
be wasteful to assign more than one interviewer to a school. The
situation is, of course, much more flexible with regard to coders.

4.   THE ESTIMATION OF THE VARIANCE COMPONENTS

Denote by $\hat{y}_{hp}$ the unbiased estimator of the true primary
mean $\bar{\eta}_{hp..}$ appropriate to the survey design and by $v_{hp} = v(\hat{y}_{hp})$,
the customary unbiased estimator of $V(\hat{y}_{hp})$ computed from the
y_{hpst}. It is clear from model (2) and condition (A) that $v(\hat{y}_{hp})$
will not involve the b_i and c_c but will of course involve the
e_{hpst} and ε_{hpst} whose variances $\sigma_e^2(h,p,s)$ will be estimated
by the between tertiaries within secondaries sums of squares. We
now write the estimator $\hat{y}_{hp}$ in the model form (derived from (2)):

$$\hat{y}_{hp} = \mu_h + \phi_{hp} + \delta\hat{y}_{hp} + b_i(h,p) + c_c(h,p) \qquad (3)$$

where $\phi_{hp} = \bar{\eta}_{hp..} - \mu_h$ and $\delta\hat{y}_{hp}$ denotes the deviation of $\hat{y}_{hp}$
(without the term $b_i(h,p) + c_c(h,p)$ afflicting it) from its
expectation $\mu_h + \phi_{hp}$. We now use the Minque (0) components of

variance estimation (see e.g. Hartley, Rao and LaMotte, 1978).
We write (3) in matrix form:

$$\underset{\sim}{y} = \underset{\sim}{X}\mu + \Sigma_h \underset{\sim h}{U}(\phi + \delta y) + \underset{\sim b}{U}\underset{\sim}{b} + \underset{\sim c}{U}\underset{\sim}{c} \qquad (4)$$

where $\underset{\sim}{X}$, $\underset{\sim h}{U}$, $\underset{\sim b}{U}$ and $\underset{\sim c}{U}$ are the design matrices implied by
model (3), for example $\underset{\sim h}{U}$ has an $n_h \times n_h$ identity matrix for
the $\hat{y}_{hp}$ in stratum h and all zero submatrices for the other
strata, where n_h denotes the number of sampled psu's in stratum
h $(h = 1,...,H)$.

To simplify the argument we assume that the sampling of
primaries is with probabilities π_{hp} and with replacement. The
ϕ_{hp} can then be regarded as a sample from an infinite population
of ϕ_{hp}'s with relative frequencies π_{hp} $(\Sigma_p \pi_{hp} = 1)$ and μ_h
is now defined by $\Sigma_p \pi_{hp} \bar{n}_{hp..}$. We also define

$$\underset{\sim h}{V} = \underset{\sim h}{U} - \underset{\sim\sim}{XX}' \underset{\sim h}{U}$$

$$\underset{\sim b}{V} = \underset{\sim b}{U} - \underset{\sim\sim}{XX}' \underset{\sim b}{U} \qquad (5)$$

$$\underset{\sim c}{V} = \underset{\sim c}{U} - \underset{\sim\sim}{XX}' \underset{\sim c}{U}$$

where, without loss of generality, $\underset{\sim}{X}$ has been standardized so
that $\underset{\sim}{X}'\underset{\sim}{X} = \underset{\sim}{I}$. The design matrices $\underset{\sim b}{U}$ and $\underset{\sim c}{U}$ are fixed by
the interviewer and coder-allocation design (see Section 7) but
the b_i and c_c are random variables through the random allocation
of interviewer and coder personnel.

We form the quadratic forms

$$Q_h = \underset{\sim}{y}' \underset{\sim h}{V} \underset{\sim h}{V}' \underset{\sim}{y}$$

$$Q_b = \underset{\sim}{y}' \underset{\sim b}{V} \underset{\sim b}{V}' \underset{\sim}{y} \qquad (6)$$

$$Q_c = \underset{\sim}{y}' \underset{\sim c}{V} \underset{\sim c}{V}' \underset{\sim}{y} .$$

The equating of the Q's to their expectations of the form

$$E(\underset{\sim}{Q}) = \underset{\sim}{K}\sigma^2 + E(\underset{\sim}{\Delta Q}) \qquad (7)$$

gives the estimates $\hat{\sigma}^2$, where $\Delta Q = \text{tr}(\underset{\sim}{V}\underset{\sim}{V}'\hat{\underset{\sim}{\Sigma}})$, $\underset{\sim}{\sigma}^2 = (\sigma_\phi^2(1),\ldots,$
$\sigma_\phi^2(h)$, σ_b^2 , $\sigma_c^2)'$ and $\hat{\underset{\sim}{\Sigma}}$ is the estimated conditional covariance
matrix of the $\hat{y}_{hp}$'s given the set of primaries (h,p) . The
inversion of $\underset{\sim}{Q} - \Delta Q = \underset{\sim}{K}\sigma^2$ leads to estimates $\hat{\sigma}_\phi^2(h)$, $\hat{\sigma}_b^2$ and
$\hat{\sigma}_c^2$ to which we adjoin

$$\hat{\sigma}_e^2(h,p,s) = \Sigma_t(y_{hpst} - \bar{y}_{hps.})^2/(\nu_{hps} - 1) \qquad (8)$$

where ν_{hps} is the number of sampled tertiaries in the secondary
hps. We hope to cover a change to a primary design which selects
primaries without replacement by a suitable change in the quadratic
forms (6) which would represent a departure from Minque (0).

5.   ESTIMATION OF VARIANCE OF TARGET ESTIMATE

We confine the discussion to estimators (of say the population mean or total) which are of the form

$$\hat{Y} = \underset{\sim}{\gamma}'\underset{\sim}{\bar{y}} \qquad (9)$$

where $\bar{y}$ is the vector of secondary means and γ is a vector of
coefficients which may depend on the sample selection of primaries
and secondaries. Let G denote a given set of selected secondaries and interviewer- and coder-randomisations (see Section 7).
Then

$$V(\underset{\sim}{\gamma}'\underset{\sim}{\bar{y}}) = \underset{G}{V} E[\underset{\sim}{\gamma}'\underset{\sim}{\bar{y}}|G] + \underset{G}{E} V[\underset{\sim}{\gamma}'\underset{\sim}{\bar{y}}|G] \qquad (10)$$

where $V[\cdot|G]$ and $E[\cdot|G]$ denote the conditional variance and
conditional expectation given a set G , and $\underset{G}{V}$ and $\underset{G}{E}$ denote
the variance and expectation over all possible sets G . An
unbiased estimator of (10) is then given by

$$v(\underset{\sim}{\gamma}'\underset{\sim}{\bar{y}}) = \underset{\sim}{\bar{y}}'\underset{\sim}{\Omega}\underset{\sim}{\bar{y}} - \text{tr}(\underset{\sim}{\Omega}\underset{\sim}{S}) + (\underset{\sim}{\gamma}'\underset{\sim}{D}_b\underset{\sim}{D}_b'\underset{\sim}{\gamma})\hat{\sigma}_b^2$$

$$+ (\underset{\sim}{\gamma}'\underset{\sim}{F}_c\underset{\sim}{F}_c'\underset{\sim}{\gamma})\hat{\sigma}_c^2 + \underset{hps}{\Sigma}[\gamma_{hps}^2 \hat{\sigma}_e^2(h,p,s)]/\nu_{hps} \qquad (11)$$

where $D_{\underset{\sim}{b}}$ is the design matrix for the variables b_i (with I columns and M rows where $M = \underset{hps}{\Sigma} (1)$, $F_{\underset{\sim}{c}}$ the corresponding design matrix for the variables $c_{\underset{\sim}{c}}$ and $\underset{\sim}{\Omega}$ is the matrix of the quadratic form providing an unbiased estimate of the variance of $\underset{\sim}{\gamma}'\underset{\sim}{\dot{\eta}}$ where $\underset{\sim}{\dot{\eta}}$ is the vector of (true) M secondary means so that

$$\underset{G}{E} \; (\underset{\sim}{\dot{\eta}}'\underset{\sim\sim}{\Omega}\underset{\sim}{\dot{\eta}}) \; = \; \underset{G}{V}(\underset{\sim}{\gamma}'\underset{\sim}{\dot{\eta}}) \; , \qquad (12)$$

$\hat{\underset{\sim}{S}}$ is the $M \times M$ matrix of conditional variance and covariance estimates of $\overline{\underset{\sim}{y}}$ given by

$$\hat{\underset{\sim}{S}} \; = \; (D_{\underset{\sim}{b}}D_{\underset{\sim}{b}}')\hat{\sigma}_b^2 \; + \; (F_{\underset{\sim}{c}}F_{\underset{\sim}{c}}')\hat{\sigma}_c^2 \; + \; \hat{\underset{\sim}{S}}_e \qquad (13)$$

where the matrix $\hat{\underset{\sim}{S}}_e$ will have diagonal terms $\hat{\sigma}_e^2(h,p,s)/\nu_{hps}$.

Equations (11) and (13) provide the formulas for an unbiased estimate of $V(\underset{\sim}{\gamma}'\overline{\underset{\sim}{y}})$ including all non-sampling error components. Immediate generalizations are feasible to any stratified multistage design in which both primaries and last stage units are drawn with replacement. While the latter condition represents a comparatively mild restriction the treatment of other primary sampling plans requires modifications custom made for each type of primary survey design.

6. ESTIMABILITY CONDITIONS

Hartley and Biemer (1978) give sufficient conditions for the estimability of all components of variance for the case in which it is feasible for two different interviewers to be allocated to different secondaries in the same primary. As stated above, in this paper we consider the situation in which (because of practical limitations) only one interviewer must carry the whole work load in a primary. Sufficient conditions for the estimability of all components of variance are then as follows:

(i)   The sample contains at least two primaries per
       stratum, two secondaries per primary, and two
       tertiaries per secondary.

(ii)   All tertiaries in a primary are interviewed by the
       same interviewer and coded by the same coder.

(iii) In at least one stratum there are at least two
       primaries entirely interviewed by different inter-
       viewers but coded by the same coder.

(iv)   In at least one stratum there are at least two
       primaries entirely coded by different coders.

The above are sufficient conditions. However, a more
reliable estimate of σ_b^2 based on more interviewer contrasts is
obtained if (iii) is replaced by the more restrictive condition:

(iii)' If the number of primaries in stratum h is n_h
       then in all strata there must be at least two prim-
       aries interviewed by the same interviewer and the
       remaining n_h-2 primaries (if any) by n_h-2
       different interviewers. (It is assumed that the
       number of interviewers, I , exceeds $\max_h (n_h - 1)$,
       a condition usually satisfied. There must be at
       least one stratum with $n_h \geq 3$.)

The condition (iii)' will certainly provide more within
stratum interviewer contrasts for the estimation of σ_b^2 but it
is difficult to assess the reduction in the variance of $\hat{\sigma}_b^2$
through replacing (iii) by (iii)'.

7. THE OPTIMIZATION OF THE INTERVIEWER ASSIGNMENT

By taking expectations the component of variance of $V(\gamma'\bar{y})$
which depends on σ_b^2 is given by

$$\text{Comp}_b \{V(\gamma'\bar{y})\} = \{\gamma' D_b D_b' \gamma\}\sigma_b^2 = (D_b'\gamma)' (D_b'\gamma)\sigma_b^2 \tag{14}$$

and this will be minimized (whatever the values of any of the

variance components) if the sum of squares $(D_b'\gamma)'(D_b'\gamma)$ is
minimized. Denoting the elements of γ by γ_{hps} , the $D_b'\gamma$
are the "interviewer totals" of the γ_{hps}. The latter are the
"jack-up factors" to be applied to the y_{hps}. and are therefore
predetermined by the survey design. Since all secondaries are
to be interviewed by the same interviewer, only the primary totals
$\gamma_{hp} = \Sigma_s \gamma_{hps}$ are available for allocation to interviewers.
Further the minimization is to be constrained by the estimability
condition (iii)'. Moreover, as a practical consideration we
would normally prefer to restrict the optimization by assigning
each interviewer "approximately" an equal number of primaries, a
concept which is discussed below. Since the total sum of squares
of the γ_{hp} is given, the minimization of the between interviewer
sum of squares is of course equivalent to the maximization of the
within interviewer sum of squares. In order to satisfy the first
part of condition (iii)' optimally we pool for each stratum h
the largest and the smallest of the γ_{hp} , denote these totals
by $\dot{\gamma}_{hp}$ (with a p-label corresponding to (say) the smaller of
the two p's) and assign the <u>total</u> $\dot{\gamma}_{hp}$ to some interviewer.
The remaining γ_{hp} are kept separate $\dot{\gamma}_{hp}$ totals. The number
of γ_{hp} is therefore given by $\nu = \Sigma(n_h - 1)$. The second
condition of (iii)' now stipulates that all $\dot{\gamma}_{hp}$ with the same
stratum index h must all be allocated to different interviewers.
We now specify the "approximately even" allocation to mean that
each interviewers quota of assigned $\dot{\gamma}_{hp}$ should only fractionally
differ from their average. Thus if

$$k = \sum_h (n_h - 1)/I \ , \tag{15}$$

then

$$k \leq \text{interviewer allocation of primaries} \leq k + 1 \ . \tag{16}$$

Denote by $\dot{\gamma}(h,i)$ *that* $\dot{\gamma}_{hp}$ in stratum h which is
allocated to interviewer i, then the task is to find *that*

allocation which minimizes

$$S^2 = \sum_i (\sum_h \dot{\gamma}(h,i))^2 = \sum_i \dot{\gamma}(\cdot,i)^2 . \qquad (17)$$

It is convenient for the algorithm to construct an $H \times I$ double array of $\dot{\gamma}(h,i)$ in which $h = 1,\ldots,H$ and $i = 1,\ldots,I$.

In each stratum h the $\nu-n_h+1$ 'excess positions" are all filled by "marked dummies" $\dot{\gamma}(h,i) = 0$ (see Figure 1 below in which an example allocation of $\dot{\gamma}(h,i)$ is exhibited for $H = 10$ strata and $I = 5$ interviewers.)

Figure 1:   Initial Interviewer Assignment

Inter-viewer i	Stratum h										Total $\dot{\gamma}(\cdot,i)$
	1	2	3	4	5	6	7	8	9	10	
1	4	0	7	0	0	0	3	0	0	4	18
2	0	4	3	0	0	0	0	3	0	2	12
3	0	2	0	0	0	3	2	0	1	0	8
4	0	1	0	0	2	1	0	2	0	0	6
5	0	1	0	1	0	1	0	1	0	0	4
$n_h-1 =$	1	4	2	1	1	3	2	3	1	2	

In the above example $S^2 = 18^2 + 12^2 + 8^2 + 6^2 + 4^2 = 584$, a high value in spite of the fact that precisely $k = 4$ values of $\dot{\gamma}(h,i)$ are allocated to each interviewer. The algorithm will commence with an "initial arrangement" such as given in Figure 1 satisfying (iii)' and (16) constructed as shown below. The "improving algorithm" will then attempt all possible double swaps of $\dot{\gamma}(h,i)$ with $\dot{\gamma}(h,j)$ and $\dot{\gamma}(\ell,j)$ with $\dot{\gamma}(\ell,i)$ which reduce S^2 and at the same time preserve the above conditions. It is programmed as a quadruple loop in h, ℓ; i, j and no double swap is made if

(a)   either $\dot{\gamma}(h,i)$ or $\dot{\gamma}(\ell,j) = 0$,

(b)   if $h \neq \ell$ and if either $\dot{\gamma}(\ell,i) \neq 0$ or $\dot{\gamma}(h,j) \neq 0$,

(c)   if $\frac{1}{2}\Delta s^2 = \{\dot{\gamma}(\cdot,i) - \dot{\gamma}(\cdot,j) - \dot{\gamma}(h,i) - \dot{\gamma}(\ell,j)\}$.   (18)

$(\dot{\gamma}(h,i) - \dot{\gamma}(\ell,j) \geq 0$.

Clearly with the above conditions $\Delta s^2 < 0$ with every legitimate double swap. If and when for a full quadruple loop no legitimate swap was found with $\Delta s^2 < 0$, the algorithm will terminate usually at the global minimum of s^2. However it is not possible to prove this in general. In the above example the algorithm did reach the global minimum as shown in Figure 2.

Figure 2. Terminal Interviewer Assignment

Inter-viewer i	Stratum h										Total $\dot{\gamma}(\cdot,i)$
	1	2	3	4	5	6	7	8	9	10	
1	0	4	3	1	2	0	0	0	0	0	10
2	4	1	0	0	0	1	0	3	0	0	9
3	0	1	0	0	0	0	3	2	0	4	10
4	0	2	0	0	0	3	2	0	0	2	9
5	0	0	7	0	0	1	0	1	1	0	10

The terminal value of $s^2\sigma_b^2 = 462\sigma_b^2$ is the minimum value of $\text{Comp}_b\{V(\gamma'\bar{\underset{\sim}{y}})\}$. The arrangement of Figure 1 (although balanced with regard to interviewer load) would have resulted in a variance 26.4% in excess of the optimum.

We now turn to the construction of the "initial arrangement" satisfying both (iii)' and (16) and at the same time giving the global minimum a finite probability to be selected. There are $\binom{I}{n_h-1}$ ways of selecting n_h-1 interviewers out of the I interviewers for stratum h and $C = \prod_h \binom{I}{n_h-1}$ possible selections of interviewers to fill the "positions" in the strata columns. Everyone of these selections has a probability of $1/C$ to be selected but most of them will violate (16). Denote by k_i the number of assignments to interviewer i, then if (16) is violated we have $(k_{max} - k_{min}) > 1$. Denote by i_{max} and by i_{min} the interviewers with k_{max} and k_{min} assignments respectively, then there must be a stratum in which i_{max} is assigned a position but i_{min} is not. Swap this assignment from i_{max} to i_{min}. This will result in an arrangement in which the numbers of interviewers with k_{max} assignments and the number of interviewers with k_{min} assignments are both reduced by 1. Continue the swapping process until the number of interviewers with k_{max} assignments and/or the number of interviewers with k_{min} assignments is zero. The new value of $k_{max} - k_{min}$ will then be reduced by at least 1. Continue the process until $k_{max} - k_{min} \leq 1$ and hence (16) is satisfied.

The swapping procedure is illustrated in Figure 3 below in which the initial assignments are $k_1 = 6$, $k_2 = 5$, $k_3 = k_4 = k_5 = 3$ and three swaps are made transferring sequentially the positions $\boxtimes^-$, $\otimes^-$ and $\diamond\!\!\!\times\!\!\!\diamond^-$ to positions $\boxtimes^+$ $\otimes^+$ and $\diamond\!\!\!\times\!\!\!\diamond^+$.

Figure 3. Illustration of Swapping Procedure

Inter-viewer	Stratum h										Swap #			
											0	1	2	3
i	1	2	3	4	5	6	7	8	9	10	k_i	k_i	k_i	k_i
1	X	X	$\diamond\!\!\!\times\!\!\!\diamond^-$			X	$\boxtimes^-$	X			6	5	5	4
2		X	X		X	X		$\otimes^-$			5	5	4	4
3		X		X			$\boxtimes^+$		X		3	4	4	4
4		X				X		$\otimes^+$		X	3	3	4	4
5			$\otimes\!\!\!\diamond^+$		X		X			X	3	3	3	4

This procedure will increase the probability of $1/C$ for any particular legitimate assignment to be selected to $(1/C) + q$ with $q \geq 0$. The above procedure of selecting an "initial arrangement" followed by the improving algorithm will result in a procedure in which the global minimum of S^2 is reached with a finite probability.

All of the above operations are automated in a computer program written by Howard Monroe (of the Institute of Statistics Texas A & M Univeristy) covering up to 50 strata and up to 10 interviewers. This is available on request.

REFERENCES

Bailar, B.A. and Dalenius, T. (1970). Estimating the response variance components of the U.S. Bureau of the Census survey model, *Sankhyā, Ser. B,* 341-360.

Battesse, G.E., Fuller, W.A. and Hickman, R.D. (1976). Estimation of response variance from interviewer re-interview surveys, *Journal Indian Society of Agricultural Statistics,* 28,1-14.

Biemer, P.P. (1978). The estimation of non-sampling variance components in sample surveys. Unpublished Ph.D. dissertation, Institute of Statistics, Texas A & M University.

Cochran, W.G. (1968). Errors of measurements in statistics, *Technometrics,* 10, 637-666.

Fellegi, I.P. (1974). An improved method of estimating the correlated response variance, *Journal of American Statistical Association,* 69, 496-501.

Hansen, M.H., Hurwitz, W.N. and Bershad, M.A. (1961). Measurement errors in censuses and surveys, *Bulletin of the International Statistical Institute,* 38, 359-374.

Hansen, M.H., Hurwitz, W.N., Marks, E.S. and Mauldin, W.P.(1951). Response errors in surveys, *Journal of American Statistical Association,* 46, 147-190.

Hanson, R.H. and Marks, E.S. (1958). Influence of the interviewer on the accuracy of survey results, *Journal of American Statistical Association,* 53, 635-655.

Hartley, H.O. and Biemer, P.P. (1978). The estimation of non-sampling variances in current surveys, Proceedings of the Section on Survey Research, American Statistical Association, Washington, D.C., 257-262.

Hartley, H.O. and Rao, J.N.K. (1978). The estimation of non-sampling variance components in sample surveys. In *Survey Sampling and Measurement* (N.K. Namboodiri, ed.) New York: Academic Press, 35-43.

Hartley, H.O., Rao, J.N.K. and LaMotte, L. (1978). A simple 'synthesis'-based method of variance component estimation, *Biometrics,* 34, 233-42.

Koch, G.G. (1973). Some survey designs for estimating response error model components, Technical Report #5, 21U-730, Research Triangle Institute.

Mahalanobis, P.C. (1946). Recent experiments in statistical sampling in the Indian Statistical Institute, *Journal Royal Statistical Society,* Ser.A, 109, 325-370.

Nisselson, H. and Bailar, B.A. (1976). Measurement analysis and reporting of non-sampling errors in surveys. *Proceedings of the International Biometric Conference* (Boston, 1976), 301-321.

Searle, S.R. (1971). *Linear Models,* New York: John Wiley & Sons.

Sukhatme, P.V. and Seth, G.R. (1952). Non-sampling errors in surveys, *Journal Indian Society of Agricultural Statistics,* 4, 5-41.

SHOULD THE CENSUS COUNT BE ADJUSTED FOR ALLOCATION PURPOSES? - EQUITY CONSIDERATIONS

I. P. Fellegi

Statistics Canada

Census population counts play a determining role in
Canada in the computation of the amount of federal transfer payments to provinces. Given the fact that the census
count is subject to error (underenumeration) whose magnitude at the province level can be approximated, the question
is raised whether the census count should be adjusted for
the estimated underenumeration when computing the amount
of transfer payments. Two types of transfer payment are
considered: 1) a fixed amount per capita; and 2) a fixed
total amount distributed in proportion to population. A
mathematical model of legislative intent is established
on the basis of which objective tests can be derived to
the question above for each type of transfer payment.
The tests are in terms of measurable statistics derived
on the basis of explicit assumptions. The paper concludes
by describing the methodology used in Canada to estimate
census underenumeration rates at the province level and
the optimum sample allocation for this measurement process
is derived if the objective is to optimize the sensitivity
of the test statistics.

1.   INTRODUCTION

This paper examines a very special kind of census data use:

its legislated utilization as input to formulae on the basis of

which funds are allocated from one level of government to another

(so-called formula allocations). To the extent that the census

counts are subject to underenumeration, their use for this purpose

represents a deviation from the legislated intent which (implicitly)

assumes the counts to be free of error.

CURRENT TOPICS IN SURVEY SAMPLING

Estimates of the undercount are often available as part of the evaluation of the census. These estimates are themselves subject to sampling and other errors. Should they nevertheless be used to adjust the census counts *for the purpose of legislated intergovernmental allocation of funds?* The present paper concentrates on the narrow question of adjustment for this single purpose. However, it is important to keep in mind that in the real world of statistical policy a number of other questions must also be answered before deciding on a specific course of action. If the counts are adjusted for one purpose (but not others), can users cope with more than one "official" set of population figures? Should intercensal population estimates also be adjusted? Should current surveys use adjusted intercensal estimates in their ratio estimation procedures? (Some of the current surveys have themselves formula allocation uses!) What would be the impact on electoral redistricting - another legislated use of census data which, however, requires considerably more geographic detail (at least in Canada) than the counts needed for intergovernmental fund allocations? None of these related questions is examined in the present paper. Even within the narrow context of a single application, i.e., legislated allocation of funds, there are several issues which have to be considered: the intent of the legislation; the danger of "politicizing" statistics or, more precisely, whether the danger of political pressures on the census increases or decreases when the counts are adjusted for underenumeration; and the long-term feedback effect identified by Nisselson (1979), i.e., whether adjusting the count diminishes the incentive, particularly for minority groups, to work with the statistical office to improve the census the next time around. Again, in this paper most of these considerations will be largely set aside, concentrating on the notion of legislative intent or "equity".

Finally, allocation formulae seldom use only the census as their data source. However, for the sake of simplicity, we will examine only the impact of errors in the census counts on fund allocations.

2. A MEASURE OF INEQUITY

Formula allocations of funds from one level of government to another take a variety of forms. Many such payments can be broadly characterized as follows.

(a) The national government (federal, for the sake of specificity) provides funds directly to the next level of government (provincial, to be specific).

(b) The legislation implies, explicitly or implicitly, the calculation of a per capita amount in province i, say X_i. The total payment intended for province i is then calculated as

$$T_i = P_i X_i , \qquad (1)$$

where P_i is the total population of province i, or that of a subgroup of the province, such as the number of university students or the number of persons below the poverty line. The quantity X_i may depend on P_i.

In the presence of some underenumeration, the quantity P_i is estimated as p_i. Applying the legislated allocation formula, but using the known quantities p_i instead of the unknown P_i, the quantity T_i would be estimated as t_i. Therefore the realized per capita payment in province i is no longer X_i but rather

$$x_i = t_i/P_i . \qquad (2)$$

Note that in (2) above the denominator is P_i, not p_i, since the actual population in province i is P_i, not p_i, so the *de facto* per capita payment in a province is equal to the actual payment (as computed using the census estimates) divided by the actual population of the province.

The per capita deviation between the amount intended by

legislation and the amount actually received is

$$d_i = X_i - x_i \ . \qquad (3)$$

We shall define the *notion of equity* as a numerical index of the extent to which legislative intent is complied with. Thus, this paper proposes as the measure of inequity due to the census under-count the square of the deviations d_i averaged over the total population given by

$$I = \Sigma \ P_i \ d_i^2/P \ . \qquad (4)$$

where $P = \Sigma \ P_i \ .$

Another way of interpreting (4) would be to think of d_i as the per capita underpayment or overpayment received by a province. In this case I is the weighted average over all provinces of the square of the per capita under or overpayments, the weights being the provincial populations. Indeed,

$$d_i = X_i - x_i$$

$$= (T_i - t_i)/P_i \ . \qquad (5)$$

Note that I does not measure only the extent to which provinces are "shortchanged" - overpayments or underpayments are given equal weight in that they both have the same effect on the legislated intent of equity.

The index I will be used to determine whether the adjusted or unadjusted census counts lead to less inequity. It is similar to the measure proposed by Jabine (1976).

3. TWO TYPES OF ALLOCATIONS

For the sake of presentation, we must simplify the large variety of allocation formulae actually used. We will focus on two prototype, or model, formulae.

3.1   Fixed per capita payment

According to this formula the amount paid to a province is directly proportional to the number of persons in the target group. If   c   is the intended per capita payment, then

$$T_i = P_i \, c \, ,$$

$$t_i = p_i \, c$$

$$X_i = c$$

$$x_i = p_i c / P_i \, ,$$

$$d_i = \frac{P_i - p_i}{P_i} \, c \quad \text{and}$$

$$I_1 = (c^2/P) \, \Sigma \, P_i \, \left(\frac{P_i - p_i}{P_i}\right)^2 .$$

Letting

$$u_i = \frac{P_i - p_i}{P_i} \tag{6}$$

denote the proportionate underenumeration in province   i   (which, of course, may theoretically be negative in the case of over-enumeration), we get

$$I_1 = c^2 \, \Sigma \, P_i \, u_i^2 / P . \tag{7}$$

A reasonable approximation of this allocation would be that legislated by the U.S. Elementary and Secondary Education Act as described in Nisselson (1979).

Note that since equity was defined as compliance with legislative intent, $I_1 \neq 0$ even if u_i is a constant over all provinces. In this case every province is short-changed by the same per capita amount so that while there is no *differential* short-changing of provinces, nevertheless the legislative intent is violated to the extent that the *de facto* per capita payment differs from the legislated one.

3.2 Fixed total payments

Under this model the federal government distributes to provinces a fixed amount, the total received by a given province being proportional to its population. Then, if C is the total to be distributed,

$$T_i = \frac{P_i}{\Sigma P_i} C \; ,$$

$$t_i = \frac{P_i}{\Sigma P_i} C \; ,$$

$$X_i = \frac{1}{\Sigma P_i} C \; ,$$

$$x_i = \frac{P_i}{P_i \, \Sigma \, P_i} C \quad \text{and}$$

$$d_i = (\frac{1}{\Sigma P_i} - \frac{P_i}{P_i \, \Sigma \, P_i}) \, C \; .$$

Using the notation of (6), it is easy to see that the corresponding inequity measure becomes

$$I_2 = \frac{c^2}{P^3 (1 - \bar{u})^2} \, \Sigma \, P_i (u_i - \bar{u})^2 \; , \qquad (8)$$

where

$$\bar{u} = \Sigma \, P_i \, u_i / P \; . \qquad (9)$$

A reasonable approximation of this model occurs in one of the payments under the Canadian Fiscal Arrangements Act. According to this Act, the revenue capacity of each province is quantified in a fashion independent of the Census. Let this measure be M_i for the ith province. Then province i receives the amount

$$\frac{P_i}{P} (\Sigma \, M_i) - M_i$$

whenever the amount above is positive and nothing if the amount is negative. Since M_i is independent of the population count P_i,

the effect of underenumeration on *receiving* provinces can be
studied within the fixed total payment model since the critical
factor is the *proportion* of the total population residing in a
province, as opposed to the absolute number applicable to the
fixed per capita payment case.

4. TESTS APPLICABLE TO THE FIXED PER CAPITA PAYMENT MODEL

We assume that estimates $\hat{u}_i$ of the undercount proportions
u_i are available which can potentially be used to adjust the
census counts. Let

$$E(\hat{u}_i) = e_i \tag{10}$$

so that

$$b_i = e_i - u_i \tag{11}$$

is the bias of $\hat{u}_i$. We also assume that the population counts
p_i have a negligible variance.

If we were to adjust the population estimates, the adjusted
counts would have a residual underenumeration (which could be
negative) equal to $u_i - \hat{u}_i$. If the adjusted census counts are
used for the fund allocation, the measure of inequity can be
obtained from (7) by substituting $u_i - \hat{u}_i$ for u_i. This yields

$$I_1^A = c^2 \Sigma P_i (u_i - \hat{u}_i)^2 / P . \tag{12}$$

Ideally, one would like to find an adjustment which
minimizes I_1^A . This is unlikely to be possible. A more modest
but still very relevant objective is to find an adjustment which
reduces the inequity, i.e., for which the difference

$$\delta I_1 = I_1 - I_1^A = \frac{c^2}{P} \{ \Sigma P_i u_i^2 - \Sigma P_i (u_i - \hat{u}_i)^2 \}.$$

is positive. Leaving out the positive factor in front of the
curly brackets, we will examine

$$\Delta I_1 = \frac{P}{c^2}(I_1 - I_1^A) = \Sigma\, P_i u_i^2 - \Sigma\, P_i(u_i - \hat{u}_i)^2. \qquad (13)$$

The expression (13) above may not be estimable directly if no unbiased estimates of u_i exist. However, after some manipulation, we obtain using the notation of (11)

$$\Delta I_1 = \Sigma\, P_i e_i^2 - \Sigma\, P_i(\hat{u}_i - e_i)^2 - 2\Sigma P_i b_i \hat{u}_i. \qquad (14)$$

All terms of (14) are estimable except the last one. The last term is equal to zero if the estimates of underenumeration are unbiased, i.e., if $b_i = 0$ for all i. If this is not the case, a simplifying assumption is needed.

Assumption A: If the estimates $\hat{u}_i$ of underenumeration have a non-negligible bias, assume that $\Sigma P_i b_i \hat{u}_i \leq 0$.

Assumption A is unverifiable but might be thought to be satisfied since the estimates $\hat{u}_i$ are typically positive and are usually underestimates of the unknown underenumeration u_i (i.e., $b_i \leq 0$). Even if the quantities b_i are not negative for all i, assumption A may be satisfied if the estimated underenumeration tends to be low $(b_i \leq 0)$ for those subgroups which the census finds difficult to enumerate (even if $b_i > 0$ for groups which the census finds easier to enumerate). In other words, assumption A is likely to be satisfied if high values of $\hat{u}_i$ are accompanied by negative values of b_i, even if low values of $\hat{u}_i$ are accompanied by positive b_i. Of course, assumption A can always be satisfied if a sufficiently conservative set of estimates $\hat{u}_i$ is used. Most methods used in practice to provide estimates of u_i, including post-enumeration surveys, dual method estimates, and reverse record checks (see section 6 for a brief discussion of the latter) are likely to satisfy the assumption. However, this assumption may not hold for so-called analytic estimates of u_i.

Denote $\Delta I_1' = \Sigma\, P_i e_i^2 - \Sigma\, P_i(\hat{u}_i - e_i)^2$. Under assumption A

$$\Delta I_1 \geq \Delta I_1' .$$

We will construct a test of the positivity of $\Delta I_1'$ which will therefore serve as a conservative test of the positivity of ΔI_1.

In order to estimate the first two terms of $\Delta I_1'$, we note that if var $\hat{u}_i$ is an unbiased estimate of Var $\hat{u}_i$, i.e., if

$$E(\text{var } \hat{u}_i) = \text{Var } \hat{u}_i \tag{15}$$

then

$$E(\hat{u}_i^2 - 2 \text{ var } \hat{u}_i) = e_i^2 - 2 \text{ Var } \hat{u}_i \tag{16}$$

so that

$$\hat{\Delta I}_1' = \Sigma \; P_i (\hat{u}_i^2 - 2 \text{ var } \hat{u}_i) \tag{17}$$

would be an estimator of $\Delta I_1'$ if the quantities P_i where known.[*] However, since their role is only to provide weights to the averaging process, (17) is not likely to be sensitive to even reasonably significant variation in the weights P_i. Therefore in place of P_i we can use census population counts p_i , or alternatively the adjusted census estimates $\hat{p}_i = p_i (1 + \hat{u}_i)$. We thus get our first test.

Test 1: Adjust the census counts if

$$\Sigma \; p_i (\hat{u}_i^2 - 2 \text{ var } \hat{u}_i) \geq 0 . \tag{18}$$

A more conservative test (and one that does not require the values P_i) can be constructed as follows.

Test 2: Adjust the census counts if

$$\hat{u}_i^2 - 2 \text{ var } \hat{u}_i \geq 0 \quad \text{for all } i. \tag{19}$$

Inequality (19) is, of course, equivalent to the condition that the estimated relative variance of all provincial underenumeration rates should be less than or equal to 1/2. It can also be

---

[*] It should be noted that $\Delta I_1'$ is itself based on sample estimates so the expected value of $\hat{\Delta I}_1'$ is not $\Delta I_1'$. However, $E(\hat{\Delta I}_1' - \Sigma I_1') = 0$.

looked upon as a form of significance test for the hypothesis of $E(\hat{u}_i) = 0$.

Test 2 is suggestive of an alternative approach whereby those provincial populations for which (19) is positive would be adjusted, leaving the remaining ones unadjusted. If s_1 is the set of provinces for which (19) is positive and s_2 is the set of all other provinces, this alternative adjustment strategy results in the following positive quantity

$$\sum_{i \in s_1} P_i(\hat{u}_i^2 - 2 \text{ var } \hat{u}_i)$$

being the change in inequity.

The application of Test 1 (or the stronger Test 2) does not *guarantee* that the adjusted census counts will in fact reduce the inequity of allocation, since the inequity measures themselves are based on sample estimates. All one can really say is that, if the sample size is large enough so that the sampling distribution of the left hand side of (18) is reasonably symmetric, it is more likely that the adjusted counts will result in less inequity rather than the unadjusted ones. Put differently, if there are no penalties attached to adjusting the counts, one could certainly use Test 1 or Test 2 as a sufficient condition for adjusting. A much more conservative test results if the basic strategy is to adjust only when the evidence, in some sense, is overwhelming that the adjusted estimates would reduce inequity. Under this strategy one would adjust the estimates only if one were reasonably certain that the unadjusted estimates lead to a higher measure of inequity.

Normally, if one wanted to construct a test for the positivity of $\Delta I_1'$, one would construct a test statistic based on its estimate $\hat{\Delta I_1'}$ and the standard error of the latter. However, since $\Delta I_1'$ itself is a mixed expression involving both population parameters (e_i) and sample estimates $(\hat{u}_i)$, that approach would lead to a test of the positivity of the common expected value of $\Delta I_1'$ and $\Delta \hat{I}_1'$, not of $\Delta I_1'$ itself. We will therefore consider the standard error σ of $\Delta I' - \Delta \hat{I}_1'$. Under the usual assumption of approximate normality, we have

$$E(\Delta I_1' - \Delta \hat{I}_1') = 0 \quad \text{and} \tag{20}$$

$$\text{Prob } (\Delta I_1' - \Delta \hat{I}_1' > -2\sigma) \doteq 0.975. \tag{21}$$

Thus, *if* we also have

$$\Delta \hat{I}_1' > 2\sigma , \tag{22}$$

it then follows from (21) and (22) that

$$\text{Prob } (\Delta I_1' > 0) \doteq 0.975 . \tag{23}$$

Therefore (22) provides the desirable test of (23). We need to estimate, however, the standard deviation of

$$\Delta I_1' - \Delta \hat{I}_1' = 2 \Sigma P_i (e_i \hat{u}_i - \hat{u}_i^2 + \text{var } \hat{u}_i). \tag{24}$$

Now, it is easy to show that up to terms of order $1/n_i$ (where n_i is the sample size in province i)

$$\text{Var}(e_i \hat{u}_i - \hat{u}_i^2 + \text{var } \hat{u}_i) = e_i^2 \text{ Var } \hat{u}_i \tag{25}$$

which, to the same order of approximation, is estimated by $\hat{u}_i^2 \text{ var } \hat{u}_i$. To the same order of approximation, it can be shown that

$$\text{Cov}(e_i \hat{u}_i - \hat{u}_i^2 + \text{var } \hat{u}_i, e_j \hat{u}_j - \hat{u}_j^2 + \text{var } \hat{u}_j) = e_i e_j \text{Cov}(\hat{u}_i, \hat{u}_j), \tag{26}$$

so that the sign of the covariance terms in the variance of (24) is equal to those of $\text{Cov } (\hat{u}_i, \hat{u}_j)$. We now make the following assumption.

Assumption B: Assume that the estimates $\hat{u}_i$ are either independent or not positively correlated with one another.

Under assumption B an overestimate of the variance of

$\Delta I_1' - \Delta \hat{I}_1'$ is given by

$$4 \sum_i P_i^2 \, \hat{u}_i^2 \, \text{var} \, \hat{u}_i \ .$$

Thus we obtain the following test.

Test 3. Adjust the census counts if

$$\sum_i p_i (\hat{u}_i^2 - 2 \, \text{var} \, \hat{u}_i) - 4 \sqrt{\sum_i p_i^2 \, \hat{u}_i^2 \, \text{var} \, \hat{u}_i} \geq 0 \qquad (27)$$

When this test indicates that adjustment is required, the probability is at least 0.975 that the adjusted counts will result in less inequity than the unadjusted ones (given assumptions A and B).

It should be noted that assumption B is satisfied for post-enumeration surveys whose stratum boundaries respect provincial boundaries. (In this case the terms of (24) are in fact independent.) It is likely to be satisfied in the sense of nonpositive correlations among the $\hat{u}_i$ if the provincial estimates $\hat{u}_i$ are based on domain estimators. (This is because the sum of the variances of estimates prepared for different domains is usually larger than the variance of the estimate prepared for the union of the domains.) At any rate, assumption B is not nearly as important as assumption A since the covariance terms are likely to be very small so long as the sampling ratios used in the survey to estimate $\hat{u}_i$ are small. In fact, they are of the order $1/P$ so that neglecting them is equivalent to neglecting the finite population correction.

Nevertheless, if assumption B cannot be accepted an even more conservative test results if the principle applied to the right hand side of (18) is applied separately to every term there. We then obtain

Test 4. Adjust the census counts if

$$\hat{u}_i^2 - 2 \, \text{var} \, \hat{u}_i - 4 u_i \sqrt{\text{var} \, \hat{u}_i} \geq 0 \quad \text{for all} \quad i. \qquad (28)$$

Based on Test 4, a conservative adjustment strategy might
involve adjusting only those provincial census counts for which
the left hand side of (28) is positive.

In conclusion, two points may be noted. First, if the
estimates $\hat{u}_i$ are <u>not</u> based on sample data (such as is the case
for analytic estimates),then $\hat{u}_i = e_i$ and Var $\hat{u}_i = 0$. Hence,
from (14), a condition for the adjusted counts to result in lower
inequity is that

$$\Sigma p_i \hat{u}_i^2 - 2 \Sigma p_i e_i b_i \geq 0 , \qquad (29)$$

Assumption A provides a sufficient (though clearly not necessary)
condition for (29) to hold.

Second, the entire development of this section is valid
(with obvious modifications) if the fixed per capita payment c
is constant within each province as opposed to its being constant
for the whole country. In other words, c can be replaced by a
set of constants c_i so long as each of them are determined without reference to the census population estimates p_i. This is, in
fact, the case with the U.S. Elementary and Secondary Education
Act mentioned above. More generally, similar results can be
derived (with P_i or p_i replaced, respectively, by $c_i P_i$ and
$c_i p_i$) under the allocation formula

$$T_i = c_i P_i + b_i ,$$

where c_i and b_i are provincial constants which do not depend
on P_i or p_i .

5. TESTS APPLICABLE TO THE FIXED TOTAL COST MODEL

If we were to adjust the census counts for the present
allocation model, the measure of inequity obtained by substituting
$u_i - \hat{u}_i$ for u_i in (8) is given by

$$I_2^A = (C^2/P^3(1-\bar{u}+\hat{\bar{u}})^2) \sum_i P_i (u_i - \hat{u}_i - \bar{u} + \hat{\bar{u}})^2 \; ,$$

where $\hat{\bar{u}} = \sum_i P_i \hat{u}_i / P$. Now if $\hat{\bar{u}} \geq 0$ (which should certainly be the case), we get

$$I_2 - I_2^A = \frac{C^2}{P^3(1-\bar{u})^2} \sum P_i (u_i - \bar{u})^2 - \frac{C^2}{P^3(1-\bar{u}+\hat{\bar{u}})^2} \sum P_i (u_i - \hat{u}_i - \hat{u} + \hat{\bar{u}})^2$$

$$\geq \frac{C^2}{P^3(1-\bar{u})^2} \{ \sum P_i (u_i - \bar{u})^2 - \sum P_i (u_i - \hat{u}_i - \bar{u} + \hat{\bar{u}})^2 \; . \tag{30}$$

As in the previous section, we define

$$\Delta I_2 = \sum P_i (u_i - \bar{u})^2 - \sum P_i (u_i - \hat{u}_i - \bar{u} + \hat{\bar{u}})^2 \; .$$

After some manipulation, we obtain

$$\Delta I_2 = \sum P_i (e_i - \bar{e})^2 - \sum P_i (\hat{u}_i - \hat{\bar{u}} - e_i + \bar{e})^2$$

$$- 2 \sum P_i (b_i - \bar{b})(\hat{u}_i - \hat{\bar{u}}) \; . \tag{31}$$

As in the previous section, the last term is zero if the quantities $\hat{u}_i$ are unbiased estimates of u_i. Failing that, we need an assumption analogous to assumption A.

Assumption C. If the estimates $\hat{u}_i$ have a non-negligible bias, assume that $\sum P_i (b_i - \bar{b})(\hat{u}_i - \hat{\bar{u}}) \leq 0$.

Assumption C should be satisfied if it is predominantly true that wherever the estimated underenumeration is above average, its measurement is also worse than average $(b_i < \bar{b})$, and conversely. Although unverifiable, this is likely to be the case at least for most survey derived estimates of underenumeration. People whom the census finds most difficult to count are, typically, also more difficult to enumerate in the evaluation survey. Note that assumption C is likely to be somewhat stronger than assumption A of the previous section since the former is equivalent to

$$\Sigma P_i b_i \hat{u}_i \le P \bar{b} \hat{\bar{u}} \ ,$$

where the right hand side is very likely to be non-positive. Assumption A on the other hand requires only that the left hand side be non-positive.

Under assumption C we readily obtain

$$\Delta I_2 \ge \Delta I_2' = \Sigma P_i (e_i - \bar{e})^2 - \Sigma P_i (\hat{u}_i - \hat{\bar{u}} - e_i + \bar{e})^2$$

$$= \Sigma P_i e_i^2 - \Sigma P_i (\hat{u}_i - e_i)^2 - P\bar{e}^2 + P(\hat{\bar{u}} - \bar{e})^2. \tag{32}$$

Excepting the P_i , all the quantities on the right hand side of (32) admit unbiased estimates. As before, if one accepts the biased estimates p_i as serviceable (their role being that of a set of weights), we get our first test.

Test 1.　Adjust the census counts if

$$\Delta \hat{I}_2' = \Sigma p_i (\hat{u}_i^2 - 2 \ \text{var} \ \hat{u}_i) - p\hat{\bar{u}}^2 + 2p \ \text{var} \ \hat{\bar{u}}$$

$$= \Sigma p_i [(\hat{u}_i - \hat{\bar{u}})^2 - 2 \ \text{var} \ \hat{u}_i + 2 \ \text{var} \ \hat{\bar{u}}] \ge 0. \tag{33}$$

An analogue to Test 2 of the previous section could immediately be deduced, but it would not be useful. In the present case it is not valid to contemplate adjusting the census count for only some of the provinces since the payment to *any* province depends on the population of *all* provinces.

The above test is designed to ensure that the census counts are adjusted if the adjusted counts are more likely to lead to a more equitable allocation of funds. As in the previous section, we can construct a test which will lead to an adjustment only if the adjusted counts are almost certain to lead to a more equitable allocation (under Assumption C). In order to do so, we need to derive the variance of $\Delta I_2' - \Delta \hat{I}_2'$, where

$$\Delta \hat{I}_2' = \Sigma P_i [(\hat{u}_i - \hat{\bar{u}})^2 - 2 \ \text{var} \ \hat{u}_i + 2 \ \text{var} \ \hat{\bar{u}}] \ .$$

It is easy to verify that

$$\Delta I_2' - \Delta \hat{I}_2' = 2\Sigma P_i [e_i \hat{u}_i - \hat{u}_i + \hat{\bar{u}}^2 - \bar{e}\hat{\bar{u}} + \text{var } \hat{u}_i - \text{var } \hat{\bar{u}}] . \quad (34)$$

The following formula, which is correct to order $1/n$, can then be obtained after some algebra:

$$\text{Var}(\Delta I_2' - \Delta \hat{I}_2) = 4\Sigma P_i^2 [e_i^2 \text{ Var } \hat{u}_i + \hat{\bar{u}}^2 \text{ Var } \hat{\bar{u}} - 2e_i \bar{e} \text{ Cov}(u_i, \bar{u})]$$

$$+ \sum_{i \neq j} P_i P_j e_i e_j \text{ Cov}(\hat{u}_i, \hat{u}_j) - e_i \bar{e} \text{ Cov}(\hat{u}_i, \hat{\bar{u}})$$

$$- e_j \bar{e} \text{ Cov}(\hat{u}_j, \hat{\bar{u}}) + \bar{e}^2 \text{ Var } \hat{\bar{u}}]$$

$$= 4 \text{ Var } [\Sigma P_i (e_i \hat{u}_i - \bar{e}\bar{u})]$$

$$= 4 \text{ Var } [\Sigma P_i (e_i - \bar{e}) \hat{u}_i]$$

$$= 4\Sigma P_i^2 (e_i - \bar{e})^2 \text{Var } \hat{u}_i + 4 \sum_{i \neq j} P_i P_j (e_i - \bar{e})(e_j - \bar{e}) \text{Cov}(\hat{u}_i, \hat{u}_j) .$$

$$(35)$$

The sign of the last term of (35) is not easy to guess. If the estimates $\hat{u}_i$ were based on simple random sampling, $\text{Cov}(\hat{u}_i, \hat{u}_j)$ would be negative and proportional to $e_i e_j$. In that case it is easy to verify that the last term of (35) is negative. In the case of more complex designs this will still hold if the design effects applicable to the covariance terms are approximately equal. Although likely to be true, the negativity of this term cannot be established in general. If the last term is negative, dropping it would lead to a conservative estimate of the variance. At any rate, for small sampling fractions usually encountered in practice, the covariances will be of the order of $1/P$ so that neglecting them is analogous to dropping the finite population correction.

Proceeding as in the previous section, we get the following test.

Test 2. Adjust the census counts if

$$\Sigma P_i [(\hat{u}_i - \hat{\bar{u}})^2 - 2 \text{ var } \hat{u}_i + 2 \text{ var } \hat{\bar{u}}] - 4 \sqrt{\Sigma P_i^2 (\hat{u}_i - \hat{\bar{u}})^2 \text{var } \hat{u}_i} \geq 0 \quad (36)$$

If Test 2 is satisfied, then under assumption C and given moderate sampling ratios in all provinces, the probability is at least 0.975 that the adjusted counts result in less inequity than the unadjusted ones.

Finally, we may note that if $\hat{u}_i$ are not sample estimates (i.e., $\hat{u}_i \equiv e_i$), var $\hat{u}_i = 0$. Hence from (31) we get

$$\Delta I_2 = \Sigma P_i (\hat{u}_i - \hat{\bar{u}})^2 - 2\Sigma P_i b_i (\hat{u}_i - \hat{\bar{u}}). \tag{37}$$

A sufficient condition for (37) to be non-negative is that assumption C holds. In this case one should adjust the census counts whenever that assumption holds.

As in the case of the fixed per capita payment model, much the same development applies under the slightly more general allocation model

$$T_i = C(P_i/P) + b_i ,$$

where C and b_i are constants which do not depend on P_i or p_i.

6.   MEASURING THE CENSUS UNDERCOUNT IN CANADA

In the Censuses of 1961, 1966, 1971 and 1976 the main vehicle used to measure the undercount has been the so-called Reverse Record Check (RRC). The methodology of this vehicle, as used in the 1976 Census, is briefly described below. Additional detail is available in Gosselin, Théroux and Lafrance (1979) and Théroux and Gosselin (1979).

For census purposes, persons are to be enumerated at their usual place of residence. There are, however, two groups which are included in the final census count but which are not counted at their usual residence in Canada. The first group consists of diplomatic, military and other personnel (e.g., merchant marine) living abroad. The second group corresponds to persons who were enumerated on a special form at a temporary address (hotels,

motels, etc.) at the time of the census and who were missed at
their usual residence.

If a sample of persons could be drawn from sources independent of the current census, and if the current address for each
selected person were determined, one could directly search the
current census file. Those who could not be found there would
represent a sample of persons not enumerated at their regular
address. The weighted total based on such a sample conceptually
provides an estimate of the number of persons enumerated away
from their regular residence, plus those missed by the census.
Since the former count is independently known, the latter is
readily estimated.

The sample frame for the 1976 RRC comprised four non-over-
lapping and exhaustive sources *:

a)   persons enumerated at their regular residence in the 1971
     Census (census frame);

b)   intercensal births (birth frame);

c)   intercensal immigrants (immigrant frame); and

d)   persons not enumerated at their regular residence in the 1971
     Census (missed frame).

The "missed frame" in its totality exists only conceptually.
However, as explained above, the RRC of the *previous* census
resulted in a sample of such persons. The first time one carries
out a RRC the frame (d) might not be available. The second time
around it would fail to include persons missed in both of the *two*
preceding censuses. Generally, frame (d) fails to include only
persons missed by each of the n preceding censuses the nth time
around. Since the chances of being missed are significantly
affected by age, the proportion of persons missed by two or more
consecutive censuses is probably of rapidly diminishing significance.

---
*
  Some very small groups are missed by the union of the four
  frames. Their discussion goes beyond the present exposition.

Having selected a sample of persons from the four sources, an extensive effort was mounted to locate their current address (tracing). The most dogged determination resulted in 97% of selected persons being traced, including those for whom it could be confirmed that they died or emigrated since 1971. The failure rate of tracing, however, varied from frame to frame: 2.9, 2.6% 8.5 and 5.1% for the census, birth, immigrant, and missed frames respectively. The untraced persons were eventually treated as non-respondents with an appropriate ratio adjustment used within each frame in a fashion analogous to other sample surveys. It is important to note, however, that persons listed in frames for which tracing was more difficult had a higher chance of being missed by the census, suggesting that those not traced were likely missed at an even higher rate. Thus it is very probable that, even after weighting for this "non-response" separately for the different frames, the RRC estimates have a higher bias (under-estimating the underenumeration) whenever the underenumeration itself is higher (assumptions A and C).

After completion of tracing, a thorough search of census records was carried out to determine whether the selected persons were enumerated at their respective traced addresses. In cases where the tracing indicated that the selected person died prior to the census, a search of the death register was carried out. All persons not found in either the census or the death register were further followed up. The objectives of this follow-up were twofold:

a)   to confirm that the traced address was correct or to obtain other addresses where the person may have been enumerated in the census;

b)   to obtain at the same time a number of census characteristics for the persons concerned which, should they turn out to have been missed by the census, would enable us to provide basic tabulations on the characteristics of persons missed by the census.

Persons who could not be traced in the follow-up operation
were added to the untraced (non-response) category. This raised
the untraced proportion to 4.8% overall, but still leaving considerable differences as between the four frames: 4.0, 7.6, 10.6 and
9.6% in the census, birth, immigrant, and missed frame, respectively. Undoubtedly, the approach of following up all those traced
persons who could not readily be found as enumerated in the census
is correct, but it also contributes to the eventual estimates of
underenumeration being conservative.

Table 1 shows the (unweighted) number of selected persons in
each final status category (enumerated, missed, deceased, emigrated
and tracing failed) for each of the four frames. The overall proportion of missed persons is higher than the final estimate for
Canada primarily because of two reasons: the percent "missed" is
actually the proportion of persons not enumerated at home (i.e.
unadjusted for persons enumerated at a temporary residence); and
the sample design used included oversampling of young males in the
census frame (where underenumeration was expected to be high), so
the unweighted proportion is an overestimate. Weighted and adjusted
estimates of the undercount for various population and household
groups are available in Théroux and Gosselin (1979).

It should be noted that the RRC does not measure overenumeration, thus no estimates of *net* underenumeration are available.
Overenumeration is, however, judged to be small.

7. SOME NUMERICAL RESULTS

Table 2 provides the basic results of the RRC by province,
while those same results are given in Table 3 by age and sex.
Breakdowns by other classifications are available in Théroux &
Gosselin (1979), including the proportion of households missed by
their characteristics. The sample size used was slightly over
33,000 persons nationally, from the four frames combined.

For purposes of allocating fixed per capita amounts
to the ten provinces, and under the appropriate assumptions,
the overall Tests 1 and 3 both indicate that adjustment

Table 1. Number of Cases in Each Final Status Category by Frame

Result	Census Frame		Birth Frame		Immigrant Frame		Missed Frame		All Frames	
	No. of cases	%	No. of cases	%	No. of cases	%	No. of cases	%	No. of cases	%
Enumerated	24,890	89.2	2,871	88.0	869	74.3	584	76.1	29,214	88.2
Missed	645	2.3	66	2.0	80	6.8	33	6.9	844	2.5
Deceased	978	3.5	47	1.4	1	0.1	36	4.7	1,062	3.2
Emigrated	272	1.0	29	0.9	95	8.1	20	2.6	416	1.3
Tracing failed	1,128	4.0	249	7.6	124	10.6	74	9.6	1,575	4.8
TOTAL	27,913	100.0	3,262	100.0	1,169	100.0	767	100.0	33,111	100.0

Table 2. Estimated Population Undercoverage By Province [+]

Province	Population Undercoverage Rate		Number of Persons Missed		1976 Census Population Total	1976 Census Population[2] Adjusted for Undercoverage[1]	1976 Census Distribution[2] (%)	1976 Census Distribution[2] Adjusted for Undercoverage (%)
	Estimated Rate (%)	Standard Error(%)	Estimated Number[2]	Standard Error				
Newfoundland	1.10*	0.39	6,200*	2,200	557,725	563,915	2.43	2.41
Prince Edward Island	0.38*	0.25	445*	292	118,230	118,675	0.52	0.51
Nova Scotia	0.86*	0.34	7,215*	2,870	828,570	835,785	3.61	3.57
New Brunswick	2.16	0.37	14,170	2,615	677,250	692,210	2.95	2.96
Quebec	2.95	0.25	189,655	16,225	6,234,445	6,525,095	27.19	27.45
Ontario	1.52	0.17	127,155	14,170	8,264,465	8,391,625	36.05	35.85
Manitoba	1.07*	0.33	11,080*	3,455	1,021,510	1,032,590	4.46	4.41
Saskatchewan	1.33*	0.34	12,440*	3,175	921,320	933,760	4.02	3.99
Alberta	1.49	0.26	27,790	4,855	1,838,035	1,865,830	8.02	7.97
British Columbia	3.13	0.31	79,775	7,770	2,466,605	2,546,385	10.76	10.88
All Ten Provinces	2.04	0.10	476,715	23,890	22,928,155	23,404,870	100.00	100.00

[+] Excluding Yukon and Northwest Territories

[*] Estimate with high relative standard error

[1] The standard error figure for the corresponding estimated number of missing persons also applies to these totals

[2] The marginal totals or percentages may differ slightly from the sum of individual totals or percentages due to rounding

Table 3. Estimated Population Undercoverage By Age and Sex [+]

Age and Sex	Population Undercoverage Rate		Number of Persons Missed		1976 Census Population Total	1976 Census Population[2] Adjusted for Undercoverage[1]	1976 Census Distribution[2] (%)	1976 Census Distribution[2] Adjusted for Undercoverage (%)
	Estimated Rate (%)	Standard Error(%)	Estimated Number[2]	Standard Error				
Male 0-4	2.53	0.46	22,950	4,240	884,750	907,695	7.75	7.76
5-14	1.14	0.21	24,520	4,590	2,123,480	2,148,005	18.60	18.35
15-19	1.93*	0.48	23,480*	5,995	1,192,600	1,216,075	10.45	10.39
20-24	5.99	0.52	67,680	6,270	1,062,345	1,130,025	9.31	9.66
25-34	3.64	0.46	68,670	9,035	1,816,765	1,885,430	15.92	16.11
35-44	2.33*	0.48	31,300*	6,580	1,310,970	1,342,265	11.48	11.47
45-54	1.63*	0.41	20,310*	5,240	1,223,510	1,243,820	10.72	10.63
55-64	1.28*	0.34	12,000*	3,260	926,530	938,530	8.12	8.02
65 & over	1.90*	0.44	16,925*	3,990	874,430	891,355	7.66	7.62
Total	2.46	0.17	287,830	19,990	11,415,370	11,703,200	100.00	100.00
Female 0-4	2.07	0.36	17,770	3,180	839,645	857,420	7.29	7.33
5-14	1.26*	0.27	25,865*	5,640	2,025,445	2,051,305	17.59	17.53
15-19	2.05*	0.51	23,990*	6,060	1,146,135	1,170,135	9.96	10.00
20-24	4.62	0.48	51,605	5,660	1,064,770	1,116,375	9.25	9.54
25-34	2.03	0.38	37,090	7,045	1,791,680	1,828,770	15.56	15.63
35-44	0.72*	0.24	9,335*	3,085	1,278,930	1,288,260	11.11	11.01
45-54	0.81*	0.38	10,135*	4,820	1,244,735	1,254,865	10.81	10.72
55-64	0.58*	0.25	5,815*	2,495	995,295	1,001,110	8.65	8.56
65 & over	0.64*	0.38	7,280*	4,305	1,126,150	1,133,430	9.78	9.69
Total	1.61	0.10	188,885	12,085	11,512,785	11,701,670	100.00	100.00

[+] Excluding Yukon and Northwest Territories
[*] Estimate with high relative standard error
[1] The standard error figure for the corresponding estimated number of missed persons also applies to these totals
[2] The marginal totals or percentages may differ slightly from the sum of individual totals or percentages due to rounding

is appropriate.   This is also the case for Test 2 (the
term-by-term version of Test 1) for every province.   Test 4 is
positive only for the four largest provinces (Ontario, Quebec,
British Columbia, Alberta) and for New Brunswick (because of the
high value of $\hat{u}_1$).   Test 1 is very "comfortably" satisfied in
that the sample could be reduced by a factor of 36 and adjustment
would still be indicated (i.e., even with an overall sample size
of less than 1,000 distributed over ten provinces!).   Test 3 would
be positive even with the sample size reduced by a factor of 4
(less than 8,000 selected persons nationally).

In the case of the fixed total cost case, Tests 1 and 2 are
both positive.   Test 2 would be positive even with a 60% reduction
of the sample size actually used.   Test 1 would be positive with
a sample size one-fifth as large as actually used, i.e., with
about 6,600 persons.   If one were to construct a term-by-term
equivalent of Tests 1 and 2 (as in the case of the fixed per
capita allocation), the corresponding tests would all be positive
in the case of Test 1, except for New Brunswick (whose estimated
underenumeration is very close to the national average).   In the
case of Test 2, *none* of the term-by-term equivalents are positive.

It is interesting to consider one of the major Canadian
federal-provincial transfer payment schemes.   Omitting the detailed
analysis of the legislation, it turns out that the relevant formulae
to use are provided by the fixed total payment tests, but extending
the summation over only seven provinces (Newfoundland, Prince
Edward Island, Nova Scotia, New Brunswick, Quebec, Manitoba,
Saskatchewan), while still retaining the national average $\hat{u}$.
Tests 1 and 2 of the fixed total cost model are still positive
when applied in this fashion, but Test 2 just barely so.   (An
overall proportional reduction of the sample by even 1% would
render the sign of Test 2 negative.)

8.   OPTIMUM ALLOCATION OF SAMPLE

The question naturally arises how to allocate the sample designed to measure the undercount proportions u_i so as to make the adjustments with maximum precision.

If the objective is to maximize the reduction in inequity after the adjustment, then one wants to maximize the test statistic in Test 1 (in both the fixed per capita allocation and fixed total cost cases).

Let us assume, for the sake of simplicity, that the variance of $\hat{u}_i$ decreases inversely proportionately with n_i. Then denote

$$\text{var } \hat{u}_i = \frac{v_i}{n_i}$$

where v_i includes all applicable design effects. In the case of the fixed total cost allocation we will neglect the effect of $\text{var } \hat{u}$ in the formulae (33) and (36). (Its numerical impact is extremely small and most unlikely to influence optimal sample allocation.)

It is now easy to verify that the optimal allocation (with fixed overall sample size n) which maximizes the test statistic in Test 1 is

$$n_i = n \frac{\sqrt{p_i v_i}}{\sum \sqrt{p_j v_j}} . \qquad (38)$$

The same result applies in the case of both fixed per capita and fixed total cost allocations. If the differences in the provincial design effects and in the proportions $\hat{u}_i$ are not large, a reasonable approximation is to allocate the sample in proportion to the square roots of the provincial populations. This provides a reasonable compromise between designing the sample for obtaining the best national estimate of underenumeration (n_i proportional to p_i) and obtaining equal reliability for every u_i (n_i equal for all i).

If we want to maximize the chance of being able to adjust the census counts with *security*, we would want to allocate the sample to maximize the value of Test 3 in the case of fixed per capita allocation and of Test 2 in the case of fixed total cost allocation. It can be verified that the following allocation

$$n_i = n \frac{a_i}{\Sigma a_i} \tag{39}$$

would achieve this second objective, where

$$a_i^2 = tp_i^2 x_u^2 v_i + p_i v_i \ ,$$

and,

$$x_i = \hat{u}_i \quad \text{in the fixed per capita allocation case}$$

$$= u_i - \hat{u}_i \text{ in the fixed total cost allocation case}$$

and t is the positive root of the equation

$$t^2 = \frac{n}{\left(\Sigma \ \dfrac{p_i^2 x_i^2 v_i}{a_i} \right) (\Sigma a_i)} \ .$$

The right hand side above involves t , but the equation can be solved iteratively, starting with an initial value of $t = 0$. In fact the convergence appears to be very rapid (the first computed value of t in several examples was correct to three significant digits).

If t were equal to zero, the allocation of (39) would reduce to that of (38). For large values of t (and if the values x_i and v_i do not vary much among the provinces) the allocation would be close to proportional to the provincial populations p_i .

The following example illustrates the differences between the two allocations. Let

x_i = 0.01 (approximately the average value of $|\hat{u}_i - \bar{\hat{u}}|$
          in the Canadian Reverse Record Check; see Table 2).

and

v_i = 0.02 (approximately the average value of $\hat{u}_i(1 - \hat{u}_i)$
          in the Reverse Record Check; see Table 2).

Then $t = 7.14 \times 10^{-4}$ when n = 33,000 (the approximate Reverse
Record Check sample size). Table 4 below illustrates the proportion of the sample to be allocated to each of ten provinces
when n_i is proportional to $\sqrt{p_i}$, a_i or p_i .

Table 4. Alternative Sample Allocations

Province	n_i proportional to			Rev. Rec. Check (actual)
	$\sqrt{p_i}$	a_i	p_i	
Newfoundland	0.057	0.039	0.024	0.053
Prince Edward Island	0.026	0.016	0.005	0.051
Nova Scotia	0.070	0.050	0.036	0.051
New Brunswick	0.063	0.044	0.030	0.052
Quebec	0.191	0.238	0.272	0.213
Ontario	0.220	0.307	0.360	0.224
Manitoba	0.077	0.057	0.045	0.056
Saskatchewan	0.073	0.053	0.040	0.055
Alberta	0.104	0.087	0.080	0.093
British Columbia	0.120	0.109	0.108	0.152

As expected, the allocation which is proportional to a_i falls
between the allocations proportional to p_i and $\sqrt{p_i}$.

Table 5 shows the realized values of Tests 1 and 2 with the
parameters as indicated above (but with the value of var $\hat{u}$
neglected in the case of Test 2).

Table 5. Realized Values of Tests 1 and 2 Under Alternative
          Allocations (Fixed Total Payment Model)

Test value	n_i proportional to			Rev. Rec. Check (actual)
	$\sqrt{p_i}$	a_i	p_i	
Test 1	2,085	2,068	2,015	2,076
Test 2	1,293	1,339	1,301	1,296

As usual, the optima appear to be broad. The allocation proportional to $\sqrt{p_i}$ maximizes, of course, Test 1, while the allocation
proportional to a_i maximizes Test 2. There is not much to choose
among them. The actual Reverse Record Check allocation was close
to the $\sqrt{p_i}$ allocation, but adjusted so as to provide acceptable
estimates of underenumeration for every province (this resulted in
the significant upward sample size adjustment for Prince Edward
Island) and to take into account past values of $\hat{u}_i$ and
v_i (this resulted in the significant upward adjustment in British
Columbia where both $\hat{u}_i$ and v_i were forecast, and turned out to
be, significantly higher than elsewhere).

    In concluding this section, it is perhaps worth once again
pointing out the somewhat curious phenomenon that the optimal
sample allocation which maximizes Test 1 (the estimated reduction
in the inequity measures between the unadjusted and adjusted
allocations) is different from the optimal sample allocation which
maximizes Test 2 (which, roughly speaking, ensures that the
adjusted census counts actually decrease the inequity with a high
probability).

9. CONCLUDING REMARKS

    As emphasized in the introduction, the present paper does
not purport to provide a tool for the definitive determination of

the answer to the question whether the census counts should be adjusted - even for purposes of fund allocation and even if the appropriate tests are positive. The tests deal only with the issue of equity, and then only for formulae which depend on census counts in a direct and untransformed fashion. This is not the case regarding the Canadian Fiscal Arrangements Act payments, except in census years. In all other years the formula uses intercensal population estimates. It is yet to be determined whether suitable intercensal population estimates can be prepared starting with the adjusted census counts. Also, the validity of Assumptions A and C can only be assessed subjectively.

Some legislated fiscal transfers depend on counts for subgroups of the population (e.g., students, or children in low income families). While one of the advantages of the Reverse Record Check methodology, as described in the previous section, is its ability to produce breakdowns of the number of persons missed by the census by their characteristics (as such it is different from, for example, analytic estimates of the underenumeration rate), the currently used sample sizes may not be adequate for adjusting the census counts for relatively small subgroups of the population.

Intercensal population estimates are also used in the weighting of current household survey data through ratio estimates. These survey estimates may, in turn, be used to determine large transfers of funds. This is the case, for example, in the Canadian Labour Force Survey whose estimates of the unemployment rate (in each of some 50 regions) determine unemployment insurance benefits. Since unemployment is significantly higher in the category where the census undercount is also worst, i.e., young males (see Table 3), it is conceivable that if the ratio estimate could be based on adjusted intercensal population estimates, the benefits to be received by some Canadians could be affected. However, the appropriate adjustment depends not only on the methodology of producing adjusted intercensal estimates but also on being able to adjust (even for the census year) by province *and* age-sex

groups, something which the current RRC sample size may not support.
Research is currently underway to investigate the feasibility of
producing synthetic adjustment factors from separate sets of
estimates of underenumeration by province and by age-sex groups.

Another problem raised by Nisselson (1979) is that by
adjusting the census counts we may remove an incentive which
otherwise might motivate national associations of difficult to
enumerate groups, such as certain minorities, to urge their
members to cooperate with the census.

The above is only a very partial list of policy issues
which have to be addressed before a decision can be made on
whether the census counts should be adjusted and, if so, for what
purpose.

REFERENCES

Gosselin,J.F., Théroux, G. and Lafrance, C. (1979). Reverse
      record check - evaluation report. Statistics Canada,
      Ottawa.

Jabine, T. (1976). Equity and the allocation of resources based
      on sample data. Statistical Working Paper 1. Office of
      Federal Statistical Policy Standards, Washington, D.C.

Nisselson, H. (1979). Comment on'Information and allocation: two
      uses of the 1980 census'by Nathan Keyfitz. *The American
      Statistician*, May 1979, 50-52.

Office of Federal Statistical Policy Standards (1978). Report on
      statistics for allocation of funds. Statistical Working
      Paper 1. Office of Federal Policy Standards, Washington,
      D.C.

Théroux, G. and Gosselin, J.F. (1979). Reverse record check -
      Basic results on population and household undercoverage
      in the 1976 census. Statistics Canada, Ottawa.

NON-SAMPLING ERRORS: GENERAL DISCUSSION

W. MADOW, *Committee on National Statistics*:   This isn't
really a question, but I thought that it might supplement Dalenius's
very interesting comments on some of the early history of non-
sampling error research.   In 1945, Fred Stephan and I tried to
begin work in this area and we felt we were failing because we
didn't feel that we had arrived at an appropriate model for non-
sampling error.   Some time later, Billy Hurwitz, and I am sure it
was with Morris Hansen, came up with the very bright idea that an
interviewer created a cluster of up to a thousand, as in the census, so
that if the interviewer was creating a non-sampling error it could
have an enormous effect on the variance.   This was one of the
precursors of the census getting up the nerve to go on a mail basis
in the 1950 census.

I am making these remarks to say that it isn't as simple
as saying you want to do response error research or total sampling
error research.   I do not know that we have appropriate models in
mind for errors of response, models that take into account the
correlations and so on that exist, and when this is done on a
within survey basis we will collect very little data that will
help us.   When it is done by comparing with outside records, the
biases are often enormous and no one knows how to use the information.   I think the bureau has done more in the way of nonsampling
error research than probably anyone else and certainly much more
than those not engaged in sampling, but I still don't think we
know what are the appropriate models and I doubt very much that
they will be general.   I think they will depend very much on the
subject matter and the survey technique, and I think therefore,
that this problem is difficult.   I am delighted with the development of models but maybe we will have better approaches than

77

we've had so far.

      T. DALENIUS, *Brown University*: I don't interpret what you
have said as a question. I agree with you entirely but my emphasis,
or what I tried to say, is that I think we should do more. Even
if we don't have a model now which we can recommend, we should at
least emphasize that there is a need for one. You can't deal with
measurement problems isolated from sampling. They have to be
somehow dealt with in conjunction, and we should pay a little more
attention to this.

      J.N.K. RAO, *Carleton University* (to H.O. Hartley): My
question is related to both Tore Dalenius's talk and Dr. Hartley's
talk. Tore raised a good point in that the sampling text books do
not give adequate treatment to response errors and other non-sampling errors. I think one of the reasons for not giving adequate
treatment of response errors is that the papers which have been
written are so dry and full of pedestrian algebra, it's pretty
hard to put that in a text book. Let me say that there is nothing
wrong with the papers. It is just that the approach used is not
terribly exciting for the students to be taught in the university.
      I think one way to overcome this problem is to formulate
the response error problems in term of linear models. There is a
lot of literature on linear models now, like the minque theory and
what Dr. Hartley has suggested to us. I am not saying that we
have solved all the problems of model formulation - there are
certainly problems there - but you could do a reasonable model
formulation. You link up the sample design with the linear model
formulation for the response errors.
      Now, mind you, this has got nothing to do with the so called
super population models for estimating sampling errors - that is a
different story. Whether we should include that in the text books
or not we can discuss tomorrow in the session on superpopulation
models! But I think even the hard core conventional samplers agree

that there are models for response errors. I think if you avoid
all the irritating finite population corrections and bring in a
model nicely and link it up with the design, you could probably
explain this in the beginning of the text books making students
aware that there is something called response error and a way to
handle the problem.

Also for the problem Dr. Hartley raised regarding the design
of the survey, we have two aspects. One is the design for sampling
errors and the other one is the design for non-sampling errors. I
think we survey samplers have been very good in designing for
sampling errors, but have completely ignored the problems of
designing for non-sampling errors. Here again the literature on
variance components might help us in the arriving at suitable
designs. For instance, my good friend R.L. Anderson has spent
many years on optimal designs for variance components. We can
learn something from Andy on this and combine the two together.

H.O. HARTLEY, *Duke University*: I would like to respond to
this as a co-author of our dry paper, Jon. I would like to point
out that the sample survey practitioners, who come perhaps originally from theory or methodology, rightly point out to us that the
area of non-sampling and measurement errors is full of practical
considerations that we probably don't know too much about. As an
apology for failing to teach some of these considerations that
Tore was asking for, I would like to point out here that as an
academic teacher I would find it very hard to examine students on
points that involve non-statistical questions, such as how you
should train interviewers, which are very important organizational
problems. Failure for the academic types to go into this is due
to the fact that it is possibly unfair to the students to set
examination questions in these areas. It's extremely difficult,
and so we face up to the fact that you have to learn a lot when
you go out into the fierce world of reality.

T. DALENIUS, *Brown University*: What Professor Hartley just
said reminded me that I missed one point in my paper. We not only
need new text books, we need new professors too!

R.L. ANDERSON, *University of Kentucky* (to H.O. Hartley):
I had to let Jon Rao talk about variance components before I did
so that it wouldn't appear too obvious what was going on. It is
very striking to me that the problems that Dr. Hartley is facing
here are the same type of problems that I face in trying to design
the type of experiments that I talked about yesterday. It is very
interesting to see that his problems of estimation are very similar
to the ones that we face in getting variances and covariances of
the components that you're estimating and then coming up with some
sort of an estimation procedure. I think that once we get away
from the finite population problem, which I think is so trivial in
comparison with many of the others, the two fields of experimental
design and sample surveys will overlap. I think that many of us
who work in both areas will appreciate that.

I want to make a few other comments. One is about students.
In the last sampling class I taught, I ran a survey in Lexington.
It was a telephone survey which was a complete flop because 50%
of the people could not be found. I don't know how you put that
into your model HOH, I haven't figured that one out yet. But also,
of those that we interviewed, two-thirds of them were the ladies
of the house rather than the men of the house and adjusting for
that difference is another problem, unless you do some sort of a
random selection of your telephoning and that is integral in this
problem.

But the main thing that comes out of this is that the
students are so appreciative of the fact that they did something
that their text books hadn't told them about, and they realize
that if they ever went out into the real world they were going to
have to think about a lot of things we hadn't talked about in
lectures. I think that this is an absolutely essential part of

any sampling course - to make students become appreciative of
these other problems, which are probably going to kill them if
they don't take care of them.

Now, one more question that I would like to put into this
game. How do you set up in a sampling program of this type that
you've done HOH, to take care of the fact that the person you are
interviewing will give a different answer on different occasions
or to different interviewers? That is a problem which has worried
me very much because I can see this happening when we run telephone interviews. Certain of the interviewers can get people to
answer differently from what other interviewers would. That is
not an interviewer bias, it is the fact that the person being
interviewed will respond differently to different interviewers.
There is an interaction taking place here which I don't know how
to take care of and I think it needs to be addressed.

H.O. HARTLEY, *Duke University*: The model that I have
discussed has a respondent error that is allowed to vary from
occasion to occasion. It does not have an interaction term with
the interviewer directly. However, if you look at it carefully,
if you allow a respondent error, it can include the interaction
between an elementary interviewer effect and a respondent effect.
It cannot include interaction between the systematic interviewer
error and the elementary error effect, but the δ_{rt} as it is in
our model does include at least that effect.

B. BAILAR, *U.S. Bureau of the Census* (to I. Fellegi): I
am absolutely fascinated by this topic as you might imagine. We
used to be able to say that as far as the question of equity was
concerned, this wasn't a statistical problem, but a political
problem, so we didn't have to think about it. Ever since I read
Ivan's paper, it has made our statistical problem somewhat larger.
One thing I did like about this was that the problems of the bias
in the way that we estimate our undercount, are taken into

consideration in some respects. What Ivan said about the fact that
in the areas where we have the largest problem with the undercount,
we probably do the poorest in estimating it, is I would imagine
quite true. This is taking into account the fact we are probably
going to be making estimates of our undercount per State and for
the Nation. I imagine that if we have to get into the business of
adjusting - and I don't want to sound like I am against it or for
it, I'm not either - I can't imagine that we will be able to stop
at the State level. This means that we will not have good sample
estimates to use at a level below that, so in case we do go below
that, we will be making use of some kind of synthetic estimate, a
regression estimate or another kind of estimate, and I am not
sure whether the bias in those estimates will still continue to
follow that same kind of pattern. I don't have any evidence about
that, one way or the other.

Everyone seems to be very eager for adjustments, and yet
with a lot of the evidence that is available it looks like there
will be many more losers than winners. I am sure that as soon as
that becomes known there will also be put into some kind of legislation a whole harmless clause that no one can be a loser and that
you will at least have to be a winner. Maybe those tests will have
to also be adjusted so that they will only go in one direction.

Then it seems to me there is another question of equity
that comes up. People have the feeling that they're losing mil-
lions of dollars because the Census has an undercount in it, so
they want this kind of adjustment as quickly as possible. But to
make a good undercount estimate takes a long time. We are currently
being sued by the city of Detroit which wants us to give undercount
estimates at the same time that we give the population counts to
the President on January 1, 1981. If we have to give an estimate
of undercount at that time, it's not going to be a good estimate.
We also have to give the population counts for each State by
April 1 for redistricting purposes. We feel that the best timing
that we can do to give a good estimate of the undercount (and

even then it isn't as good as we can do if we have more time)
would be by September of 1981. To make it even a better one,
which means using administrative records, would make it go into
'82 probably late in the year. So it seems to me there is another
equity problem. If you are going to adjust, should you do it well
and take a long time or should you do it poorly but quickly?

        I. FELLEGI, *Statistics Canada*: Well you are raising some
absolutely relevant questions, as of course an insider would.   I
don't know if I should answer fully now, but I will at least
address the questions. First of all, so far we in Canada are
fortunate that our legislators, at the Federal level at least,
have written laws which require fund allocation only explicitly
from the Federal to the Provincial level. I know in the U.S. the
Federal fund allocations go down to something like 40,000 local
areas, depending on the particular legislation, directly from the
Federal Government to the local areas. This is one instance at
least where the strong Provincial autonomy, or spirit of strong
Provincial autonomy, that prevails in Canada works in our favor.
There would absolutely be a rebellion of Provincial premiers, if
the Federal Government were even thinking of directly allocating
money to a sub-Provincial level of government. So thank God for
small mercies. I know that this doesn't help you but it helps me.
The problem is very serious and, without trying to be facetious,
trying to estimate the undercount for some 40,000 local areas,
would be a pretty daunting problem, and obviously the presentation
would have to be re-thought.

        One of the surprising things that came out of my own study
(surprising to me) is that pretty thin samples would still provide
estimates that would guide you strongly to adjust if there are in
fact sufficient differences between the areas to which the allocation is to be made.

        I address a second problem, where again we are more fortunate
I think, in our legislators' serendipity (not wisdom necessarily),

than in the U.S. What happens in the Federal fund allocations
that are legislated in Canada, is that legislation prescribes an
initial payment, a sort of downpayment, which is based on the best
available estimate, with the clear understanding that when better
estimates come about, the allocation will be re-computed. There
may be, in fact there were, instances where money was re-claimed
from the provinces - after some heated correspondence needless to
say - but it was re-claimed and re-paid. So the timing of course
is an absolute issue and it boils down to somehow trying to pene-
trate the legislators' minds, or in fact the civil servants who
advise the legislators, about the fact that that kind of an escape
hatch (being able to re-compute later when better data become
available) is an absolute necessity. It is a bind but it is a
no-win situation without it. Everybody loses, including first of
all the statisticians, but also the Federal Government and the
Provinces.

Finally, addressing the question that you raised about
possibly some provinces or some local governments being losers and
the potential for re-writing legislation, well we are trying to
deal with that problem by in fact not retroactively setting the
rules of the game, but prospectively. Fortunately just about 1982,
the present fiscal arrangement act expires and new ones will have
to be put in their place. So we are in a position to advise the
Department of Finance and the Provinces (those are the players in
the scene) before there are any census data, let alone evaluated
census data, as to what the rules of the game would be if we were
to adjust, and what might be the legislative implications if they
want to take those legislative implications into account. In
other words, if they want to re-write the legislation to eliminate
the possibility of losers, well, that's their prerogative.

III CURENT SURVEY RESEARCH ACTIVITY

TWENTY-TWO YEARS OF SURVEY RESEARCH AT THE
RESEARCH TRIANGLE INSTITUTE: 1959-1980

David L. Bayless

Research Triangle Institute

Research Triangle Institute is a not-for-profit
organization performing research in many disciplines
for government, industry and other clients throughout
the United States and abroad. The paper describes
the organization of the survey research centers at
the Research Triangle Institute, survey research
projects and selected characteristics of the staff
that conduct survey research. In most projects, the
survey research functions of developing the measurement instruments, data analysis and interpretation of
the survey findings for policy implications, and final
report writing are carried out by the professional
staff of the Institute. The projects are assessed by
tabulating the number of projects, percentage of
projects, average cost per project (in dollars), total
cost (in dollars), and the percentage of the total
project cost for each category of each project
characteristic.

In recent years, large research projects
included studies on drinking habits and patterns
among U.S. teenagers, fire protection, and on drug
abuse. The paper also discusses the design, findings
and policy implications of two large national surveys,
the Safe School Study and the National Longitudinal
Study of the High School Class of 1972.

1. INTRODUCTION

This paper is an assessment of the survey research projects

conducted at the Research Triangle Institute (RTI) for the past

twenty-two years from its inception in 1959 to the current year

87

of 1980. The survey research projects have been classified into two categories: *methodological projects* and *survey projects*. A project that assesses the quality or cost-effectiveness of a sample survey technique is termed a *methodological project*. A project that implements at least one of the sample survey functions of questionnaire development, sampling, data collection, data processing, and data analysis is termed a *survey project*.

Survey research is carried out primarily through observation and measurement of populations as they exist, rather than by controlled trials or experiments. Research projects which are not experiments, but which select and measure samples from populations of interest and use sample survey techniques to provide the requisite data or evaluate the quality of sample survey techniques, are called survey research projects (Horvitz, 1979).

The project characteristics assessed in this paper are the type of project (*survey* and *methodological*) and type of agency sponsoring the project. All of the projects assessed used a probability sampling design. These characteristics will be analyzed for all the 245 *methodological* and *survey* projects completed by RTI in the past twenty-two years. In addition, discussion is provided that assesses the changes in the survey design or data collection methodologies of the survey projects over time. Some of the data collection issues in the transition from the face-to-face personal interviews to the telephone interviews are addressed.

Characteristics that will be analyzed for the 212 *survey projects* are the year the project was completed, subject area, and the geographical coverage (nationwide, statewide, and local).

Other project characteristics are indeed of interest, such as response rate, cost per respondent, cost per survey design component, success of methods to improve response rate, methods used to assess nonsampling errors, and sources of bias in the survey results. The resources devoted to the preparation of this paper did not, however, permit the assessment of these important characteristics. In addition, the projects could have been categorized by type of survey design such as: cross-sectional, longitudinal, or surveys conducted on one occasion or multiple occasions, and intended uses of the survey results (descriptive versus analytical). The information available for each project did not permit classification for these characteristics.

The paper first describes the organization of the survey research at the Research Triangle Institute and selected character-istics of the survey research staff. The paper then describes and interprets the results of the assessment of the survey research projects in terms of the project characteristics. The purpose, summary of the methodology, certain results, and policy implica-tions of selected survey projects are also described. The projects are assessed by tabulating the number of projects, percentage of projects, average cost per project (in dollars), total project cost (in dollars), and the percentage of the total project cost for each category of each project characteristic.

2. ORGANIZATION OF SURVEY RESEARCH AT RTI

The Research Triangle Institute is a not-for-profit organ-ization performing research in many disciplines for government, industry, and other clients throughout the United States and abroad. Current research volume is more than $40.9 million annually. RTI was incorporated at the end of 1958 by joint action of the University of North Carolina at Chapel Hill, Duke University in Durham, and North Carolina State University at Raleigh. RTI is a separately-operated affiliate of the three schools with its own staff and facilities. Close working ties at many levels are maintained with the founding universities. RTI's staff of more than 1,100 occupies 323,500 square feet in 14 laboratory and office buildings on a campus of 180 acres in the 5400-acre Research Triangle Park centered near Raleigh, Durham, and Chapel Hill. A separate organization, the Research Triangle Foundation, promotes

and develops the resources of the Park which has become the site
for over 30 other governmental, industrial, and university-related
research installations. RTI research operations are organized
into major groups covering the social sciences, statistical
sciences, chemistry and life sciences, and energy, engineering
and environmental sciences.

Survey research projects at RTI are generally carried out
jointly by the social sciences and statistical sciences centers.
The associated survey tasks of designing and selecting the sample,
developing and implementing the measurement process or data
collection procedures, editing, coding, processing, and analysis
of the data are basically statistical. In most survey projects,
the social sciences centers carry out the functions of developing
the measurement instruments, analyzing the data, interpreting the
survey findings for policy implications, and final report writing.

The social sciences consist of four centers:
1. Center for Population and Urban-Rural Studies
2. Center for Educational Research and Evaluation
3. Center for the Study of Social Behavior
4. Center for Health Studies
and the statistical sciences consist of four centers:
1. Sampling Research and Design Center
2. Survey Operations Center
3. Computer Applications Center
4. Statistical Methodology and Analysis Center.
Each center has a director who reports to a vice president and
has overall responsibility for the management of the center.

Table 1 contains the number of professional and support
level staff in the statistical and social sciences centers. Over
the years, the growth of staff at RTI in survey research has increased considerably. In 1962, seven people were employed in the
sample survey section of what was then called the Statistics
Research Division. Today, a total of 361 people are conducting

survey research projects and of the 58 professionals who have a
doctorate, 21 are in the Statistical Sciences Group.

Table 1. Research Triangle Institute Number of Professional Staff
By Educational Degree and Support Level Staff of the
Statistical and Social Sciences

(February 29, 1980)

Educational Degree	Statistical Sciences	Social Sciences	Group I Total
Doctorates	21	37	58
Masters	63	31	94
Bachelors	70	33	103
Total Professional	154	101	255
Total Support	80	26	106
Total	234	127	361

3. ASSESSMENT OF THE SURVEY RESEARCH PROJECTS

3.1 All Survey Research Projects

During the twenty-two year period from 1959 to 1980, a
total of 245 projects costing a total of $66 million have been or
are being conducted by survey research staff at Research Triangle
Institute. Table 2 gives the number of survey research projects,
their percent of the total number of projects, and total and
average costs for both methodological and survey projects. Thir-
teen percent of RTI's survey research projects have been "methodological." These projects represent two percent of the total cost
of the survey research projects during the past twenty-two years.

The first methodological project was an evaluation of the
1959 Census of Agriculture. This led to other methodological
projects in agriculture, one of which was a study of multiframe
sample designs. Other methodological projects were for research

on nonsampling errors in sample surveys, and a methodological
study completed in 1969 evaluating the randomized response technique to estimate abortion rates. A current methodological study
at RTI is evaluating the content and methods of four national
health surveys for application to local health planning areas.
Such methodological studies are an effort on the part of RTI
survey researchers to improve the quality of the statistical
surveys conducted for RTI's clients.

    A large percentage of survey research projects at RTI in
the past twenty-two years have been funded by the Federal Government. Table 3 characterizes the survey research projects by the
type of sponsoring agency. Forty-four percent of the total survey
research projects were funded by Federal Government agencies,
which amounted to 89 percent of the total funding for survey
research projects during RTI's first 22 years. State and local
government agencies were responsible for 27 percent of the projects,
amounting to 6 percent of the total revenue of all projects. The
other survey research projects funded by universities, foundations,
and commercial agencies accounted for the remainder of the projects
(29 percent), and a small share of total funding for the survey
research projects (5 percent).

Table 2. Number of Survey Research Projects, Percent of Total
          Projects, and Average Cost per Project by Type of Project

Type of Project	Number of Survey Research Projects	Percent of Total Projects	Average Cost Per Project (In Dollars)	Total Cost of all Projects	Percent of Total Cost: of All Projects
Methodological	33	13	45,490	1,501,156	2
Survey	212	87	305,136	64,688,871	98
Total	245	100	270,163	66,110,027	100

Table 3. RTI Survey Research Projects the Past Twenty-Two Years
          By Type of Sponsoring Agency

Type of Sponsoring Agency	Number of Survey Projects	Percent of Survey Projects	Average Cost Per Survey Project in Thousands of Dollars	Total Cost of Survey Project in Thousands of Dollars	Percent of Total Cost of Survey Projects
Federal Government	106	44	554	58,775	89
State & Local Government	67	27	63	4,198	6
Universities	45	18	48	2,162	3.5
Foundations	8	3	44	351	.5
Commercial	19	8	37	694	1
Total	245	100	270	66,180	100.0

Over the years, the survey designs or data collection methodologies used by RTI survey researchers has varied very little. In
most applications, face-to-face personal interviews have been the
primary method used to collect the survey data; only in a small
number of projects was the mail or telephone method used. In the
period 1959 to 1971, three RTI survey research projects were
mail surveys. A very small number of RTI's survey research projects have been telephone surveys. In the time period 1959 to
1979, for example, three projects were telephone surveys.

It is expected that the telephone will replace personal
interviews in future household surveys, primarily to reduce cost
and to improve or maintain acceptable response rates. With the
advent of the Computer-Assisted Telephone Interviewing (CATI)
system, currently under development at the Research Triangle
Institute, this goal should be attainable. CATI offers an

opportunity to cut survey costs by combining interviewing and data
entry. The interviewer asks each question as it comes up on the
terminal screen, along with codes for the set of possible responses.
As soon as the interviewer keys the code appropriate to the
respondent's answer, the next question consistent with prior
responses comes up on the screen and the process is repeated. CATI
is also expected to improve the quality of interview data, primarily
because the interview is completely programmed. This should
minimize interviewer errors which ordinarily occur through failure
to ask the correct set of questions for each respondent (i.e., use
the correct skip patterns in the questionnaire). In addition, the
interview program can recall previously recorded information for
the interviewer's use in subsequent questions. The survey research
challenge of the 1980's is to adapt the methods of existing surveys
being conducted with the more expensive fact-to-face personal
interview method to telephone interviews using CATI.

3.2.   The "Survey" Projects

Of the 212 survey projects which included at least one of
the survey techniques of instrument development, sampling, data
collection, data processing, and data analysis, nearly half (48%)
have been in the field of education with an average cost per
project of approximately 36 thousands of dollars (see Table 4).

The most significant educational survey undertaken by RTI
in its first ten years was the National Assessment of Educational
Progress (NAEP). NAEP measured educational achievement levels in
various subject matter areas, including science, math, reading,
social studies, and writing, among 9-, 13- and 17-year-olds.
Approximately 25,000 students in each age group were selected to
participate each year from national probability samples of schools.
Response rates among schools ranged from 91 percent to 95 percent,
and among students from 78 percent to 85 percent. In addition,
NAEP also measured achievement levels among young adults 26-35 who
were identified in national samples of households. RTI continued

Table 4. Number of Survey Projects, Percent of Survey Projects, and Average Cost per Survey Project by Substantive Area

Substantive Area	Number of Survey Projects	Percent of Survey Projects	Average Cost Per Survey Project in Thousands of Dollars	Total Cost of Survey Projects in Thousands of Dollars	Percent of Total Cost of Survey Projects
Health	49	23	440	21,538	33
Agriculture	4	2	8.7	35	.1
Education	102	48	360	36,754	57
Economics[1]	16	8	29	464	1
Aging	6	3	65	392	.6
Planning[2]	13	6	63	823	1
Population[3]	5	2	138	690	1
Behavior[4]	11	5	242	2,658	1
Defense	3	1.5	73	218	.3
Fire	3	1.5	372	1,116	2
TOTAL	212	100	305	64,689	100

[1] Economics includes projects in the subject areas of marketing research and commerce.

[2] Planning includes projects in the subject areas of recreation, population, and health.

[3] Population includes projects in the areas of population studies, roadside litter, and other projects, and health effects.

[4] Behavior includes projects in drug and alcohol abuse and mental health.

after more than ten years as both the sampling and the data
collection contractor for NAEP under contracts totalling $20
million. The extensive experience gained by many RTI staff
members from the NAEP surveys, both in-school and out-of-school,
has been an important factor contributing to RTI's survey expertise
in other areas.

There was also a NAEP spinoff, as individual states began
to undertake their own assessments. RTI has completed statewide
educational assessment surveys for North Carolina, Maine, Minnesota,
Connecticut, Texas, and other states. These national and state
assessments account for the majority of the educational projects
conducted by RTI.

Health-related projects account for the second largest
category of 440 thousands of dollars per project (see Table 4).
A significant large-scale health survey has been the National
Medical Care Expenditure Survey (NMCES), which is an assessment
of the personal health care expenses of the U.S. population.
NMCES has included six national surveys of 13,500 families, two
national surveys of physicians, a national survey of hospitals
and clinics, three national surveys of hospitals and clinics,
three national surveys of employers, and a national survey of
health insurance companies. RTI's other significant surveys in
health and health-related subjects include a national survey of head
and spinal cord injuries, a national survey of dental practices,
a national survey of teenage sexual behavior, a statewide health
survey in Virginia, and a pilot study for a national survey of
epilepsy. Health surveys at RTI are most expensive. They are
more costly than educational surveys (see Table 4) because they
often require national geographic coverage and personal face-to-
face household interviews.

Two large national surveys that illustrate RTI survey
research experience which combine behavior/education, not just
education, are the Safe School Study and the National Longitudinal

Study (NLS) of the High School Class of 1972. The Safe School
Study was mandated by Congress to determine the frequency, serious-
ness, and incidence of crimes in public schools, the costs of
vandalism, and effective preventive measures. The study consists
of three phases. Phase I was a mail survey which collected data
from 4,014 school principals with a response rate of 73 percent
for nine months in the period from February 1976 to January 1977,
regarding incidents occurring at their schools and their recommendations of how the problems of school violence and vandalism could
be solved. Phase II collected the same data as that in Phase I
from 642 principals, with a response rate of 76 percent, and
asked 23,895 teachers and a sample of 31,373 students, with a
response rate of 76 percent and 81 percent respectively, about
their background, attitudes, and experiences as victims of crimes
in the schools using the victimization survey technique. In
Phase II student and teacher data was collected by RTI field staff
using anonymous self-report questionnaires and face-to-face
interviews of 6,813 students at a response rate of 83 percent.
Some of the Safe School Study findings were as follows (National
Institute of Education, 1977):

- The typical secondary school student: a 1 chance in 9
  of having something stolen in a month; a 1 chance in 80
  of being attacked; and a 1 chance in 200 of being robbed.
- For the typical teacher in the nation's secondary schools:
  she/he has around 1 chance in 8 of having something
  stolen at school in a given month, 1 chance in 167 of
  being robbed, and 1 chance in 200 of being attacked.
- The best estimate of the yearly replacement and repair
  costs to the nation's schools due to crime is around
  $200 million.
- Of the attacks with injury recorded for the survey, only
  one-sixth were reported to police. Even when serious
  violence is involved, as with attacks requiring medical
  treatment, only about one-third of the offences are

reported.

- Of all offences taken together, about one-third are
  reported to police.

Some of the policy implications of the Safe School Study were:

- Consideration should also be given to ways of decreasing
  the impersonality of secondary schools and increasing
  the amount of continuing contact between students and
  teachers.

- Schools and their communities should recognize the key
  role of the principal in troubled schools, and give
  special attention to recruiting and training principals
  for the schools seriously affected by crime and disrup-
  tion.

- Security measures can also be helpful in reducing
  violence and property loss in schools, provided they are
  not used as a substitute for effective governance.
  School systems with serious problems of violence and
  vandalism can benefit from hiring additional security
  personnel with training in interpersonal skills as well
  as security functions.

The National Longitudinal Study of the High School Class of
1972 (NLS) is designed to collect and report data on what happens
to young adults after high school (National Center for Education
Statistics, 1977). The major data were collected by a student
questionnaire about plans, aspirations, attitudes, and experiences.
The 1972 base-year survey national probability sample included
over 23,000 students from 1,200 public, private, and church-
affiliated high schools; about 21,000 seniors from over 1,000
schools participated. Follow-yp surveys of the 1972 class begin
in October of each year: first followup, 1973; second, 1974;
third, 1976; and fourth, 1979. There were 21,350 participants who
completed a First Followup Questionnaire - 69 percent by mail and
31 percent by personal interview. Participants were asked where
they were and what they were doing with regard to work, education,

and/or training in October 1972 and 1973 to facilitate tracing of progress and defining factors that affected progress since high school. The second followup completed by RTI had 20,872 participants who completed questionnaires - 72 percent by mail and 28 percent by personal interview. Data were obtained on October 1973 and 1974 activities of 93 percent of the sample members. The third followup provided data on October 1975 and 1976 educational and work experiences since the last survey.

Some of the major findings of the National Longitudinal Study were that women have better academic credentials than men, women still have a somewhat lower college attendance rate than men; the difference, however, has decreased over the past decade. It was also found that, although whites, in the aggregate, are more likely to attend college than blacks, this difference is wholly a function of class background and academic preparation. In addition whites, on the average, are more likely than blacks to enter colleges of higher quality as indicated by higher selectivity level. The data show that, while students in academic high school programs are better prepared for college education even after class background, ability, and aspiration are considered, financial aid programs do not seem to have a substantial impact of college persistence. The data also show that proximity to a two-year college has an important impact on college attendance, primarily for middle class students.

The growth trend of survey projects at RTI is shown in Figure 1. The four years 1977-1980 have accounted for 37 percent of RTI's survey projects and 59 percent of the total survey project costs (see Table 5). Figure 1 also contains the trend of the average cost per project for the past twenty-two years. The decline in average cost for the period 1974-76 is partially explained by the fact RTI conducted a large number of lower-cost surveys during this period.

Early years of survey research at RTI included a survey of

Table 5. Number of Survey Projects, Percent of Survey Projects,
          and Average Cost per Survey Project by Time

Year Groups	Number of Survey Projects	Percent of Survey Projects	Average Cost Per Survey Project in Thousands of Dollars	Total Cost of Survey Projects in Thousands of Dollars	Percent of Total Cost of Survey Projects
1959-1961	7	3	6.6	46	.1
1962-1964	11	5	8.4	92	.1
1965-1967	11	5	49	537	.8
1968-1970	17	8	217	3,692	6
1971-1973	32	15	374	11,977	19
1974-1976	55	26	188	10,136	16
1977-1980	79	37	484	38,210	59
TOTAL	212	100	305	64,689	101

gasoline consumption by registered motorboat owners to determine
an approximate allocation of gasoline tax revenues to a water
safety program; a survey of child day-care facilities across the
State of North Carolina to assist in the preparation of regulatory
legislation; and a survey of the cost of care for persons in homes
for the aged and nursing homes to determine the daily allowance
to be paid for welfare patients.

In later years, surveys were carried out to provide an index
of the economic well-being of rural families, to measure the
relationships between drug usage and crime, to evaluate programs
designed to change drinking and driving patterns, and to estimate
the consumption of alcohol by teenagers.

In recent years, large survey projects were completed on
drinking habits and patterns among U.S. teenagers, on fire
protection, on the children of migrant workers, on drug abuse
rehabilitation programs, on unemployed youth in North Carolina,

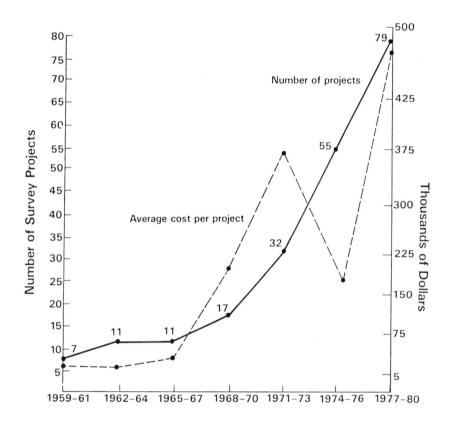

<u>Figure 1.</u>   <u>Trend of Survey Projects at the Research Triangle</u>
<u>Institute from 1959 to 1980.</u>

on individualized education for handicapped children, and on
environmental health effects.

The geographical coverage of the survey projects was
analyzed.   Each project was classified into one of three categories
depending on the geographical area the sample of the survey represented.   These categories were: nationwide, statewide, and local.

In terms of number of projects, survey projects conducted
at the national and state levels were about the same (85 projects
versus 76 projects, see Table 6).   As expected, the percent of
projects and total survey project cost and average cost per survey
project varied considerably.

In summary, survey research at the Research Triangle
Institute during the past twenty-two years has been an important
and growing research enterprise.   The survey research projects
have ranged considerably in size and have generated several
significant policies.

Table 6.   Number of Survey Projects, Percent of Survey Projects,
           and Average Cost per Survey Project by Geographical
           Coverage.

Geographical Coverage	Number of Survey Projects	Percent of Survey Projects	Average Cost Per Survey Project in Thousands of Dollars	Total Cost of Survey Projects in thousands of Dollars	Percent of Total Cost of Survey Projects
Nationwide	85	40	673	57,218	88
Statewide	76	36	65	4,927	8
Local	51	24	50	2,545	4
TOTAL	212	100	305	64,689	100

ACKNOWLEDGEMENT

The author acknowledges the effort of Mr. David Moazed for assisting in the development, coding, and editing of the project data file and his computer programming assistance in conducting the data analyses of the project characteristics. In addition, special thanks is given to Ms. Linda Higgs and Ms. Thelma Stone for their assistance in developing the project data file and typing this paper, Mrs. Sophie Burkheimer for her efforts in graphics, Dr. Jay Wakeley for assistance in editing the paper, and Mr. Richard Platek of Statistics Canada for his suggestions with respect to the final version of the paper. The content of the final paper, however, is the responsibility of the author.

REFERENCES

Horvitz, D.G. (1979). Survey research. *Hypotenuse,* Research Triangle Institute, Research Triangle Park, North Carolina.

National Center for Education Statistics (1977). National Longitudinal Study of the High School Class of 1972 (Review and Annotation of Study Reports). Department of Health, Education and Welfare, Washington, D.C.

National Institute of Education (1977). Violent Schools - Safe Schools (The Safe School Study Report to the Congress). Department of Health, Education and Welfare, Washington, D.C.

COST BENEFIT ANALYSIS OF CONTROLS IN SURVEYS

R. Platek and M. P. Singh

Statistics Canada

The purpose of the paper is to identify and
relate errors in surveys to control programs. These
programs are discussed in relation to their objectives,
usefulness and cost. An attempt is made to provide
an index of survey performance and to identify the
conditions under which this index would be operational.

1. INTRODUCTION

In surveys, as in many other areas, controls are of para-
mount importance. To disregard them could be expensive in the
long run and in some cases could seriously damage the credibility
of the survey results. For the purpose of this paper, the overall
control system will comprise of a number of control activities
which ensure that the actual survey performance conforms to predetermined targets, such as cost, timeliness and reliability of
the statistics produced. These control activities will be discussed in the context of large scale continuous surveys involving
several complex stages of operations each of which are likely to
contribute to the total survey error. The main thrust of the
paper is to identify the major sources of errors and to discuss
corresponding control activities. The paper further emphasizes
that to be effective, the control activities should be considered
as an integral part of survey planning and operations.

Most of the large scale continuous household surveys use
complex designs and estimation procedures and incorporate such

features as deep stratification, multistage sampling with unequal
probabilities, sample rotation and ratio estimation procedures.
The discussions on survey errors and corresponding control
programs, although presented in the context of the Canadian Labour
Force Survey, which is a monthly survey of 55,000 households, are
also relevant to other similar surveys.

Three main aspects of the survey, namely design, collection
and data processing are discussed separately in sections 2, 3 and
4 from the point of view of identification of sources of errors.
In order to facilitate the understanding of the relevance and the
importance of different types of errors, operational steps
involved in the LFS are briefly mentioned in each of these
sections. A more detailed description of various phases of this
survey is available from Statistics Canada (1977).

Section 5 presents an outline of different control programs
used in the Canadian Labour Force Survey which aim at minimizing
the impact of errors discussed in the earlier sections. The
control programs of both types, namely diagnostic and preventive,
are discussed. Diagnostic controls are usually built-in special
procedures which examine the performance of a survey at any given
time. In the majority of cases, diagnostic controls provide
numerical measures indicating the performance of various aspects
of the survey operation whereas the preventive controls are aimed
at minimizing the effect of errors from various sources. The
effect of some preventive controls can be measured; for others
it is not measurable. Classification of the control programs
into these two types is made in section 6.

It is our experience that in the case of continuous surveys,
there are times when one has to introduce new controls while at
other times, some of the existing controls have to be removed
altogether or reduced in size. This is often due to restraints
imposed from time to time on the available resources as well as
the fact that it is neither possible nor necessary to insist on a
fool-proof survey instrument in all cases. In the last section,

a survey performance index is proposed which, if calculated
regularly, would suggest what type of control program should be
given priority at a particular point in time. The resources used
in major LFS operations are briefly discussed.

2. SAMPLE DESIGN

2.1 Frame

The most common frame for household surveys is an area
frame with census enumeration areas (EAs) or some other well-
defined areas as the building blocks. The basic frame for such
surveys generally consists of several subsampling frames depending
upon the stages of clustering and the randomization scheme involved.
This basic frame is frequently supplemented by list frames (listing of address) whenever such lists can be easily procured and
inexpensively updated. For example, in the Canadian Labour Force
Survey (LFS), Primary Sampling Units (PSUs) in rural areas are
formed by grouping the desired number of EAs, whereas in cities,
clusters which are usually city blocks serve as the first stage
sampling units. In addition to the area frame, list frames of
large apartment buildings are used in certain selected metropolitan areas. There are surveys where a list frame (also referred
to as an address register) obtained from different sources has
itself become the basic frame with respect to large sampled areas;
however, we shall not concentrate on this situation since the
control mechanisms are very much dependent on the source of the
list frame. The following are two major sources of errors that
may be introduced at the planning (for the sampling operations)
stage itself leading to deficiencies in the sampling and subsampling frames:

E_1:   inadequacy and incompleteness in the description of
the boundaries of the selected areas on the map as
well of the definitions of units to be included from
the selected areas at any stage of sampling.

E_2:   Choice of dwelling list, its coverage and outdatedness.

Errors occurring at the data collection stage are mentioned
in section 3 (see E_8).

2.2 Stratification

Surveys conducted on a national scale usually involve
several stages of stratification. In the Canadian Labour Force
Survey, Economic Regions (ER) within the provinces are treated as
primary strata and estimates are produced regularly for these
areas or combinations thereof. Within each ER, further stratification is carried out using data on certain important stratification variables, information on which is obtained from the most
recent census. In the context of a continuous survey the changes
in the relative importance of these variables over a period of
time would lead to deterioration in the efficiency of stratification. This deficiency induced in the estimate could be attributed
to the following:

E_3:   Outdatedness of the stratification variables over a
period of time.

2.3 Sample Allocation

Generally, for large scale sample surveys, the overall
size of the sample is determined by the amount of resources
allocated to the program, leaving a rather limited scope for
technical input in this regard. However, once the total sample
size is fixed, its allocation to various provincial or subprovin-
cial regions is largely a matter of technical and operational
considerations in order to satisfy the demands for reliable data
at the national and subnational levels at the time of designing
the survey.

Population growth at the national level would result in a
corresponding increase in the sample and if the growth is uneven
among provinces, this could create an imbalance in the original

allocation, causing a decrease in efficiency per unit cost. This
gives rise to the following:

E_4:   Changes in the population distribution at various
         levels of aggregation would adversely affect the
         reliability as well as increase the cost of the survey.

2.4   Degree of Clustering

Economical designs often include unequal probabilities of
selection, stratification and several stages of clustering
requiring several stages of randomization. While normally
stratification and unequal probability selection result in reduced
variances, clustering usually results in increased variances due
to positive intra-class correlations. Various forms, sizes, and
stages of clustering are quite prevalent in area sampling since
they help reduce the cost of collecting data. In continuous
surveys, the increase in variance due to clustering may vary over
a period of time, affecting the performance of the design. This
is because variance is a function of the correlation, which may
vary from one time to another depending upon the changes in
population structure in the clusters. Therefore, for continuous
surveys:

E_5:   Efficiency of the estimates would be affected by
         changes in the degree of intra-class correlation.

2.5   Sample Selection

Probability proportional to size (PPS) sampling is
freuqently used in large scale surveys at various stages of
sampling and the measure of size is usually related to the
characteristics of the units. PPS sampling results in reduced
variances for various characteristics as long as these characteristics remain highly correlated with the measure of size. However,
a continuous survey is subject to structural changes over time.
One of the important changes is faster growth of population in

larger cities than in rural areas. Furthermore, population gravi-
tates to different parts of cities due to job opportunities and
other facilities in those areas. The combined effect of these
events is an uneven growth in different sections of the cities
and a resultant decrease in the correlation between the characteristics and the size measure. Thus, another source of error for
continuous surveys is due to:

   E_6: outdatedness of the original selection probabilities
           resulting from an uneven population growth within
           a stratum.

3.   DATA COLLECTION METHODS

3.1 General

     Data collection for the Canadian LFS, including such
aspects as recruiting and training of interviewers, data trans-
mission and respondent relations are the responsibility of the
Regional Operations Division within the Statistics Canada. There
are eight Regional Offices and about 1,000 interviewers engaged
in collecting LFS data from about 56,000 household during a
special field week (survey week) of each month for information
pertaining to the previous week (reference week).

     The questionnaires filled in by the interviewers are sent
to Regional Offices where data are captured and transmitted to
the Head Office for detailed editing and processing. Some
important steps that affect the quality of LFS data which may also
be relevant to a number of other surveys are discussed below.

3.2 Training of Interviewers

     Training of interviewers is a continuous process involving
initial training, ongoing training, group training, self study
and special training based on observation and reinterview programs.
These latter two programs are discussed in the section on quality

controls. Any weakness in the proper understanding of the interviewing procedure and the conceptual and technical issues at this
stage would result in:

E_7: Error component (often referred to as the correlated
response variance) that may be attributed to the
interviewers themselves. Generally, lack of understanding of the conceptual and technical issues would
lead to higher bias and the lack of understanding and
adherence to the interviewing procedure would increase
the correlated response variance.

3.3 Listing and Interviewing

Listing of dwellings is the process of identifying and
recording of dwellings within the boundaries of specified clusters.
Each dwelling is given a unique serial number and an address and
description on the listing schedule. All dwellings which can be
determined as being habitable (whether vacant or occupied)
within the boundaries as outlined on the cluster map are to be
listed.

Upon identifying sampled dwellings, the first contact by
the interviewer with the persons in the selected dwelling is
always by personal visit. The interviewers are allowed to accept
response about all eligible members of the household from any
responsible member (15 years of age and over) of that household.

In larger cities, the interviews in the subsequent months
are conducted by telephone, except when the respondent prefers
personal interviews. In rural areas all interviews are personal.

In spite of all efforts, there are always certain households for which data are not available in a given month. This
may be due to no one at home, temporarily absent, refusal, etc.
These households are coded non-respondents and are imputed for
at the weighting stage.

Some primary sources of errors related to either the

listing, interviewing or data entry stages are mentioned below.
Note that certain components of these errors may be in common with
the sources mentioned earlier because the approach used in the
identification of source is based on the *operational stages used
in a survey* and the same type of error may occur in more than one
stage of operation.   The possible overlaps are therefore identified.

E_8:   At the data collection stage factors such as incorrect
inclusion or exclusion of dwellings from the sampled
clusters, incorrect identification of cluster boundaries, outdated cluster maps and cluster lists, missed
dwellings within clusters, failure to list all members
of the household would cause coverage errors.   Normally
these lead to undercoverage (slippage) of the population
in the surveys (see E_1 and E_2).

E_9:   Proxy responses (as they are allowed in the LFS) may
lead to response errors of varying magnitude for
different subgroups (age-sex) of population.   Initia-
tives to increase non-proxy responses may lead to an
increase in the non-response rate unless sufficient
funds are made available for the resulting increase
in the number of call-backs.

E_{10}:   Instead of asking the questions as worded, the interviewers may try to interpret the questions and guide
the respondents according to their own understanding
of the concepts and general perception of the respondents.   In addition, the respondents themselves may
accidentally or intentionally make an error in providing
a response.   This could lead to errors in data, causing
an increase in response bias and variance of the
estimate (see E_7).

E_{11}:   In general, any change in the survey conditions, which
may result from legal requirements, for example, could

affect response errors.

E_{12}: High non-response rates may lead to complete distortion of statistics due to increased imputation variance and bias.

E_{13}: Errors may be introduced at the data entry stage. These errors may vary from one Regional Office to another, depending on the quality of the operators employed.

4. DATA PROCESSING METHODS

Detailed editing and imputation in LFS are carried out at the Head Office, with the edited data then being weighted. The final weight comprises four main factors, namely, the design weight, an urban-rural factor, balancing for non-response, and an age-sex factor. Although the last three factors are aimed at increasing the efficiency of the usual design-based estimate, they could also be the sources of errors in the final estimate. For example, the urban-rural factor (the ratio of urban or rural population in the sampled PSUs in a province to that of all strata in the provinces) may become out of date and cause reduction in efficiency. Similarly, balancing for non-respondents may induce bias in the estimate if the characteristics of the respondents and non-respondents differ significantly and if characteristics and response probabilities are correlated. Also, the fact that the population projections used in the determination of age-sex factors are not adjusted for the census undercoverage of the population may induce a certain amount of bias. Thus,

E_{14}: Various steps involved in processing and weighting may result in imputation bias, coverage bias and may thus increase the mean square error of the estimates.

5. QUALITY CONTROL PROGRAMS

Every statistic derived from a census or a survey is subject to various sources of error contributing to the total survey error. Attempts are therefore made to control and minimize the impact of these errors in the final estimates, In this section, the control programs used in the Canadian Labour Force Survey are briefly discussed. Some of these programs aim at ascertaining or to minimizing the impact of error arising from a single source (e.g. data entry quality control) while others are multipurpose in the sense that they aim at more than one source of errors, (e.g. observation program, see 5.14). These programs are outlined below.

5.1 Sample Maintenance

This activity related primarily to the upkeep or maintenance of the sampling frame. In any rotation design, rotation of the units takes place not only at the dwelling level (ultimate sampling unit) but also at the earlier stages of sampling over a period of time. Various measures (such as control in the duplication of units and in the use of random starts) are taken to ensure proper selection and rotation of units at various stages. Other activities include identification and definition of the selected units by means of maps and descriptions, proper conduct of sampling operations in the field such as counting, listing, updating.

5.2 List Maintenance

There are two measures adopted to ensure that the cluster lists are up-to-date. Firstly, thorough special checks are carried out by the interviewers from time to time, for example, at the time of introduction of the cluster or changes in the interviewer's assignments. In addition to these checks by interviewers, listing is also checked by senior interviewers on different occasions such

as at the time of interviewer induction, observation or reinterviewer (see 5.14, 5.15).

5.3 Slippage

In spite of the implementation of the preceding two measures
on a regular basis to ensure adequate coverage of the population
to be surveyed, certain households or persons within the households
are bound to be missed in any survey of a human population. In
the LFS, therefore, a program of slippage evaluation is undertaken
on a regular basis. The slippage rate is defined as the percent
difference between the population estimated from the sample and a
census-based estimate of the population. An acceptable range for
the slippage rate is established on the basis of past data and
any significant increases are monitored closely and corrective
action taken to ensure proper use of the field procedures. At
the national level, the slippage rate in the LFS is normally about
four percent, with significant variation by age-sex groups and by
regions.

5.4 Cluster Yield Check

The actual number of dwellings is compared with the expected number of dwellings at the time of design. When the difference
exceeds a predetermined maximum, reasons for the difference are
established. Often the causes of the difference are due to a
construction or demolition of dwellings, but a significant portion
is attributed to listing errors followed up with corrective action.

5.5 Sampling Variance

A regular program of variance-covariance analysis is undertaken in the LFS. The coefficients of variation of about ninety
major characteristics are produced and analyzed every month.

5.6 Components of Variance

For the purpose of obtaining measures of reliability of

estimates, it is seldom necessary to split up the total variance
into components and to estimate each component. In a multi-stage
design, however, questions frequently arise as to the contribution
of various stages to the total variance. Examination of the
relative magnitude of the variance components may suggest more
efficient sample designs which may be realized through changes in
the sizes of the units or changes in the allocation of the sample.

5.7 Design Effects

This measure (sometimes also called a Binomial Factor) is
employed to assess the overall performance of the design strategy
as compared to simple random sampling (SRS). It is computed from
the variance-covariance system as the ratio of the variance of
the estimate obtained in the survey using a particular design to
the variance of an estimate assuming SRS design. Cost considerations aside, the higher the design effect, the worse is the design
used in relation to SRS. Regular monitoring of this measure is
one of the least expensive means of detecting the areas that would
benefit from partial or complete redesign.

5.8 Stratification Index

As mentioned in section 2.2, stratification may become
inefficient over time in a continuous survey such as LFS and
result in an error E_4. The decrease in the efficiency of
stratification may be reflected in an increase in the design
effect although one cannot distinguish deterioration in the
stratification from deterioration in the sample design within
strata by observing the design effect alone. From the variance-
covariance system, stratification indices (Gray, 1966), are
calculated in different economic regions each month for a few
major characteristics and stratifying variables as well as for
other super-strata to monitor possible changes over time. The
index measures an approximate reduction in the variance from that
which would occur if the economic regions and other super-strata

were no further stratified.

5.9 Sample Size Stabilization

Irrespective of how cost-efficient a design (or redesign) is at the initial stages of introduction, for a continuous survey the natural increase in the population resulting in an increase in total sample size would lead to an overall increase in cost without a corresponding gain in efficiency. If left unchecked, the LFS sample size of 62,800 dwellings (56,000 households) per month would grow, due to the fixed sampling ratios, at the rate of about 2.3% per year as a result of natural population growth and the general phenomenon of declining household size. In order to check this growth, a system of sample size stabilization was introduced on a regular basis during 1977. Obviously, there are several ways of stabilizing the size of the sample. Each of the methods has a different cost and variance impact on the monthly estimates. The method chosen for the LFS is aimed at retaining maximum efficiency of the sample on a regular basis. The general methodology (Drew, 1977) adopted is briefly outlined below.

Let n_b denote the base (or desired) sample and let n_s be the number of dwellings actually selected in a given month. Note that n_b and n_s may be defined at any level of geographical detail. For a given month, in all those areas where $n_b \geq n_s$, no dwelling is dropped and the stabilization weight to be applied to the records would be unity.

For areas where $n_b < n_s$, $(n_s - n_b)$ dwellings are sub-sampled systematically from n_s dwellings using the inverse sampling ratio: $R = n_s/(n_s - n_b)$ and these sub-sampled dwellings are identified and dropped from the survey. A stabilization weight is applied to all non-dropped dwellings. The procedure for dropping the dwellings has been automated and is carried out every month in all areas with only certain exceptions due to operational constraints.

5.10 Program of Updating Selection Probabilities

As mentioned in section 2.5, the outdatedness of the original selection probabilities introduces a major component of
error (E_6) into the monthly estimates for larger cities having
high population growth. In order to deal with the growth problems,
the primary criterion for the choice of design in cities was its
suitability to incorporate changes in the selection probabilities.
The random group method (Rao, Hartley & Cochran, 1962) used in
the larger cities is found to be quite suitable for incorporating
changes in the selection probabilities (Platek and Singh, 1976;
Drew, Choudhry & Gray 1978). A program of updating the selection
probabilities using the new size measures (dwelling counts by
clusters) first introduced during 1977 in the provinces of Alberta
and Saskatchewan on an experimental basis showed a gain in
efficiency of the order of 50% in certain strata. As a result,
the LFS sample is updated in a number of growth areas on a regular
basis.

In order to realize the benefits of new size measures as
soon as they become available, an additional step is included in
the weighting system on an interim basis, where the new size
measures are utilized with considerable gain in efficiency in
the form of a ratio estimation procedure (Singh, 1978). This
ongoing program controls the possible increases in the variance
resulting from outdated size measures.

5.11 Data Capture Quality Control

During the data capture operation (carried out through the
mini computers in the Regional Offices), it is necessary to
maintain a program of quality control to ensure that the errors
introduced during the data entry operation do not affect the
data significantly. A sample verification of records is carried
out each month by qualified data entry operators. The work of
each operator is verified on a sample basis, using a sampling

plan based on the operator's earlier performance. The sample
verification scheme used for the LFS ensures that over a period
of time, at least 97% of the documents are free of data entry
errors.

5.12 Edit and Imputation

The essential purpose of the editing system is to ensure
completeness and logical consistency of data collected. Imputation is performed using data from responses in the previous month
or from respondents with similar characteristics in the same
month. The frequency of such imputations and edit failure rates
are monitored regularly by means of the Field Edit Module (Newton,
1980).

5.13 Field Edit Module

This system is designed to produce reports relating to
edit changes made during the head office processing. These
reports can be produced at various levels of aggregation and
serve as part of the feedback procedure to the Regional Offices.
This may also be used as an important component in the calculation
of the interviewers' performance indicator.

5.14 Observation Program

Observation generally occurs at the time of the actual
interviewing. In the LFS, the observation program is undertaken
with the aim of evaluating and improving the performance of the
individual interviewer. It provides an opportunity for training
(or retraining) and motivating the interviewers, updating the
listings and identifying field problems. All regular interviewers
are observed at least once a year. In addition, special observation takes place in a few cases depending upon the performance
of the interviewers.

5.15 Reinterview Program

A reinterview is an interview with the households that
have already been interviewed during the survey week, conducted
by a Senior Interviewer in the week immediately following the
survey week and in which the same questions as previously asked
by the interviewer are repeated by the Senior Interviewer. The
differences observed between the two sets of responses are
attributable to several sources, such as the respondents, the
interviewer, and the reinterviewer. In addition, the wording and
sequencing of the questions and some general survey conditions
will have a different effect on the two interviews. In the LFS,
this program has become a regular feature and has as its dual
objectives, the provision of measures of the components of the
response errors and use as a management tool for the evaluation
of interviewers' performance.

The sample is split into two parts: with and without
reconciliation. The purpose of reconciliation is to provide a
measure of response bias whereas the unreconciled sample is used
to measure the simple response variance. A measure which aims
at assessing the quality of statistics over time is referred to
as an 'index of inconsistency'. The index measures the proportion
of the total variance accounted for by the simple response variance.

Another measure to indicate the relative stability of
various characteristics in a given period could be the coefficient
of response variation defined as the ratio of the square root of
the simple response variance to the average of the responses in
the two interviews. This measure, being independent of scale,
is more appropriate for the comparison of stability between the
characteristics.

The details for the methodology of the reinterview program
in the Canadian Labour Force Survey have been described by Tremblay,
Singh & Clavel (1976) and Ghangurde (1979).

5.16 Non-Response Rates

Every month non-response rates within various categories are computed, analyzed and communicated to the Regional Offices as a means of monitoring the operation of the survey. Effects of high nonresponse on small area estimates are monitored.

6. ANALYSIS OF CONTROL PROGRAMS

6.1 Classification of the Program

In the earlier sections, we have discussed possible sources of errors and various quality control programs in the context of large scale continuous surveys with special reference to the LFS. These programs may be divided into two types: preventive and diagnostic.

The primary purpose of preventive measures (controls) is the minimization of errors rather than their assessment or evaluation. In some cases more than one program is needed to control errors arising from a single source while in others a single program may help minimize errors from more than one source. An example of the latter is the observation program which is aimed at reducing various components of the response error as well as coverage error. While the actual effects of preventive programs are not measurable directly from a given data set, in most cases the relative performance of one program over another may be measured from various data sets. The choice of a particular program may be made under the assumption of the similarity of general survey conditions. As an example, it is generally not possible to measure the effect of an ongoing training program on a continuous survey, however, in a specially designed test it is possible to measure the relative performance of one type of training program over another.

The main purpose of diagnostic measures is in assessing and evaluating the performance of different aspects of the survey program. In the majority of cases, diagnostic controls result in

measurable statements. Some programs no doubt serve dual purposes,
in that they are used as preventive controls as well as to provide
measurable statements of the performance of the specific aspects
of the survey instruments. Such programs of necessity require
much more careful planning and execution.

The measures that have been considered above are listed
below with the appropriate references and labelled preventive
and/or diagnostic. The errors associated with the various
measures are also included.

It is recognized that the list of the sources of errors
discussed earlier is by no means complete. Two important omissions
are, for example, the specification errors introduced at the
planning stages of the survey and the interpretation error that may
occur at the final stage of analysis and use of the data. A more
complete enumeration of errors reference has, however, been
prepared by the U.S. Bureau of Census (1978).

Just as it is not necessary to evaluate the impact of a
preventive program on the data, it is not necessary that a diagnostic program should lead to a preventive action. Diagnostic
measures generally are meant to be used for evaluating the performance of different aspects of the survey. Since perfection can
never be achieved in surveys of human populations, certain
margins of error are allowed at various stages of the operation,
and over a period of time, it becomes possible to establish a
range for the magnitudes of the errors. As long as the measures
lie within the acceptable range, no action is required. For
example, the design effects are computed every month in the LFS
and no action is taken unless there is a significant increase over
a period of time, in which case certain preventive programs such
as updating of selection probabilities may be installed. Thus,
the diagnostic measures are simply the predictors of the performance of the survey instrument. They themselves do not improve
the performance whereas preventive measures are meant to improve
the performance, although their impact on the quality of data

Table 6.1. Classification of control measures

Measures	Reference	Pre-ventive	Diag-nostic	Control Evaluation of Errors
Sample Maintenance	5.1,2.1	✓		E_1,E_6
List Maintenance	5.2	✓		E_1,E_2,E_8
Slippage	5.3		✓	E_2,E_8,E_{14}
Cluster Yield Check	5.4		✓	E_1,E_8
Sampling Variance	5.5		✓	E_4,E_5,E_6
Components of Variance	5.6,2.3		✓	E_5
Design Effect	5.7		✓	E_3,E_4,E_5,E_6
Stratification Index	5.8,2.2		✓	E_3
Sample Size Stabilization	5.9	✓		E_4
Updating Selection Prob.	5.10	✓		E_6,E_2,E_4
Data Capture Control	5.11	✓	✓	E_{13}
Edit and Imputation	5.12	✓		E_{13}
Field Edit Module	5.13	✓		E_{13}
Observation	5.14	✓		*
Reinterview	5.15	✓	✓	E_7,E_{10},E_2,E_8
Non-Response	5.16		✓	E_{12},E_{14}

*   The observation program is a preventive control for many errors
including E_1, E_2, E_7, E_8, E_{10}, E_{11}, E_{12} .

cannot be assessed. Evidently, the two types of measures are complementary to each other. However, because of the general interest in producing better quality data rather than determining the quality of the data produced, the preventive measures do get higher priority in becoming an integral part of the survey instrument.

We have purposely omitted any references to survey specifications, which may under certain conditions give rise to survey errors. The reason for this omission is that maximum care should be taken to minimize such errors at the planning stage and in the majority of cases, it would be difficult to assess the impact of such errors.

7. SELECTION OF CONTROLS BASED ON SURVEY PERFORMANCE

7.1 Survey Performance Index (SPI)

Quite often survey practitioners face the problem of comparing the overall performance of different surveys or the same survey over a period of time. In large scale surveys where there are several diagnostic measures available and in the case of their conflicting performances, the problem of comparison becomes very complex and subjective decisions are taken in most cases. We propose a composite SPI with the hope that further theoretical and empirical evaluation of the model and the parameters involved in it would make it useful for assessment of the overall survey performance. Further, as indicated in the following subsection, it may also be used in setting up priorities among the control programs in order to maximize survey performance for a given cost.

In general, the SPI may be defined as a function (F) of total error and cost, that is,

$$I = F(TMSE, C_1, C_2, C_3) \qquad (7.1)$$

where C_1 is the cost of survey operations excluding the cost of
control programs, C_2 is the cost of preventive controls and C_3
denotes the cost of diagnostic controls. The Total Mean Square
Error (TMSE) is used as the measure of total error. First, the
components of TMSE and their association with various sources of
errors are given below. Special forms of the index are then discussed.

Total mean square error may be expressed as

$$TMSE = NSV + SV + TB^2 \qquad (7.2)$$

where

NSV = [RV + IV] is Non-Sampling Variance,

TB = [RB + IB + SB] is Total Bias,

SV = Sampling Variance,

RV = Response Variance,

IV = Imputation Variance,

IB = Imputation Bias,

SB = Sampling (or coverage) Bias, and

RB = Response Bias.

It may be noted that the response error (i.e. RV and RB) in this
context includes all sources of errors (see Table 7.1) in the
survey operation except the effect of nonresponse and undercoverage. The sampling variance in the same way considers surveys
with complete response and coverage. Thus, these quantities when
estimated from the sample should be adjusted for full response
and coverage. The imputation variance and bias on the other hand
are the results of nonresponse and undercoverage as well as the
methods of imputation adopted. For a detailed discussion of these
concepts and their evaluation, reference is made to Platek,
Singh & Tremblay (1978) and Platek & Gray (1980).

In order to carry our a cost benefit analysis of
the control programs, the index in (7.1) should ideally incorporate not only the various components of TMSE as stated above but
also detailed components of C_2 and C_3. This would allow
association of costs (C_{ij}) incurred in controlling and evaluating

Table 7.1. Components of TMSE by possible sources of error

Error Components (i)	Sources of Errors (j)
SV	E_3, E_4, E_5, E_6, E_{14}
RV	E_7, E_9, E_{10}, E_{11}, E_{13}, E_{14}
IV	E_8, E_{12}, E_{13}, E_{14}
SB	E_1, E_2, E_8
RB	E_7, E_9, E_{10}, E_{11}, E_{14}
IB	E_8, E_{12}, E_{13}, E_{14}

individual sources of errors (j) with the corresponding reduction
(δ_i) in the respective components (i) of the TMSE. It would
then be possible to establish the magnitude of reduction per unit
cost (δ_i/C_i) by individual components of TMSE, which will permit
identification of programs that are less cost effective for a
particular survey. However, in the majority of large scale
continuous as well as ad hoc survey operations, the cost (C_3) of
diagnostic controls as identified in the previous section in
Table 6.1, are negligible. As regards the cost of preventive
controls (C_2) which are normally a significant proportion
of the overall cost, it is virtually impossible to
isolate the cost of certain control measures as they tend to
become, as they should be, part and parcel of the production
process. In other cases where the cost estimates are available,
they may be confounded either with the cost of operations (C_1)
or with the cost of some other control programs themselves.
Both these factors are discussed further in the following subsection in the context of the Labour Force Survey. However, for
these reasons, we will discuss below a special form of the index
(1) that is a function of sampling and non-sampling error

components of TMSE, given by

$$I = \frac{NSV + TB^2}{SV + NSV + TB^2} \tag{7.3}$$

The index includes all aspects of data quality and compares it with the sampling variance which can be kept well under control at the design stage in the case of ad hoc surveys and through regular updating of the sample for continuous surveys. The index lies between 0 and 1 for a given survey with a fixed sample size. The lower the value of the index, the better is the perform-ance of a survey and thus the index will reflect any deterioration in the control programs. Lack of control causing a high non-response rate in a particular survey month, for example, would result in an increase in the components of imputation variance (IV) and imputation bias (IB); this in turn increases the magnitude of the index implying a decrease in the performance of the survey.

It may be noted that the index will generally vary from characteristic to characteristic. Therefore, its comparison must be viewed in relation to the importance of the characteristics for a particular survey occasion. Further, since the index may become small simply because of high sampling variability, strict monitoring of the coefficient of variation (CV) in case of contin-uous surveys and adjustments in CV for comparison of different ad hoc surveys would be essential. However, when this index is computed on a regular basis for one or more important character-istics over a period of time, the relative magnitude of the index for a given characteristic will indicate the relative performance of the survey instrument over time.

We now consider some special cases of the index in (7.3). For surveys where the nonresponse and slippage rates are low or the methods used for balancing of the nonresponse and the under-coverage are highly reliable, the contributions from each of the components IV, IB, and SB will be negligible, implying that

$$NSV \doteq RV \quad \text{and}$$

$$TB \doteq RB \; . \qquad\qquad (7.4)$$

Thus, if (7.4) holds, the index

$$I = \frac{RV + RB^2}{SV + RV + RB^2}$$

$$= \frac{1 + RB^2/RV}{1 + \dfrac{SV}{RV} + \dfrac{RB^2}{RV}} \qquad\qquad (7.5)$$

Further, analysis of data from the LFS reinterview program show
that the ratio RB^2/Rv during 1978-79 remained on an average
below one percent for the three important characteristics, namely,
.06% for employed 0.4% for unemployed and 0.9% for not in labour
force.   In general, it seems that for variety of situations the
contribution from this term would be negligible, and the index in
(7.5) may be expressed as

$$I \doteq \frac{RV}{RV + SV} \qquad\qquad (7.6)$$

which measures the proportion of 'total variance' (i.e. RV + SV)
accounted for by the response variance.

Note that RV in (7.6) measures the response variance from
all the sources and similarly that the sampling variance (SV)
refers to the situation of complete coverage and full response,
as mentioned earlier. Further, under the assumption (7.4), the
sampling variance as obtained from the variance-covariance system
(see section 5.5) may be used to calculate the value of 1 in (7.6).

Thus, under the very simplistic assumptions

i)   (7.4) holds,

ii) $RV \doteq SRV$, i.e. components other than simple response
variance (SRV) have negligible contributions to the RV,
and

iii) $\dfrac{RB^2}{RV} \doteq 0$,   (7.7)

the general index I in (7.3) becomes

$$I = \frac{SRV}{SRV + SV}$$

$$= \frac{SRV}{SRV + d \cdot BV} \qquad (7.8)$$

where BV is the Binomial Variance (i.e. variance of the estimate
under the assumption of simple random sampling) and

$$d = \frac{SV}{BV} \quad \text{is the design effect.} \qquad (7.9)$$

The Index I in (7.8) may also be expressed as

$$I = I^* \frac{SRV + BV}{SRV + d \cdot BV}$$

$$= I^* \left[1 - (d-1) \left(\frac{BV}{SRV + d \cdot BV} \right) \right] \qquad (7.10)$$

where

$$I^* = \frac{SRV}{SRV + BV} , \qquad (7.11)$$

is the usual 'Index of Inconsistency' as determined from the
reinterview programs (see section 5.15).

Thus if the three assumptions stated in (7.7) hold then it
would be possible to determine the performance index I as in
(7.10), based on the data from the unreconciled part of the
reinterview program which provides a measure of simple response
variance and the index of inconsistency I^*. The reinterview
program may also be used to compute I as in (7.6) which avoids
the assumption (iii) of (7.7), since an estimate of RB would be
available from the reconciled part of the reinterview sample.
Assumption (ii) of (7.7) is a serious one and seems to be crucial,
as its relaxation would require knowledge of correlated response
variance.

Due to the small size of the reinterview program of the
LFS the values of the index of inconsistency I^* are computed as
an average based on six months data. The value of d , the design
effect was also determined for the same period. The value of I
as given (7.10) was accordingly computed for the LFS at Canada
level for the characteristic unemployed. The index has shown a
gradual decline over a period of time, stabilizing about 14%
during 1979. This decline in the index is indicative of gradual
improvement in the performance of the survey and that the non-
sampling errors are well under control. The index when computed
regularly by the Regional Offices would indicate the relative
performance of the survey over a period of time.

7.2 Priorities of Control Programs

In this section, a brief analysis of the resource allocation primarily for the control programs in the LFS is presented.
Several comments are provided on the use of the survey performance
index in establishing priorities among the control programs.

We present in Table 7.2, the distribution of cost of field
operations by some major components in two different periods.
Major deviations in terms of the cost allocations between the
periods are discussed.

The shift in the distribution of resources between the two
periods should be studied in the light of two major changes that
were introduced in the LFS during the period, namely that (i) the
redesigned sample was introduced during 1975/76 and (ii) the LFS
sample size increased from about 35,000 households to 56,000
households during 1976/77. In spite of these significant changes,
it is interesting to note a great deal of similarity in the
distribution of the field operations resources.

Note that items 2 and 3 are the control programs of the
preventive type while items 1 and 4 include regular production
type activities. A detailed distribution of cost, although
available for certain major subcomponents of item 1 and 4, is not

Table 7.2. Distribution of field operations cost for the LFS

Components of Field Operations	Percent Distribution of Cost	
	1974/75	1979/80
1. Basic Operations (hiring, basic training, listing, interviewing)	59	65
2. Interviewer's Quality Control (Observation, reinterview, ongoing training)	11	9
3. Sample Updating (field counts in growth areas)	-	2
4. Regional Office (supervision, general administration)	30	24
	100	100

presented since our main purpose here is to study the changes in the priorities of the control programs. Under item 2, several control programs are grouped together in order to ensure comparability of resources and control programs between the two periods. Further, it is difficult to determine precisely the resources required by individual programs. Also, certain portions of the resources spent on the preventive control program, (e.g. the observation program) may in fact be reported under supervision in item 4. This however, should not affect the comparison of the distributions between two periods as it will affect both distributions equally. The following remarks may be made about the distributions.

1. The shift of 6% of the total resources from item 4 to item 1 may have resulted from the combined effect of significant increase in overall size of the sample and better management of resources included in item 4 under Regional Office expenses.

2. During 1974/75, no resources were allocated under item 3

because the new redesigned sample was to be introduced a year
later. A total of 11% of the resources was therefore allocated under item 2 which included preventive controls aimed at
minimization of errors contributing to various components
(i.e., NSV, IB, RB and SB) of the total mean square error
(see 7.2), excluding the sampling variance component. It may
be noted that the situation remained basically the same in
this respect until after two years of the introduction of the
new sample during 1975/76.

3. Regular monitoring of the diagnostic measures indicated a
    deterioration of the design due to uneven population growth
    in certain parts of the country. On the basis of a study
    conducted to examine the effectiveness of a regular program
    of updating the sample in growth areas (Drew and Singh, 1978)
    a decision was made to shift a part of (2% of the total) of
    the resources from item 2 to a new preventive control designed
    to update the sample on a regular basis. As mentioned in
    section 5.10, efficiency gains of up to 50% were noted in
    certain strata as a result of this program.

    Remarks 2 and 3 above, suggest that diagnostic measures
separately or together serve very useful purposes at minimal cost.
Also, they suggest that the decision of reducing the intensity of
preventive control under item 2 was guided more by intuitive
considerations than an actual evaluation because, as pointed out
in section 6, the impact of reducing the intensity of preventive
control (e.g. ongoing training) is difficult to assess.

    The resources for field operations, discussed above,
comprise about 70% of the total budget of the LFS. Of the
remaining 30% distributed over methodology, systems, processing
and analysis, less than 5% are spent on various preventive controls (such as sample maintenance, data entry quality control).
The diagnostic controls such as design effects, field edit module,
slippage, mentioned earlier put together account for only about

one percent of the total budget.

Another aspect of diagnostic measures worth pointing out in the context of the LFS is that at times they have led to programs requiring changes in the allocation of the overall sample in various areas resulting in more efficient estimates for the same cost or the reduction of the sample size at predetermined levels of efficiency. For example, the sample size stabilization program discussed in section 5, and introduced in the LFS, has led to savings of about 1% of the total budget without any loss of efficiency.

On the basis of the above analysis of the control measures and their effects on the cost and efficiency on the LFS estimates, we discuss some general guidelines for the choice of control programs.

It would be unrealistic even to attempt to prioritize the control programs for surveys in general as the choice of programs should depend upon the type of survey instrument and availability of resources. It is, however, safe to assume that the needs of ad hoc surveys are quite different from those of the continuous surveys and in that respect, the priorities of the control programs could well be different. Ad hoc surveys generally command higher priorities for preventive controls but some diagnostic measures that are intended to evaluate the data quality should also be included.

With respect to continuous surveys, it is generally an accepted practice to include some preventive controls at the time of development of surveys. However, for the efficient development of the survey instrument, it is essential that *both* preventive and diagnostic type programs should be considered as an integral part of each of the three phases of the survey, namely, design, collection and processing. Since diagnostic controls are intended to evaluate and monitor rather than improve the performance of a survey, there is a general tendency to attach lower priority to these measures. But on account of the lower relative cost for

diagnostic controls and high potential benefits particularly with
respect to continuous surveys, it is highly desirable that such
control should be given very high priority. Further it should be
noted that data released without appropriate indicators of quality
may leave the users involved in planning and policy decisions
faced with unknown risks.

Having decided on various diagnostic measures to be used
in a survey, the choice among the preventive controls which are
generally much more expensive may be made on the basis of expert
opinion. Also, the survey performance index suggested earlier
could be used in establishing priorities for various measures at
the time of designing the survey or in changing the priorities
as time goes on. It should be possible to establish a lower (L_1)
and upper (L_2) limit for the index (I) on the basis of earlier
experience on the same or similar surveys. As long as for the
current survey $L_1 \leq I \leq L_2$, it may be safely assumed that the
survey performs as planned. Changes would be needed only when
the index does not lie within these limits. If in other similar
surveys I has been found generally to be less than L_2 , and
in the current survey $I > L_2$, this points to the need to consider
whether extra or stronger preventive controls should be implemented.

In such a situation, programs relating to interviewers'
quality control such as observation, training, reinterview list
maintenance, etc. should be given higher priority. Some of these
may even be made more intensive at the cost of reducing the overall
size of the sample or elimination of such programs as updating of
selection probabilities, in order to bring the non-sampling errors
under control. On the other hand, if I has been found to be
greater than L_1 in similar surveys and in the current surveys
$I < L_1$, then the measures should be introduced to reduce the
sampling errors. However, since the index may become low either
due to (a) an increase in sampling error or due to (b) a decrease
in nonsampling errors, additional checks using diagnostic measures
(e.g., design effects) would be needed to confirm the course of

action. Irrespective of the reason, a low index would simply indicate that there are too many preventive measures to control the non-sampling errors. In case of (b), some of these measures may simply be dropped or made less intensive, resulting in a saving of the resources. However, if (a) holds, then these resources should be redirected towards such preventive controls as updating the sample in order to reduce the sampling error. An example of this course of action taken in respect to the LFS was mentioned earlier in this section.

In conclusion, one of the essential features of the control programs should be their flexibility, so that it is possible to change their intensity in order to maximize the performance of the survey instrument at a fixed cost. We would like to stress that the controls should be considered as an integral part of the survey and it should be recognized that a similar relationship exists between the cost of controls and the non-sampling errors as it does between cost and sampling error.

ACKNOWLEDGEMENTS

We wish to express our thanks to Dr. I. P. Fellegi, Dr. B. A. Bailar and several staff members of the Census and Household Survey Methods Division for their very useful comments and discussions on an earlier draft of this paper.

REFERENCES

Drew, J.D. (1977). LFS sample size stabilization. Technical memorandum, Household Surveys Development Division, Statistics Canada, Ottawa.

Drew, J.D., Choudhry, G.H. & Gray, G.B. (1978). Some methods for updating sample survey frames and their effects on estimation. *Survey Methodology* 4, 225-263.

Drew, J.D. & Singh, M.P. (1978). An integrated program for continuous redesign and sample size reduction - A proposal for self-representing areas. Technical memorandum,

Household Surveys Development Division, Statistics Canada, Ottawa.

Ghangurde, P.D. (1978). Estimation, interpretation and use of measures of response errors based on reinterview data. Technical memorandum, Household Surveys Development Division, Statistics Canada, Ottawa.

Gray, G.B. (1976). Stratification index: methodology and analysis. *Survey Methodology* 2, 233-255.

Newton, F.T. (1980). An overview of the field edit module. Technical memorandum, Census and Household Survey Methods Division, Statistics Canada, Ottawa.

Platek, R. & Gray, G.B.(1980). Imputation methodology(total survey error). Prepared for the Panel of Incomplete Data set up by the National Academy of Science, awaiting publication.

Platek, R. and Singh, M.P. (1978). A strategy for updating continuous surveys. *Metrika* 25, 1-7.

Platek, R., Singh, M.P. and Tremblay, V. (1978). Adjustments for nonresponse in surveys. In *Survey Sampling and Measurement* (N.K. Namboodiri, ed.) Academic Press, New York. 157-174.

Rao, J.N.K., Hartley, H.O. & Cochran, W.G. (1962). On a simple procedure of unequal probability sampling without replacement. *Journal of the Royal Statistical Society* B24, 482-490.

Singh, M.P. (1978). Alternative estimators in PPS sampling. *Survey Methodology* 4, 264-280.

Statistics Canada (1977). Methodology of the Canadian Labour Force Survey. Catalogue No. 71-526, Statistics Canada, Ottawa.

Statistics Canada (1978). A compendium of methods of error evaluation in census and surveys. Catalogue 13-564E occasional, prepared by J.-F. Gosselin, B.N. Chinnappa, P.D. Ghangurde and J. Tourigny, 1-25.

Tremblay, V., Singh, M.P. & Clavel, L. (1976). Methodology of the labour force survey reinterview program. *Survey Methodology* 2, 43-62.

U.S. Bureau of the Census (1978). An error profile: employment as measured by the current population survey. Statistical Policy Working Paper 3, prepared by C.A. Brooks, B.A. Bailar, et.al. U.S. Bureau of the Census, Washington,D.C. pp.1-89.

SURVEY SAMPLING ACTIVITIES AT THE SURVEY RESEARCH CENTER

Irene Hess

University of Michigan

The Survey Research Center's primary concern is
with the application of sample survey methods to a variety
of social problems. The design, selection and servicing
of a national sample of counties and households for
personal interview surveys is a major sampling activity.
The design of the county sample is discussed and its
multipurpose uses illustrated with brief descriptions
of several classes of household and nonhousehold
sample surveys. Other types of sample designs developed
at the Center are presented and some of the more unusual
ones described.

1.   INTRODUCTION

1.1 Historical Notes

In 1946 members of the professional staff of the Division
of Program Surveys, Bureau of Agriculture Economics, United
States Department of Agriculture, moved to The University of
Michigan to create a new type of survey organization, the Survey
Research Center (SRC). Two years later, when the senior staff of
the Research Center for Group Dynamics (RCGD) at the Massachusetts
Institute of Technology joined the SRC in Ann Arbor, the Institute
for Social Research (ISR) was formed. In 1964 the Center for
Research on the Utilization of Scientific Knowledge (CRUSK) was
organized and in 1970 the Center for Political Studies (CPS) was
established. The last two centers were outgrowths of two programs

of research within the SRC: the Program of Human Relations and
Social Organization, and the Program in Attitudes and Behaviour
in Public Affairs.

The Sampling Section, an organizational unit within the
SRC, may provide sampling services and consultation for each
center within the Institute, although its first responsibility
is to the SRC.

From its beginning, the SRC's primary concern has been with
the application of sample survey methods to a variety of social
problems: psychological, sociological, economic and political. In
the early years the Center undertook projects that met four
criteria (Survey Research Center, 1952):

- A project must clearly be in the public interest;
- It must give promise of contributing to important
  scientific goals;
- It must be directed toward the common good rather
  than for the benefit of one organization;
- The Center must be permitted substantial freedom in
  the design of a study and the publication of the results.

Within those limits, the Center undertook studies with the
sponsorship of government agencies, private business organizations
and research foundations. In some cases studies were instituted
at the request of an outside agency, and in other cases members
of the Center's research staff took the initiative.

There was no intention to become competitive with government agencies or with other research organizations, nor were there
any plans to offer survey research services to groups outside of
the Institute. Requests for proposals from federal and state
government agencies not having developed to current volume, much
of the funding was in the form of grants.

Within the Center, research programs have evolved around
the special interests of the research staff. The oldest and
largest program, largest in the sense of number of staff members
and volume of work, is in Economic Behavior. Other current

programs include: Organizational Behavior; Urban Environmental
Research; Social Indicators; Youth and Social Issues; Family and
Sex Roles; Group Identification, Interrelations, and Achievement;
and Survey Methodology. To a large extent, these programs determine the sampling activities at the SRC.

1.2 The Sampling Section Organization

From a group of five or six in the early years of the
Center, the sampling staff has grown to 15 or 20 members, including professional, secretarial and supporting staff, and graduate
student assistants. About half of the staff are full-time
employees, while others have fractional appointments. Some are
employed on a temporary basis at times when the volume of work is
unusually large. Hence the size of the staff varies from time to
time.

The section has neither fixed funding nor a fixed schedule
of projects, since both depend on the success that the several
programs may have in developing proposals and receiving grants or
contracts. Members of the sampling staff participate in sample
design and related aspects of proposal development and prepare
cost estimates for sampling activities included in the proposed
research.

When a low volume of program related research is anticipated, the sampling staff may accept from organizations outside
of the Institute requests for consultation or sampling services.
Or the sampling section and field office may jointly accept an
outside request for sampling and data collection services if the
area of research is of interest to either group and supplements
rather than competes with other SRC projects. Occasionally in
the past and more frequently during the last two or three years
since the sampling staff has expanded, members of the section
have been successful in negotiating funding for research in their
fields of interest.

1.3 Defining Survey Sampling Activities

Survey sampling activities at the SRC as here discussed relate primarily to sample design and selection. The activities of methodological research, consultation, teaching and training are omitted. The design, selection and servicing of a national sample of counties and households, a major sampling activity at the Center, is described and the multipurpose uses of the county sample illustrated with brief descriptions of several classes of both household and non-household surveys occurring over the life of the Center. Next, other national designs are presented. These are followed by examples of state and local area surveys. Attention is given to the designs for telephone surveys, particularly those conducted from a central location. Brief comments about weighting, adjustment for nonsampling errors, estimation and sampling error calculations conclude the remarks on survey sampling activities.

2. A NATIONAL SAMPLE OF COUNTIES AND ITS MULTIPURPOSE USES

2.1 Description of the County Sample

The predecessor of the first national sample at the SRC was developed at the Department of Agriculture to obtain factual information about the distribution and uses that people expected to make of liquid assets accumulated during the war years. After the Survey of Liquid Asset Holdings, Spending and Saving was conducted in 1946 for the Board of Governors of the Federal Reserve System, the sample, as well as the professional staff, was transferred to Ann Arbor.

The multistage, area probability sample had been designed to have 66 strata and 66 primary sample areas, one from a stratum. Twelve of the strata, corresponding to the 12 most populous metropolitan districts, had only one primary area. Consequently, each of the 12 areas entered the sample with certainty. Each of the remaining 54 strata, approximately equal in adult population, contained two or more primary areas only one of which was selected for the sample. The primary areas were single counties or groups of counties. The sample was designed to yield approximately 3,000 completed, personal interviews for the Federal Reserve Board survey.

Although that first sample was not planned for long-term use at the SRC, the sample size of 66 primary areas proved to be well suited for the type of research to be conducted at the Center. Since a national sample of counties has been the foundation for the majority of the SRC sample surveys, acquaintance with the basic design is a necessary first step to the understanding of many of the survey sampling activities. Furthermore, the data collection organization must be considered. The interviewing staff is composed of local people, usually two per sample area except in major metropolitan areas where there are more. These interviewers are trained and supervised by 10 to 12 travelling supervisors each of whom is responsible for the staffing in several of the sample areas.

Just as many survey samples begin with first and even second and third-stage sampling units already determined, most of the interviewing staff has already been engaged and has had some experience with the SRC interviewing practices.

In the second year at the SRC, a complete redesign and reselection of primary sample areas was undertaken. The 12 major areas and the remaining metropolitan districts were redefined to conform strictly to the federal government's definitions of metropolitan districts which did not always follow county lines. Nonmetropolitan primary areas were sometimes made more heterogeneous by including more than one county in a primary area or by combining counties and part counties to form sampling units. Selection probabilities were changed from a base of adult population to one of total population. Explicit strata were formed within each of the four geographic regions to give them proportionate representation and to permit separate estimates and analysis by region. The probability technique of controlled selection was developed and applied to control the sample selection within a region by degree of urbanization and groups of states beyond the control that could be expected from simple stratified random sampling alone (Goodman and Kish, 1950).

After each decennial census, the more than 200 metropolitan areas have been redefined to conform to the federal government's latest definitions of metropolitan areas. Although adjustments to such definitional changes disrupt many strata, it is thought that a clear distinction between Standard Metropolitan Statistical Areas (SMSA's) and non-SMSA's, each with representation in proportion to its population, has advantages in estimation and analysis, in making comparisons with census data and with data from other organizations, and in the sampling of subgroups of the population for special purpose studies.

The 1960 sample changes included expansion of the 54 areas to 62 in order to increase the number of strata in the South and in the West where the largest growth had occurred during the 1950's. The sample revisions in 1960 and 1970 were made using techniques to maximize the likelihood of retaining former sample areas after changing strata and measures of size (Keyfitz, 1951; Kish and Scott, 1971).

The distribution of the 74 areas in use during the 1970's appears in Table 2.1 along with the distribution of the total population by region, certainty and noncertainty areas, and by SMSA and non-SMSA classifications. It can be seen that the 12 certainty areas contain nearly 30 percent of the total population. These areas vary in population from two million to 16 million. The other SMSA's, with 40 percent of the population, must have a minimum size in excess of 50,000 since the presence of a city of 50,000 is one of the qualifications for SMSA status. The remaining 31 percent of the population is located in non-SMSA counties which are grouped to achieve a minimum size of 2,000 population.

Table 2.1.   Distribution of the SRC Sample of County Areas by
Geographic Regions, Certainty and Noncertainty
Selections, Compared with Conterminous United
States Population Distribution, 1970

Item	All Regions	North- east	North Central	South	West
All sample areas	74	14	22	26	12
Certainty selections[1]	12	4	4	2	2
Noncertainty selections	62	10	18	24	10
SMSA's[2]	32	6	8	12	6
Non-SMSA's	30	4	10	12	4
Conterminous United States					
Population					
Number (millions)	202.1	49.1	56.5	62.7	33.8
Percent	100.0	24.3	28.0	31.0	16.7
In certainty areas					
Number (millions)	59.3	26.6	16.2	4.9	11.6
Percent	29.3	13.2	8.0	2.4	5.7
In remaining areas					
Number (millions)	142.8	22.5	40.3	57.8	22.2
Percent	70.7	11.1	20.0	28.6	11.0
SMSA's					
Number (millions)	80.4	13.8	21.2	30.2	15.2
Percent	39.7	6.8	10.5	14.9	7.5
Non-SMSA's					
Number (millions)	62.4	8.7	19.1	27.6	7.0
Percent	31.0	4.3	9.5	13.7	3.5

[1] The 12 certainty selections are the New York - New Jersey, and
the Chicago - Northwestern, Indiana, Consolidated Areas, and
the 10 largest SMSA's outside of the consolidated areas.

[2] The February, 1971 SMSA definitions remain in current use.

In the table it can be seen that the 62 sample areas are
distributed in even numbers by SMSA classification across the
regions. This forcing to even numbers keeps the regions and
classes self-contained when a paired difference technique is used
to calculate measures of sampling variability, or when 31 sample
areas, one-half of the 62, are sufficient for research needs.
The approximate geographic locations of the 74 areas in current
use are shown in Figure 2.1.

Alaska and Hawaii which entered the Union in 1959 are not
included in the SRC sample of counties. Most frequently household
surveys are in the range of 1,500 to 3,000 interviews. The small
number of interviews to be taken in Alaska and Hawaii as self-
representing areas become disproportionately costly. Occasionally,
however, for special purpose studies data collection may be
conducted in those states.

Figure 2.1.   Geographic Distribution of the 74 Primary Areas in
the SRC Sample of Counties

2.2 Surveys of Household Populations

2.2.1 *Household Samples and respondent selection procedures*

By far the majority of SRC national sample surveys have
required personal interviews with household members. Just as
the samples of counties had their origin with the Surveys of
Consumer Finances, so did the samples of households within sample
counties. Developing and maintaining household samples for
continuing use through personal interviewing was and is a major
sampling activity. In addition to Census publications, many
Enumeration District lists and corresponding maps are purchased
from the Census Bureau. Sources for current county, city and
town maps must be located and purchases made. Local areas are
contacted to obtain information on major new residential construction which then is incorporated into the sampling process.

Within primary units, area probability sampling continues
through several stages to control the household selections by
urbanization, geographic location, economic level and racial
composition. Household samples large enough for six to 12 surveys,
depending on individual survey needs, are selected and subsampled
without replacement to choose the specific area segments and
households to be contacted for a particular survey. Usually
sample dwellings are selected in clusters of approximately four.
But there can be variations. For example, when searching for a
rare population, larger clusters are used. In the early survey
years, an average of only two sample dwellings were selected from
urban blocks, but blocks were selected in pairs so that an
expected four dwellings came from two nearby blocks.

As a general practice, for a particular survey, households
are given equal selection probabilities over all stages of
sampling. But again there may be research demands that justify
disproportionate sampling.

The specifications for respondent eligibility may define a
unique individual: the major wage earner; the person with major

financial responsibility for the household; the only member with
certain qualities or characteristics. Or two or more individuals
may meet eligibility criteria but only one interview is desired
from the household. In that event an objective procedure for
choosing one member is used with no substitution of respondents
permitted (Kish, 1949). Consequently, respondents may have
unequal chances of selection.

2.2.2 *Surveys of Consumer Finances*

The Survey of Consumer Finances continued annually from
1947 through 1959 with Federal Reserve Board sponsorship. For
these surveys as well as for the Liquid Assets Survey of 1946, it
was considered necessary to give higher selection rates to the
households in the higher income groups in order to obtain a
larger proportion of high income families, and more reliable
estimates of dollar values, than would have been obtained with
equal selection probabilities. After the survey at the Agriculture Department, data were analyzed to learn what improvements
might be made in the sample for the following year. While
sampling blocks at differential rates was beneficial, it was
thought that a more efficient procedure would be disproportionate
sampling of dwellings based on interviewer ratings assigned from
observation as residential housing in sample blocks was listed.
Three ratings and three sampling rates were used. That procedure,
used first in 1947, led to the selection of annual samples of
approximately 6,000 dwellings which were then subsampled disproportionately to yield approximately 3,000 completed interviews
with heads of spending units. (Spending units were family members
who pooled their incomes for their major items of expense. There
could be more than one spending unit in a household.) Although
the Survey of Consumer Finances as operated through the 1950's
has been discontinued, the financial studies continued into the
1970's with some changes in research design. The sample sizes
became smaller than formerly and disproportionate sampling of

dwellings by economic rating was discontinued.

2.2.3. *Interim or omnibus surveys and the studies of consumer sentiment*

Beginning in the early years at the SRC and continuing into the 1970's the Economic Behavior Program organized two or three personal interview surveys each year in addition to the Surveys of Consumer Finances, always conducted in January and February. These interim, quarterly or omnibus surveys, from which stem the consumer sentiment reports, have had variations in sample design, sample sizes (usually 1,250 to 1,500 interviews) and respondent selection procedures. The surveys, combining a series of economic related questions with blocks of questions from other subject areas, demonstrated that one survey could collect data that cut across several social science disciplines.

These surveys have frequently served as vehicles for exploratory work in new subject areas and for methodological studies in questionnaire design and question wording. The experimental work sometimes challenged the sampling staff to devise procedures to ensure that each experimental question form, or block of questions was administered to replicate samples of respondents. The last personal interview survey of this series was conducted in the spring of 1976. Since that time data collection has continued by telephone.

2.2.4 *Surveys of Political Behavior*

One of the most successful explorations into a new subject area was the addition of a few questions concerning vote intention to the omnibus survey in the fall of 1948. The accuracy of the returns, not entirely due to scientific sampling and research for there was a large element of luck involved, prompted a hastily organized post-election study. From this small beginning, the studies of political behavior developed. It must be clearly understood that the samples are not designed to predict election returns nor is that the goal of political science research.

Beginning with the political study in 1952, these surveys have recurred biennially, sometimes with and sometimes without researchers in other subject areas as collaborators. Respondents have been citizens of voting age. Sample sizes have varied from 1,000 to 2,600 completed interviews, frequently a combination of current and panel studies.

For the purpose of studying Representatives, the constituency they represent and the linkages between these two groups, it was recognized that counties as first stage sampling units were unsatisfactory because of the large variations in numbers of designated respondents by congressional districts with constituents in the sample. The 1976 survey was the last political study conducted within the SRC's 74-area sample of counties. In 1978 a new series of studies began with the congressional districts as primary areas.

2.2.5 *Other sample surveys of household populations*

To illustrate the variations in sample design that occur within the county and household framework, four other household surveys are mentioned:

- The family planning studies had their beginning in 1955 with approximately 2,600 interviews conducted with white married women age 18-39 with husband present or in military service, and another 400 interviews with single women age 18-24 years (Freedman, Whelpton and Campbell, 1959). A second family planning study in 1960 included in the study population all married women age 18-30 years with husband present or in military service, and added two other age groups of white women with specific marital qualifications. After these pioneering experiences, the series of family planning studies have been continued quinquennially by other survey organizations.

- The panel study of family economic progress began in 1968 and continues annually. The research design combined two household samples; (1) a cross-section sample of about 3,000 families selected from the SRC's 74-area sample; (2) a supplemental sample of approximately 2,000 low-income families chosen at disproportionate rates from a larger sample selected two years earlier and twice interviewed by the Census Bureau (Survey Research Center, 1972). Sample families were requested to sign releases to permit transfer of identication information to the SRC. The supplement was drawn from the SMSA's of the four geographic regions and also from non-SMSA's of the South. This survey has the most complex weighting system of any of the SRC household samples.

- The largest sample selected for one household survey was a sample of 28,000 drawn for a researcher at another university interested in the study of the nutritional status of preschool children ages 1 through 5 (Owen,Krom, Garry, Lowe and Lubin, 1974). About one household in six had eligible children. The data collection included physical examinations by a mobile medical team. Since the cost of moving the team around the country ruled out a random assignment of interviewing dates, over a two-year period the team moved back and forth across the country in a pre-designated pattern that led to the completion of half of each region's sample each year, with data collection during at least two different seasons.

- During a 55 week period, December 1976 through December 1977, data were collected on the mail flow to and from about 5,500 households in the United States, approximately 100 per week. The assignments of households to a data collection week were randomized to form 55 replicates that could be combined to give quarterly, semi-annual and annual estimates of the household mailstream (Kallick,1978).

An initial personal contact with each household was
followed by daily telephone calls during the week of data
collection. These data were needed by the Mail Classification Research Division of the United States Postal
Service.

2.3 Surveys of other populations

Three samples constructed with the 74 areas as primary
selections illustrate how the sample of counties can be used as
first-stage units for surveys of government officials, clients of
service organizations, and schools. Often it might appear desirable to ignore counties and begin with organizations as first-
stage sampling units. On the other hand, there may be no complete
list of organizations, or the number may be so overwhelmingly
large that one prefers to begin with clusters of organizations,
the county being one type of cluster or group. A primary consideration is that the study population be distributed approximately
as the total population, the base for the probability selection
of sample counties. If the Center's interviewing staff is
involved, it is desirable that the sample be in or near the areas
where interviewers are located. Also both the cost and time
required to develop a sample may be reduced if the sample selection
can begin with first-stage units already defined and those
probabilities calculated.

2.3.1 *State, county, township and municipal governments and the general revenue sharing program*

A study designed to aid policy-makers in their evaluation
of the federal government's General Revenue Sharing Program
required personal interviews with government officials in each of
the 50 state governments, in a sample of counties and townships,
and in five size classes of municipalities (Juster, 1975). The
SRC's 74-area sample required supplementation before the desired
samples of municipalities could be satisfied. It was necessary

to make a second primary selection in strata where any primary area fell below a specified proportion of the stratum population.

2.3.2 *Public assistance clients in the vocational rehabilitation program*

A similar situation arose with a study of public assistance clients in the vocational rehabilitation service system. The sample selection, undertaken for a research institute at another university, did not involve the SRC field staff but by using the Center's 74-area sample as first-stage units, there were appreciable savings in the time required to implement the sample. Again the sample was supplemented in strata where any primary area had less than a specified proportion of the stratum population. A probability selection of counselors serving sample counties was made and monthly samples of new clients drawn over a six-month period, fiscal considerations having reduced the desired time period from 12 months to six.

2.3.3 *National samples of youth enrolled in primary and secondary schools*

Sampling youth enrolled in schools is a familiar and often repeated sampling activity.

- In 1952 a sample of adolescent boys, ages 14 through 17, was chosen by sampling schools within sample counties, then selecting home rooms within schools and pupils within classes. That study sponsored by a national organization for boys was followed in 1956 by a similar study for a national organization for girls (Douvan and Adelson, 1966). The population of adolescent girls was defined as girls in grades 6 through 12, since grades within schools are readily defined and located while pupils within some specified age span may be scattered through many grades.
- In 1958-59 samples of schools were selected to serve two

studies: a survey of boys in grades 5 through 9; and a
survey of the fitness of American youth in grades 5
through 12. The latter study, initiated by the
President's Council on Youth Fitness and directed by the
Department of Physical Education at The University of
Michigan, was the first national survey of the physical
fitness of school children. Youth fitness surveys have
been repeated periodically sometimes with and again without the participation of the SRC sampling staff (Reiff,
Kish and Harter, 1968).

- Beginning in the spring of 1975 and continuing annually
to the present and into the future, a national study of
high school seniors in public and private secondary
schools monitors drug use and related attitudes of youth
(Bachman, Johnston and O'Malley, 1980). The 74-primary
area sample again was supplemented in much the same
manner as for the samples of governments and vocational
rehabilitation centers. The research design required
that there be about 120 sample schools per year, that a
school remain in the sample for two consecutive years,
not to be returned to the sample for at least two years
thereafter, and that each year one-half of the schools
should overlap the preceding years sample while one-half
enter the sample for the first time or after a lapse of
two years. When the number of seniors in a school is
less than 200, all are included in the sample. With
larger numbers there is some subsampling, usually by
classes. About 18,000 self-administered questionnaires
in five forms are completed annually.

3. OTHER SAMPLE DESIGNS

3.1 National Samples

Until 1978 the SRC had one national sample of counties,
staffed by local interviewers, where almost all of the personal

interview surveys were conducted. Most frequently the study populations would be household members, but they could be schools and pupils, units of government and other types of organizations. In 1978 and 1979, in quick succession three other national samples evolved.

3.1.1 *Sample of postal areas and nonhouseholds*

To complement the household mailstream survey, data were desired on the mailflow from any business, nonprofit organization or unit of government that receives mail or purchases postal service, or from any location that by outward sign reveals the presence of occupants other than householders - hence the term nonhouseholds. Data were needed by several types of nonhouseholds, by industry codes, class of mail, and several postal regions. A complete list of postal areas with their postal revenues being available, it was thought desirable to consider postal areas as primary units and postal revenue as measures of size. Postal areas vary in annual revenue from approximately $500 million to less than $1,000. In choosing the 153 primary selections, smaller areas were linked to achieve a minimum size (Survey Research Center, 1978). Also one of the stratification variables and a control in the sample selection was whether a postal area was in, near to, or far from one of the 74-area sample points. In Figure 3.1.1 it can be seen that nearly half of the selections are in the 74-areas. Within the 153 selected postal areas, establishment lists were compiled, and a sample selected and scheduled for 52 weeks of data collection. Data on originating mail were obtained from about 100 establishments per week.

Consideration is now being given to the feasibility of converting the establishment sample into a continuing sample to collect data on entrepreneurial sentiment, analogous to the consumer sentiment surveys.

LEGEND
● In the SRC County Sample (74)
▲ Not in the SRC County Sample (79)
2● Number of sample postal areas in SRC
County Samples

Figure 3.1.1 Geographic Distribution of the Sample of 153 Postal
Areas

3.1.2. *Sample of congressional districts for studies of policital behavior*

In 1978 a post-election survey of approximately 2,500 interviews was conducted with a sample of congressional districts as
first stage sampling units. One hundred eight of the 432 districts
were selected for the primary sample areas. In the selection
process one of the stratification variables classified districts
as being within one of the 74-area sample points, or containing
all or part of one of the 74-areas, or completely disassociated
from the 74-area sample. Congressional districts, approximately
equal in 1970 population, are most irregular and variable in
geographic area encompassed.

The sample of secondary selections, which are counties or
part counties, was controlled within the bounds of probability

sampling to maximize the overlap between the 74-area sample and
the secondary selections of the 108 sample districts. In Figure
3.1.2 it can be seen that 85 of the secondary selections are in
the 74-area sample. Insofar as the samples of cities, towns and
open country selections, and segments within those areas could be
utilized by the congressional district sample, the county and the
district samples share sampling materials, but not households
which are sampled without replacement.

LEGEND
● In the SRC County Sample (85)
▲ Not is the SRC County Sample (23)
2● Number of secondary selections in SRC
   County Sample
●─● Two Secondary Selections from one
   Congressional District

Figure 3.1.2.   Geographic Distribution of Secondary Selections in
                the Sample of 108 Congressional Districts

During 1980 about 4,500 new or first-time interviews are
being taken at definitely scheduled time periods from January to
November, plus around 1,500 reinterviews prior to election day and
an estimated 1,360 post-election reinterviews. The entire sample
of 4,500 new interviews was selected and then subsampled to
designate the several subsets to be interviewed at specific times.
Subsets are national samples varying in size from 650 to 1,000
interviews.

3.1.3 *Sample of black households*

The black population is not distributed in proportion to
total population. Therefore, a new sample of counties and households was needed to satisfy the research request for a sample of
approximately 74 primary areas to yield about the same level of
precision as the SRC 74-area sample of counties and households.
The desired sample size was 2,500 completed interviews.

Again there was the definite intention to utilize as much
as possible the materials assembled and developed for the 74-area
sample. The primary areas defined for the SRC county sample were
retained with no revisions. The measure of size was changed from
total population to black households. Eighteen primary areas,
one per stratum, were included with certainty. Fifty-eight other
strata were formed from the original 74 by reducing the number
of strata in three regions and by increasing the number of strata
in the South. One sample area was selected from each of the 58
strata using probability techniques to retain wherever possible
sample areas in the current 74-area sample. The approximate
locations of the 76-area sample points are shown in Figure 3.1.3.
The high proportion of new sample points occurs because a little
over one-half of the black population is located in the South
whereas only about one-third of the total population is in the
Southern Region.

Figure 3.1.3   Geographic Distribution of the 76 Primary Areas in
               the Sample of Counties Selected for the Study of
               Black Americans

3.1.4   *Sample of nonfederal, general medical hospitals and
        admissions*

    A sample for a national study of changing patterns of

hospital care and costs over a 15-year period was designed and

selected in cooperation with the University's Bureau of Hospital

Administration (American Medical Association, 1964).   Three

hospital samples were required from three classifications of

nonfederal general medical hospitals: (1) those in existence at

the beginning of the period;   (2) those in existence eight years

later, a mid-point in the time period;   (3) and those in operation
during the last year of the research period. A hospital selected
for the first-period sample was retained for the following periods
if it remained in operation. Likewise, a hospital entering the
sample for the second period was retained for the third period
also. Within sample hospitals a full year of hospital admissions
were sampled to obtain individual records for analysis. Since the
SRC's field staff did not participate in the study, there was no
reason to construct the sample around the SRC sample of counties.
The hospitals were the first-stage sampling units and the
admissions the second stage.

3.2   State Samples

3.2.1   *In the State of Michigan*

　　One might think that with survey research facilities located
within the State there would be a continuing demand for those
services which would justify keeping samples of Michigan households and establishments selected and ready to implement on short
notice. Such has not been SRC's experience.

-   In 1950 a survey of establishments in three major
    industrial classifications - metal products, furniture,
    and chemicals - looked at factors that attract manufact-
    urers to the State and reasons why some manufacturers
    remain in the State while others consider moving to
    another state (Survey Research Center, 1950). A sample
    of nearly 200 plants was chosen from a list that included
    the number of employees in each plant, permitting
    selection probabilities in proportion to that measure of
    size. Plant executives were contacted for personal
    interviews. Two other surveys of Michigan industry have
    been conducted at intervals of about 10 years. Each
    sample was selected independently, with variations in
    sample size and in stratification by geographic locations,

number of employees and industrial classifications.

- In 1955 the SRC's first household sample for the State
  was selected with 17 primary areas comprising 23 counties.
  Six of the areas which were included with certainty
  contained two-thirds of the State's population. Sampling
  households in cities, towns and open country areas within
  the sample counties was designed in much the same manner
  as for national samples. The State sample design provided
  for several surveys in the range of 400 to 1,000 interviews. The first study of approximately 600 interviews
  collected residents' perceptions of price changes, price
  expectations, and attitudes toward the current costs of
  consumer goods.

- In the fall of 1958, as part of the comprehensive study
  of hospital and medical economics, conducted for the
  Governor's Study Commission on Prepaid Hospital Administration, about 1,000 interviews were taken with the
  household population - consumers of medical care
  (McNerney, 1962).

- Another part of the Michigan medical study required a
  probability sample of nonfederal, general hospitals for
  short-term care and samples of patient records for the
  calendar year 1958 (Hess, Riedel and Fitzpatrick, 1975).
  The sample of 47 hospitals was used for two parts of the
  comprehensive study: Character and Effectiveness of
  Hospital Use; and Changing Patterns of Care. The design
  for the national study of changing Patterns of Hospital
  Care, which occurred a few years later, was an outgrowth
  of the state study design.

3.2.2 *Study of medical care in the State of Hawaii*

A third survey in the medical care field for which the SRC's
sampling staff provided consultation and sampling services was the
1970 study of personal medical care in the State of Hawaii

conducted by researchers from the University of Michigan School
of Medicine (Payne and Lyons, 1972). A year of hospital discharges in 16 diagnostic categories from the 22 nonfederal, short-
term hospitals in the State were sampled to obtain approximately
3,700 discharges for the research. Both hospital records and
ambulatory care records before and after hospitalization were
examined.

The Office Care Study, independent of the hospital discharge
study, was designed as a two-stage sample, the first stage being
a sample of primary-care physicians selected from lists provided
by the research staff. The second-stage was a sample of visits
logged by each of the selected physicians for 12 diagnoses or
conditions over a six-weeks period.

3.2.3 A multipurpose household sample for a midwestern state

In the spring of 1976 the sampling staff designed and
selected a master sample of counties and housing units for the
State Planning Agency of a midwestern state. A state sample
similar in design to the Michigan sample would be a fairly routine
undertaking. However, this other state had 13 planning regions
for which it was thought separate surveys would be needed in the
near future. The state sample was stratified and controlled so
each region had proportionate representation in the state sample.
Furthermore, any sample cluster in the state sample was designed
to be a sample cluster in a regional sample. Three of the regional samples were completed for data collection simultaneously with
the first state survey in the summer of 1976. The SRC staff
prepared field maps and sketches to direct state interviewers to
sample housing units for the first survey. State interviewers
were responsible for all field work, both listing of residential
units and interviewing. All work materials were transmitted to
the state. The SRC has no continuing responsibility for this
multipurpose sample.

3.3   Local Area Samples

Except for the sample size, local area household surveys may differ little in design whether the community is the entire survey area or part of a state or national survey. But not all local surveys are routine. Two with unusual designs deserve mention.

3.3.1   *Visits to hospital emergency units*

A medical foundation requested a design for a sample survey of emergency unit visits to hospitals in a metropolitan area. Since records were not kept in a manner suitable for sampling, the flow of emergency unit visits through hospitals was viewed as one might regard the flow of motor traffic through key intersections. Hospitals were assigned to seven groups, three in the city and four suburban rings that could be included or excluded, individually or collectively, according to later developments. There was no sampling of hospitals, only the sampling of visits. The medical foundation was to be responsible for employing and training personnel to record visits to emergency units at times specified by the sample design. Data collection was planned to cover a full year with a self-weighting sample of 4,000 visits allocated to represent hospital groups, seasons, days of the week, and three 8-hour shifts in proportion to the volume of emergency unit visits within groups and time periods. Data about the 4,000 cases were to be collected through interviews with certifying personnel. Patients or their family members were not to be contacted at emergency units although a subsample of 800 patients or family members were to be interviewed in their homes several weeks after the emergency unit visit. The seven regions and seven days per week suggested a series of 7×7 Latin squares to schedule the sample by weeks, regions and days. Listing sheets were designed and sampling instructions provided for each data collection period.

3.3.2 *Continuous mental health assessment*

Another local area study was conducted to demonstrate the utility of continuous community mental health assessment (Roth and Klassen, 1973). The research design required that about 25 personal interviews be taken each week for a period of 52 weeks with household members, 18 years of age or older, one per household. Each respondent was to be given three short forms to be completed and mailed to the research office one per week for three weeks following the personal interview, after which the respondent was withdrawn from the study. It required four weeks of field operations before the data collection was in full production with returns from 100 respondents per week, 25 personal interviews and 75 mail returns. By allowing for vacancies and nonresponse, it was estimated that an expected 28 sample housing units would be required each week for 55 weeks. However, the numbers 28 and 56 were better suited for sampling purposes. Therefore, the sample design specified that 56 homogeneous strata of approximately equal size be formed using as stratification variables racial composition, model neighborhoods (disadvantaged neighborhoods with special social programs), geographic location and economic level. From each of the 56 strata, two sample blocks were selected with probability proportional to the number of housing units, yielding 112 sample blocks. From each of the 112 blocks an expected 14 housing units were selected to yield an expected 1568 housing units for the sample at a constant over-all rate. For scheduling weekly samples the 56 strata were then combined in groups of four to form 14 large homogeneous strata. Each of the 14 large strata contained 8 sample blocks with 14 sample dwellings and therefore contributed two sample dwellings per week over a period of 56 weeks. Within each large stratum the four small strata formed two pairs or two half-strata. The assignment of housing units to the 56 weeks required that within a half-stratum there should be a selection from each small stratum once every two weeks, and from each sample block a selection once every four weeks. With those controls, assignments were randomized over each 4-week period. To return the sample to the required 55 weeks, dwellings scheduled for the 56th week were never contacted, and the over-all sampling rate was reduced accordingly.

3.4 Samples for Data Collection by Telephone

For a number of years the surveys of consumer sentiment used two forms of data collection: (1) personal interview surveys in which respondents were asked to supply a telephone number whereby they might be reached for reinterviewing; (2) recontacts of former respondents by telephone calls from the local interviewing staff. The subsampling of respondents who reported telephone numbers was a routine task for the sampling staff. But with the discontinuation of personal interviewing for omnibus surveys, that source of telephone numbers disappeared. However, with the advent of the wide area telecommunications service, and the development of techniques for generating samples of telephone numbers by computers, the economic studies of consumer sentiment now collect data monthly. The sample design clusters residential subscribers by telephone exchange. Each quarter a new sample of exchanges is selected. Interviewing is conducted from the central telephone facility in Ann Arbor.

Two types of national telephone samples are available at the SRC: (1) samples of telephone numbers within the 74-area sample of counties; (2) stratified samples of telephone numbers within conterminous United States (Groves and Kahn, 1979). With either design telephone numbers are clustered by exchange. One purpose of the former design is to permit screening by telephone to locate members of a rare population that are then to be interviewed in person. The full potential of this system has yet to be utilized.

Several telephone samples of Michigan residential subscribers have been selected as well as samples in some local areas.

4. OTHER ACTIVITIES

Sampling activities do not end when data collection begins. The sampling staff is available for problem solving during the interviewing period of a study. Also attention must be given to any other factors that affect the sample.

4.1 Weights for Disproportionate Sampling

If respondents have been given unequal chances of selection, the effects of such disproportionalities are examined and appropriate weights supplied by the sampling staff.

4.2 Adjustments for Nonresponse

Responsibility for nonresponse adjustments is shared by the sampling and research staffs. Adjustments may be in the form of imputation or weighting or a combination of these. Sometimes no adjustments are made.

4.3 Estimation Procedures

Estimates from sample data are generally ratios, means, indexes, correlations, regression coefficients and other multivariate analyses as decided by the research staff. If estimates of population totals are desired, the sampling staff may be requested to advise on the use of some ratio estimator since with small sample sizes a simple unbiased estimate is often imprecise.

4.4 Sampling Error Calculations

It is the responsibility of the sampling staff to provide codes so that each interview (response, record) can be identified with the primary sampling unit or cluster of which it is a member. Formulas for the calculation of sampling errors appropriate to the estimation techniques and the sample design must be provided also. Generally the existing computer programs can satisfy requests for sampling error calculations. The sampling staff may take full responsibility for the calculations or advise others in the use of the programs.

5. SUMMARY

To summarize, the purpose of survey research at the Survey Research Center is to study social problems. Survey sampling activities, primarily sample design and selection, are determined by the research investigations negotiated by the research staff with governments, business or industry, foundations or other organizations. National surveys for personal interviewing center around a probability sample of 74 county areas staffed by local interviewers, trained and directed by travelling supervisors. Surveys of household populations usually require that a specific individual be designated as the respondent rather than accepting an interview with any eligible household member. The recent development of a centrally located telephone interviewing capability has required designs for computerized sampling of telephone numbers. Many special purpose samples of a variety of national, state and local populations have been designed and selected. The Center's samples are generally small and the designs complex.

REFERENCES

American Medical Association (1964). *Changing Patterns of Hospital Care*, Report of the Commission on the cost of Medical Care IV, Chicago: The Association.

Bachman, J.G., Johnston, L.D. and O'Malley, P. (1980). *Monitoring the Future, 1976*. Ann Arbor: Institute for Social Research, The University of Michigan.

Douvan, E. and Adelson, J.(1966). *The Adolescent Experience*. New York: Wiley.

Freedman, R., Whelpton, P.K. and Campbell, A.A. (1959). *Family Planning, Sterility and Population Growth*. New York: McGraw-Hill.

Goodman, R. and Kish, L. (1950). Controlled selection - a technique in probability sampling. *Journal of the American Statistical Association* 45, 350-372.

Groves, R.M. and Kahn, R.L. (1979). *Surveys by Telephone*. New York: Academic Press.

Hess, I., Riedel, D.C. and Fitzpatrick, T.B. (1975). *Probability Sampling of Hospitals and Patients*. 2nd ed. Ann Arbor: Health Administration Press.

Juster, F.T., editor (1975). *The Economic and Political Impact of General Revenue Sharing*. Ann Arbor: Institute for Social Research, The University of Michigan.

Kallick, M. and study staff (1978). *Household Mailstream Study*. Ann Arbor: Institute for Social Research, The University of Michigan.

Keyfitz, N. (1951). Sampling with probability proportional to size: adjustment for changes in probabilities. *Journal of the American Statistical Association* 46, 105-109.

Kish, L. (1949). A procedure for objective respondent selection within the household. *Journal of the American Statistical Association* 44, 380-387.

Kish, L. and Scott, A. (1971). Retaining units after changing Strata and probabilities. *Journal of the American Statistical Association* 6, 461-470.

McNerney, W.J. and study staff (1962). *Hospital and Medical Economics*. Chicago: Hospital Research and Educational Trust.

Owen, G.M., Kram, K.M., Garry, P.J., Lowe, J.E. and Lubin, A.H. (1974). A study of nutritional status of preschool children in the United States, 1968-1970. *Pediatrics, American Academy of Pediatrics* 53, No. 4, Part II, Supplement April.

Payne, B.C. and Lyons, T.F. (1972). Episode of illness study and office care study. In: *Method of Evaluating and Improving Personal Medical Care Quality*. Ann Arbor: The Univeristy of Michigan School of Medicine.

Reiff, G., Kish, L. and Harter, J. (1968). Selecting a probability
    sample of school children in the coterminous United States.
    The Research Quarterly 39,

Roth, A. and Klassen, D. (1973). *A Report on the Planning,
    Implementation and Evaluation of the Kansas City Community
    Mental Health Assessment Project.* Kansas City: Epidemio-
    logic Field Station, The Greater Kansas City Mental Health
    Foundation.

Survey Research Center (1950). *Industrial Mobility in Michigan.*
    Ann Arbor: Institute for Social Research, The University of
    Michigan.

Survey Research Center and Research Center for Group Dynamics
    (1952). Institute for Social Research, 1952, publication
    36. Ann Arbor: Institute for Social Research, The
    University of Michigan.

Survey Research Center (1972). *A panel study of family income
    dynamics,* Vol. 1, Section 11. Ann Arbor: Institute for
    Social Research, The University of Michigan.

Survey Research Center, Sampling Section and Postal Study Research
    Staff (1978). *A Quantitative Description of the Current
    Nonhousehold Mailstream* , Task 2 report. Ann Arbor:
    Institute for Social Research, The University of Michigan.

SURVEY RESEARCH AT THE BUREAU OF THE CENSUS

Barbara A. Bailar and Gary M. Shapiro

U.S. Bureau of the Census

Some of the current research efforts at the U.S. Bureau of the Census are described. Two of them -- estimation for small areas and the redesign of recurring household surveys -- are highlighted in this paper. Other research efforts of major consequence on imputation techniques, multiple frame methodology, and developing public use samples for the 1940 and 1950 censuses are described briefly.

1. INTRODUCTION

In addition to being in the midst of carrying out the 1980 Census of Population and Housing, we, at the Bureau, are conducting a wide range of surveys in the demographic and economic areas. In the design and maintenance of these surveys, a large amount of research in all areas of sample surveys, not just sampling issues, is conducted at the Bureau. In this paper, we describe in some detail the need for research in making estimates for small areas and some of our research efforts along those lines. We also describe some of the studies being conducted as part of the redesign of the recurring household surveys. Finally, in Section 4 we briefly describe some additional topics of major research interest.

2.   SMALL AREA ESTIMATION

The Bureau is faced with a variety of situations in which
small-area data are needed but are not available. In Sections
2.1, 2.2, and 2.3 we give three examples of these needs, discussing the methodology in 2.4.

2.1 The 1978 Census of Agriculture

In the 1969 and 1974 Censuses of Agriculture, the data
collection was conducted completely by mail using a list frame.
The desired product was a complete and fully accurate count of
farms, farmland, and farm production. However, the complex
structure of agriculture made this difficult to achieve. Coverage
evaluation studies using independent area segment samples indicated substantial undercoverage.

In 1978 the census data collection was based on a list
sample developed independently from earlier censuses. In addition,
the data collection included a sample of area segments which were
canvassed for farms. All farms located in the area sample were
matched to the farms on the list sample. Those not matching any
farm were verified to make sure they were not already counted in
the census. Any that were not matched and were found to have been
missed from the list sample were used to make State estimates for
farm operations not included on the mailing list.

The sample design for the area sample provided for reliable
estimates for States but not for counties. Thus, in the publications, in the columns for State tabulations, there is a total that
is the sum from the farms reported on by mail based on the list
sample and the estimate of farms from the area sample. The farms
from the mail portion are distributed by county, but those from
the area sample are not. For county summary data tables, the area
sample data are shown as one line for the State with the stub "Not
allocated to counties."

Data users and analysts would like the data from the area sample to be allocated to counties. Research is currently underway to assess different methods of distributing the data.

2.2 Commerce Cities Project in Denver

The Commerce Cities Project is a Department of Commerce demonstration program for helping cities. The objective of the program is to demonstrate how cities can use existing Department of Commerce resources in better ways to deal with specific city needs.

Planners in the City of Denver, Colorado want to develop a data base to use in economic development planning. The data base is to consist of socioeconomic data items for the city level and, where possible, for smaller geographic areas. The Census Bureau is providing assistance to Denver by investigating the feasibility of using nonsurvey methods to estimate and project selected data items to be included in the data base. If the effort in Denver is successful, the Bureau may try to apply the methods in other cities.

The general approach is as follows. The Census Bureau and Denver, in collaboration, have selected a set of variables for which estimation and projection methods will be applied. The list of variables include population by age, race, sex, and income, and the number of persons whose incomes are below the poverty level -- for census tracts or for Denver's ten "communities", which are aggregates of tracts.

The Bureau and Denver are now developing sets of Federal, State and local data that can be used in attempting to apply estimation and projection methods. The methods are being identified and their potential use in the project is being evaluated. The methods range from the basic synthetic method to regression methods to empirical Bayes methods. Some are described more fully in Section 2.4.

2.3 <u>Distributing the 1980 Undercount to Small Areas</u>

Even though we have instituted a number of new procedures
in the 1980 census aimed at reducing the differential undercount
between subgroups of the population, there will probably be an
undercount of some size when the results are all in. The Bureau
has made no decision to adjust census counts for this undercount,
and the discussion in this paper should not be misinterpreted as
an implication that we shall do so. However, in order to decide
whether there has been an undercount and whether it has affected
subgroups of the population differently, we have to estimate that
undercount. In estimating the undercount, there is interest in
estimating at levels below the national and State levels, even if
only for the Bureau's own use in evaluating methodology.

In 1980 we will produce national estimates of the undercount by the method of demographic analysis. Essentially this
means using birth and death records as well as records on immigra-
tion and emigration.

We will also be conducting a large-scale matching study,
involving a sample of about 170,000 households nationally. There
will be a separate sample for each State and the District of
Columbia so that estimates of the total corrected population may
be made for each of these places, with an acceptable level of
reliability. The matching study also provides for a corrected
estimate of the Hispanic and Black populations at the national
level. Corrected population estimates may also be made for the
10 largest SMSA's.

The estimates from the matching study will probably be
biased because of the high probability that many people missed in
the census will also be missed in a household survey. For this
reason, these total corrected population estimates may be "raked"
to the demographic estimates. State estimates from the matching
study would then be adjusted accordingly. However, within States,

there are many other units of local government -- cities, counties, and townships -- for which estimates of the total corrected population or the undercount are of interest. Again, a method of producing local-area estimates is desirable.

2.4 Making Local-Area Estimates

In two of these three examples, we have sample data for a larger area (States) that we want to allocate to small domains. For the allocation of farms we wish to go from State data to county data; for the allocation of the undercount, we wish to go from State data to counties or municipalities or towns. The third example, the Denver project, requires the allocation of data as yet unknown -- perhaps State, perhaps regional -- to the city, to communities, and to tracts.

In all of these cases, it is possible to use the basic synthetic estimator. This estimator uses sample data for the larger area to estimate the variable of interest for different subgroups of the population. Then these estimates are scaled proportionally by the incidence of the subgroups within the domain of interest. Thus, for estimating the undercount of a county, the State estimates of undercount would be estimated by age, race, and sex. Then the estimate of undercount for each county would be estimated by scaling the estimates of the State undercount by subgroup to the incidence of the subgroup in each county. When these are added over all counties, the county estimates add to the State estimate.

There are some major reservations about this estimation procedure. First, the method will be biased if the undercount for the State by subgroups is not reflected in the subgroups' undercount in each county. If there are local factors that change the undercount for some counties, the synthetic estimator will not reflect this. It is quite noticeable in applications of this method that unless the variables defining the subgroups are highly

correlated with the variable of interest, the synthetic estimates
tend to cluster near the mean for the larger domain. It is also
evident from past studies that there is considerable geographic
variation in the estimates of undercount that would not be
accounted for in the synthetic procedure.

Another method that has worked well for estimating population is the sample regression method, discussed in several papers
by Ericksen (1973, 1974). With this method, a regression equation
is set up using selected symptomatic variables, measured for each
small domain as independent variables, but using current sample
data for the variable of interest as the dependent random variable.

There are many other methods as well, many of them discussed by Purcell (1979) in his dissertation and in his paper with
Leslie Kish (1979). It is possible that no one method is best for
every problem. Somehow data must be derived or experiments run to
make an assessment of the methods possible.

A significant problem in making small-area estimates is in
the selection of the best set of associated variables. For the
undercount estimates, we have used age, race, and sex at the
national level. For local areas we may wish to consider degree
of urbanicity, amount of imputation, or other such factors.

For estimating the number of farms not counted in the mailing procedure, we may want to consider farm sales, acreage, and
type of farm. For the City of Denver, age, race, sex, income,
educational attainment, and other economic and demographic variables will be studied.

3. REDESIGN OF THE MAJOR DEMOGRAPHIC SURVEYS

Most of the sample units for recurring household surveys
conducted by the Census Bureau are selected from the decennial
census lists of addresses. Thus, a redesign of new sample
selection for the surveys has been done after each decennial

census. If we were to go as long as 20 years between new sample selections, a very large proportion of the samples would be from the separate new construction universe. Because of shifts in the population, increased variability in the survey statistics could result.

In addition to just reselecting the sample following the 1980 census, we intend to do a total survey redesign, questioning all aspects of the design and estimation procedure for each recurring household survey. We are conducting substantive research in many areas. Redesign will be carried out for the following surveys:

- Current Population Survey (CPS), a monthly labor force survey sponsored by the Bureau of Labor Statistics.

- Health Interview Survey, mostly annual publications, sponsored by the National Center for Health Statistics.

- Annual Housing Survey, annual publications, sponsored by the Department of Housing and Urban Development.

- National Crime Survey, mostly annual publications, sponsored by the Bureau of Criminal Justice Statistics.

- Survey of Residential Alterations and Repairs, a quarterly survey sponsored by the Census Bureau.

Separate staffs, for the most part, are working on each one of these surveys. Over the remainder of this year, each staff will determine what is best for each survey. Then, an effort will be made to see if common design decisions can be made that are nearly optimal for all surveys. If this is not possible, some of the surveys may operate under substantially independent designs and procedures, although some minimal degree of coordination would be maintained to ensure efficiency and non-overlap of samples.

Because so many different redesign projects are going forward, we cannot discuss many in detail. In Sections 3.1 and 3.2
we describe two important areas of research in detail and in
Section 3.3 briefly mention several other topics.

3.1 Stratum and PSU Definitions

Major areas of concern for all the surveys are the formation of strata and the formation of primary sampling units (PSU's)
within strata. In the past, one set of stratum definitions has
been used for nearly all demographic surveys. In redesigns after
past censuses, we were conservative in changing stratum definitions mainly because we wanted to maximize the overlap of sample
PSU's between the new and old designs. This would maximize the
number of interviewers to be retained.

However, since the 1970 redesign, there have been three
major expansions of the CPS so that stratum definitions have been
modified in an unplanned manner. Thus, the conservative approach
of defining strata will not be followed. For a survey such as
the Survey of Residential Alterations and Repairs, which is in a
small number of sample PSU's, the problem of retaining interviewers is greater than in the CPS and some of the other surveys, so
that greater effort in maximizing the overlap in the new design
may be desirable.

We have decided to use a clustering algorithm for stratifi-
cations for most of the surveys. In particular, we are looking
into the Friedman-Rubin (1967) algorithm. The stratification
variables that will be used for CPS and the Annual Housing Survey
are mostly 1970 and 1980 decennial census variables. Unfortunately, the redesign will take place too soon after the 1980 census to
make use of the sample data.

In the other surveys, determining the best stratification
variables is more difficult because the important survey characteristics are not collected in the censuses. Thus, different

approaches are necessary. For the National Crime Survey, for
example, a multiple regression model will be developed to determine the relationship between crime statistics of interest and
socioeconomic variables. We will then proceed in one of two ways.
One possibility is similar to what we will do for the CPS. Having
established the important socioeconomic variables from the regression model, we will use them as stratification variables in a
clustering algorithm to determine the strata. The second possibility is more complex but has the advantage of more fully
exploiting the relationship of the socioeconomic variables to the
crime statistics in the stratification process. We will first
determine predicted crime values from the regression model for
important crime items. If the number of crime items of interest
is large, factor analysis can be used to construct a more manageable number of predicted crime "indexes." Each predicted crime
index will be a weighted index of the socioeconomic variables.
At this point the predicted crime indexes will be input to a
clustering algorithm.

The present PSU definitions were essentially defined about
30 years ago and are the same for all surveys. Except for New
England, PSU's are defined in terms of counties. In the redesign,
PSU's may vary for the different surveys. We will probably have
some sub-county divisions for defining PSU's, especially in some
of the counties in the West which cover large land areas and thus
involve high interviewer travel costs. Also, for most, if not all
surveys, we will form strata before forming PSU's, the opposite of
the present system.

To illustrate what we are doing in defining PSU's, consider
plans for the Annual Housing Survey. Some research will be done
to determine the optimal workload size per interviewer in nonself-
representing PSU's. Most of the research is planned to determine
when and how counties and subcounties should be combined to form
PSU's. Variance and cost data will be utilized to determine the

approximate optimal and upper bound for the geographic size of a
PSU. We will then attempt to combine counties in cases where a
single county is much smaller than the optimal, and will generally
have a one-county-PSU whenever the county is larger than the
optimal. We are also investigating which set of variables is best
in determining those counties to combine to form maximally heterogeneous PSU's.

3.2 Rotation Scheme

In most household surveys, we employ a rotation scheme with
sample units interviewed several times but eventually rotated out.
Because of changes in survey objectives there is a need to re-
evaluate the rotation schemes now in use.

The CPS has the most complex pattern of any of the surveys,
with units in sample for 4 months, out for 8 months, and in for 4
months. For all important CPS characteristics, correlations over
time are positive and diminish gradually over longer time periods.
Thus, the existing scheme is relatively efficient for estimating
month-to-month changes at the national level. However, with
changes in survey objectives, there is a great need to get reliable annual estimates of level for States and Standard Metropolitan Statistical Areas (SMSA's) as the basis for distributing
Comprehensive Employment Training Administration (CETA) funds.
Thus, we plan to examine rotation patterns that may be more
efficient with respect to the variance of annual estimates formed
by averaging 12 months of data.

Intuitively, the most efficient rotation pattern for
variances of annual estimates would be to have units in sample
only one time a year, but this is not necessarily the most efficient rotation pattern from the cost standpoint. If such a scheme
were used, the percentage of telephone interviewing would be
reduced from well over 50 percent now to almost nothing, thus
increasing per unit costs. It is our goal to determine the

rotation pattern that will achieve specified reliability levels
for both sets of objectives at the lowest cost.

We are now estimating variances for selected characteristics by rotation group and covariances between rotation groups both
within a given month and between different points in time. Initial
results from these calculations were surprising, leading us to
believe that our methods are yielding overestimates of variance
for self-representing PSU's and of within-PSU variance for nonself-
representing PSU's. Additional work is going forward to estimate
the degree of overstatement. Also, cost estimates will be made
for different rotation plans, and will be utilized together with
the variance/covariance estimates.

Because rotation group bias is an important phenomenon in
the CPS, we are determining whether rotation group biases differ
by region. This is important because comparability between States
and parts of States is necessary for the equitable distribution of
funds. Bias in national estimates is second in importance to
differential bias among States and sub-State areas. Indications
of substantial variation in rotation group bias patterns will be
considered in deciding which rotation pattern to use. However,
since we have no way of knowing which month in sample provides the
best estimate, we are uncertain exactly how to use this information.

There has been no sample rotation for the Annual Housing
Survey. Sample units have been interviewed once each year from
the time the survey started in 1973. By introducing the redesigned
sample in stages we will have the opportunity to determine if
there is any rotation group bias in this survey. The results of
this determination will be used in the decision about the type of
rotation pattern, if any, for the future.

The Health Interview Survey has operated exactly the
opposite of the Annual Housing Survey -- each unit is interviewed
only once. In order to improve estimates of year-to-year change

as well as change over longer periods of time, we will study
various rotation plans for this survey, although no good covariance
estimates will be available. There is also considerable interest
in obtaining good estimates of annual medical expenditures. To do
this, it would be necessary to have several interviews at the same
housing unit within a year. Thus we will consider the possibility
of interviewing a subsample of the full sample at about 3 month
intervals during the year.

The major concern with the rotation scheme in the National
Crime Survey is how it fits in with the reference period. At
present, we ask respondents about crime victimizations during the
preceding 6 months, and respondents are contacted seven times at
6-month intervals. If it is more cost efficient and not badly
biasing to the data, asking respondents about victimizations
during the preceding 12 months may be an alternative. Then a
rotation scheme requiring contact with respondents only once a
year is indicated.

3.3 Other Redesign Research

There are many other sample design considerations being
studied, too many to discuss in any detail in this paper. A
brief description of a few follow.

- *New Construction Sampling*. Partial computerization of the
  selection of construction permits for the near future and a
  comprehensive computerization for the more distant future may
  improve sampling from new construction.

- *All Area Segment Sampling*. Area samples are used only in rural
  areas for all surveys at present, but will probably be used
  everywhere for the Health Interview Survey because of the
  sponsor's wishes and confidentiality problems. There is an
  interest in increasing area sampling for other surveys as well
  to simplify sample selection and control. Since area sampling
  generally results in more variable segment sizes and is more

costly than census list sampling, there are potentially bigger
gains to be realized by research in this area. Thus, we will be
investigating alternative forms of area sampling, trying to
minimize cost, variance and bias.

- *Telephone Interviewing*. A recently completed field study for
  CPS is being analyzed to see if there is any evidence of differences in labor force data between telephone and personal
  visits. In the Annual Housing Survey, now conducted entirely
  by personal interview, a small feasibility test of telephone
  interviewing will be conducted. In the National Crime Survey,
  there is a greatly increased use of telephone interviewing this
  year. The results will be closely analyzed.

- *Proxy Respondents*. Analysis of a recently completed study for
  CPS is underway to examine differences by respondent type for
  labor force data. Since past data in the Health Interview
  Survey has shown important differences between self-respondents
  and proxy respondents, we are considering a change in the
  respondent procedure.

- *Estimation and Weighting*. Research will include improved methodology for noninterview adjustments, ratio estimation, raking,
  and composite estimation.

     Though there are many other redesign research projects
being conducted, this description gives some indication of the
kinds of studies being carried out.

4.   OTHER RESEARCH AREAS

4.1   Imputation Techniques

     In any survey in which data are collected from people,
respondents tend to leave some items unanswered. In demographic
surveys, income is frequently left blank; in economic surveys,
inventory is frequently left blank. In order to present statistics

that are the most useful, we impute for the missing data fields.
Since we can use additional information from the same record or
from administrative records, the Bureau can presumably impute for
missing data better than users can. Yet a fundamental question
arises concerning imputation. At what point should imputation not
be done and the statistic not be published because there was too
much imputation?

In the demographic area we impute data for some items in a
hot-deck procedure based on the use of suitable "donors." For
each item nonrespondent we obtain a donor (respondent) with similar
survey characteristics. The imputed value for the missing item is
taken to be the response for the item reported by the donor. In
the economic area, we base imputations on past records for the
same respondent or the same kind of business.

Some important questions regarding imputation need to be
answered. What are the ways of imputing data that will be
reflected in the estimated variances of the statistics? Should
more than one value be imputed for each missing data item, as
suggested by Don Rubin (1977)? A significant amount of research
is going on in this area at the Bureau.

4.2 Multiple Frame Methodology

The Registration and Voting (RAV) Survey in 1976 made
extensive use of multiple frames. This type of methodology was
needed because voting and registration rates, estimated from
household surveys, are generally biased upward since some people
who are not registered or do not vote report that they do. Some
kind of additional record on actual behavior seemed desirable.
In addition, the goal of the survey was to estimate voting and
registration rates for each of 73 counties, 11 towns, and 9 States
by selected minority and non-minority groups. For areas with
small percentages of specific minority groups, an additional frame
might help in providing more accurate estimates. The definition

of the minority groups varied by jurisdiction. For example, in Honolulu County, the Filipino, Chinese, and Japanese were minorities and other groups were classified as non-minorities.   In Coconino County, Arizona, Spanish and American Indians were the minorities.

Thus, to meet the needs of the survey for future years, some kind of record check with registration lists seemed needed. Also, it appeared that the registration lists could be used as frames for sampling. These lists provided information on actual registration and voting, but no information by minority group.

Thus, two frames were available from which to select samples. One was a household frame based on the 1970 census supplemented with new construction. The second frame was the county registration list, available at the county seat. It was assumed that the coverage of the second frame was contained within that of the first.

As a part of a methodological study to investigate the feasibility of a multiple frame survey, independent samples were selected from each frame. As a part of the estimation procedure, households at the addresses provided on the registration lists were included in the sample only if the selected registrant was a member of the household. The estimators of the voting and registration rates consisted of weighted averages of the characteristics over the overlap domain covered jointly by the two frames plus the additional component from the household survey covering the remainder of the universe.

Several alternative designs were considered:

- Assume the registration list is completely up-to-date so that every registrant would be found at the address on the registration list.

- Use a double sampling scheme in which the information from a subsample of households from the household frame is checked

against the records. Then the reported information for the full
sample from the household frame would be used in a difference
estimator.

- Check the records for every household in the household sample.

    Estimators based on the three different designs were
compared and the double sampling scheme performed best overall
in a cost-variance framework. A complete report on this study
carried out by Isaki, Huang, and Hogue (1980) is available.

4.3 Developing Public Use Samples for 1940 and 1950

    The purpose of this project is to provide public use
samples on computer tape for the 1940 and 1950 censuses. The
project, sponsored by the National Science Foundation, and
costing $7 million, is being carried out jointly by the University
of Wisconsin and the Bureau of the Census. It is scheduled for
completion in late 1982.

    The 1940 and 1950 census records are stored on about 11,000
microfilm reels -- about 5,000 for 1940 and 6,000 for 1950. The
records microfilmed are the original, handwritten schedules.
Since no other sources of these records are available, the 1940
and 1950 samples will have to be selected from the microfilm
records which are sometimes illegible. Also, there are problems
of duplication, omission, and records being out of sequence.

    The basic intention in generating these tapes from the
microfilm records is to make the 1940 and 1950 tapes comparable
to the 1960 and 1970 census public use tapes. For 1960 and 1970,
the tapes include 1 percent samples of households and their
members, subsampled from the households that received the "long-
form" census questionnaire. In 1940 and 1950, however, the
additional long form questions were asked on an individual basis,
rather than on a household basis. Consequently, a household in
the 1940 or 1950 census would typically have at most one of their

members providing additional "long-form" data.

This fundamental difference in the way the censuses were taken has introduced complexity in the design of the selection procedures for the 1940 and 1950 public use samples. In order to select households for the 1940 or 1950 samples that contain persons with long-form data, the initial unit of selection will be a *person* listed on a long form, rather than a household. A systematic sample of persons listed on long forms will be selected first. The household containing each selected person will be identified. Rules have been developed to determine whether or not to retain the household for the public use sample.

In 1940, the selection rules have been derived in such a way that the resulting sample of households will be self-weighting. This will be accomplished by retaining a household for the 1940 sample with probability $1/h$, where h is the number of household members. In 1950, since virtually all the "useful" census items are long-form items, all long-form persons selected at the initial stage will be retained along with their households. In addition to the basic household samples, self-weighting samples of persons with long-form data will also be available for both censuses.

A pretest for selection of the 1940 sample is being carried out in Pittsburgh, Kansas, at this time in order to compare three data transfer procedures and to test the sampling and other aspects of the operation. The pretest is being carried out for 60 microfilm reels. The sample selected from these reels will contain about 4,500 households. In addition to providing a comparison of data transfer methods, the pretest should provide needed information on (1) cost and time requirements of various operations, (2) irregularities in the microfilm records, (3) completeness of sampling specifications and (4) completeness of coding specifications. The pretest results will have an important influence on the finalization of the procedures for selecting

Barbara A. Bailar and Gary M. Shapiro

the 1940 sample.

5. CONCLUSION

Two of the research efforts that are now receiving major attention at the Bureau -- small-area estimation and the redesign of the recurring surveys have been described. In addition, brief descriptions have been provided for some other research efforts of major consequence. Major research efforts are also going forward in other areas such as questionnaire design, alternative training methods, and developing a computer-assisted telephone interviewing system, but time prohibits any further discussion. It should be evident that survey research is flourishing on a number of fronts at the Bureau of the Census.

ACKNOWLEDGEMENTS

Several people were of major help in putting this paper together. Cary Isaki, Charles Rogers, David Chapman, Arnold Reznek, Donald Malec, and Larry Cahoon were major contributors. To them, we give our thanks; to us, we reserve the responsibility for any errors.

REFERENCES

Ericksen, E.P. (1973). A method for combining sample survey data and symptomatic indicators to obtain population estimates for local areas. *Demography* 10, 137-160.

Ericksen, E.P. (1974). A regression model for estimating population changes of local areas. *Journal of the American Statistical Association* 69, 867-875.

Friedman, H.P. and Rubin, J. (1967). On some invariant criteria for grouping data. *Journal of the American Statistical Association* 62, 1159-1178.

Isaki, C., Huang, E. and Hogue, C. (1980). Comparisons of
    multi-frame with single frame sample designs using registra-
    tion and voting survey data. Internal Census Bureau
    memorandum. Washington: United States Bureau of the Census.

Purcell, N.J. (1979). Efficient estimation for small domains: a
    categorical data analysis approach. Unpublished doctoral
    dissertation,

Purcell, N.J. and Kish, L. (1979). Estimation for small domains.
    Biometrics 35, 365-384.

Rubin, D.B. (1977). Formalizing subjective notions about the
    effect of nonrespondents in sample surveys. Journal of
    the American Statistical Association 72, 538-543.

CURRENT SURVEY RESEARCH ACTIVITY: GENERAL DISCUSSION

D. DALE, *Carleton University* (to D. Bayless): Before you asked how many people here had actually designed a questionnaire, you might have asked how many people are on the academic staff of a university. (This does illustrate I think that we have to be careful how we ask our questions!)

I think there is a basic difficulty in the coordination of work that is done in the university within a mathematics or statistics department. The difficulty is that at present students are given very little experience in the handling of "dirty" data. We teach from Bill Cochran's text, or we use Mendenhall if you are starting out early, but we fail to develop the art of this kind of survey work.

I know that in your case, you are subject to time constraints. You can't wait till Thursday if they want to see you on Tuesday. You have to get something done quickly and so you develop the talent of overcoming this type of obstacle. The university doesn't have that. Based on your contacts with various universities and with the Research Triangle Institute, I'd like to know how you expect to stimulate academic operations to mesh with yours under the constraints that you do have.

D. BAYLESS, *Research Triangle Institute:* To your first comment, I wanted to applaud! To the second, let me point out that the Research Triangle Institute is owned by three universities. One of these is a private institution (Duke), the other two are state universities. We have a cadre of consultants, many of them from these universities. We use these people in various ways on actual projects. Of course they are paid, but they learn

189

by working with us on actual tasks, and I think some of that
carries back to the curriculum and back to the classroom where
they are dealing with students.

As a matter of fact, I was attending a meeting of the
National Committee of Vital and Health Statistics just last month
and this was one of the very points that was made there.

There is a demand for people with practical experience
which can be seen from the growth on my chart. Government agencies
ask for people who can come in and translate sample designs,
including measurement and data collection designs, into something
meaningful for the analysis in the final report. I think it's an
important area or I wouldn't have put it in this paper.

As a final comment, I would like to say that I wish we had
more time at the Institute to start a program, to be able to share
some of the arts, some of the practices and some of the practical
problems that we have with others. But, as you said, we have got
to have the job done on Tuesday. There certainly is insufficient
time to adequately communicate some of this practical experience
over to where it can be useful for teaching purposes.

T. Dalenius, *Brown University* (to D. Bayless): Following
up on Professor Dale's question, I would like to ask how many
professors have conducted an interview themselves, that is, taken
a questionnaire and collected the data themselves?

Teaching students how to do this work is obviously very
difficult in the university environment unless you include some
sort of internship (say at the Census Bureau) as part of their
formal training. In the PhD program for which I am responsible,
students in fact have to spend six months of their time on actual
survey work. When these students finish they know what surveys
are all about, not from reading a book but from working on real
life data and real life problems. In those schools where there
is a statistical agency close by, I think such training is most
worthwhile. There are some formal problems, but they can be solved
pretty easily.

A.R. SEN, *Environment Canada* (to D. Bayless): I have been listening since yesterday to some very interesting discussions on sampling errors and the teaching of students. I would like to know if in your own work, you come across clients who have no notion of survey sampling. In other words, when they come to you with a problem do they have a feel for survey sampling and survey design?

D. BAYLESS, *Research Triangle Institute:* The answer to your question is yes. If you do consulting in survey research you do run across people who know nothing about survey design. That's why they are asking you to come in. You are the doctor and they are the patients and you have to help them out. I think there is a thirst for more knowledge of the basic principles of statistics on their part. Part of the message I was trying to get across involves the discussion, planning and training that should take place before the survey is conducted, not after we have the data there on the file and are wondering what we are going to do with it.

B. BAILAR, *U.S. Bureau of the Census* (to R. Platek): I know you said the listing of items was incomplete but I would like to comment on one that was not mentioned. A source of a great deal of trouble is interpretation errors that have occurred during the analysis. People pick out certain points to highlight and wonder whether the inferences that they make in some cases are really justified on the basis of the data, the sampling errors, the known biases and so forth.

I also don't understand exactly how you estimate the index of survey performance. I thought you were including in it the actual sampling variance that's measured from the survey itself and the response variance that would be appropriate based on the survey design. Yet when it comes down to a simple index of inconsistency, we're measuring sampling and response variance on the basis of a simple random sample. Thus, it does not take into account some of the complexities of the design.

R. PLATEK, *Statistics Canada:* The index of inconsistency is determined using reinterview data. It is true that taking $d = 1$ (design effects) is a considerable simplication. However, in the final version of the paper we intend to point out under certain assumptions the relationship between the index of inconsistency and the performance index. This should be useful for the purposes of computation of the performance index and it will take into account the actual design effect.

D. BAYLESS, *Research Triangle Institute* (to I. Hess): Did I understand you correctly, Irene, that in your general purpose sample of 74 PSU's, you go out and you list and screen enough households for 6 to 12 surveys in the future? In other words, do you prelist and prescreen?

I. HESS, *University of Michigan:* Not all at once. We select the sample of segments, but the segments themselves may not be listed until the time they are being used. Even if they were listed, and the listing is shared by more than one survey, these would be updated for each survey.

D. BAYLESS, *Research Triangle Institute* (to I. Hess): I was concerned about the issue of respondent fatigue, that is, sample wearout or sample incompleteness. The question I was really going to ask was whether you had some rule of thumb or guidelines as to how long you leave the sample out there before you use it. I think it's a practical problem that we all face.

I. HESS, *University of Michigan:* The reasons for selecting segments for several surveys all at once is a matter of convenience and savings. But we can also avoid technical difficulties in sampling with probability proportional without replacement. It's easier I think if you select a large sample and then subsample it.

As far as putting the segments into use is concerned, this is not done until they're really going to be involved in a data

collection operation and at that time the segments are listed.
If these listings contain more housing units than would be
involved in one study, then on another study we may sample from
the remaining housing units. In this case, the listing is updated at that time, so that the housing unit list always reflects
the current situation.

H.O. HARTLEY, *Duke University* (to B. Bailar): My first
question is referring to a topic mentioned very early on in
Barbara's talk, that is, the undercount in the Agricultural Census.
I wonder whether the definition of a farm operator presents
difficulties because of modern diversification. Currently many
individuals are involved in a great variety of businesses, with
farm operations representing only part of their activities. I
know of an individual living in the city of Houston involved in a
major operation carried out by pilots as well as many other
businesses that he's conducting at the same time. It seems to be
sort of difficult to envisage how this fellow can be gotten hold
of using an area sample. Thus, diversification seems to present
ever increasing difficulty with regard to defining and counting
the number of farm operators.

A second question is a very technical one. The multiple
frame that you mentioned is a very interesting development. One
of the advantages of using multiple frame operations is that the
always difficult matching operation is only required for sample
units and not for the total population. If it were required for
the total population it would indeed be essentially unmanageable.
The question is, is it because of this that the double sampling
procedure turns out to be the most economical one?

B. BAILAR, *U.S. Bureau of the Census:* In answer to the
first question, the complexity of modern farm operations does
give us some difficulty in the Census of Agriculture, but we do
not expect to pick up large diverse operations in the area sample.
These we would have on various lists. For example, we have the

Internal Revenue Service records of people who report farm operations. Anything really large would come from that. We also have
lists from the Department of Agriculture. We find that we get a
lot of small farms in the area sample that are not on any of our
lists. So though we are perhaps adding a lot to the count of the
number of farms, we aren't adding a whole lot to the volume of
farm land or farm production coming from this.

As a matter of fact, somebody said to me yesterday that
since the census girl was obviously already making corrections
for the undercount to the farm census, we must agree in principle
that we should be making adjustments to censuses. I was very
quick to point out that we think differently about our Agriculture
Census than our decennial census.

Although I'm really not very clear on your second question,
I may state that there is a very large report on this study which
I have not yet gone through.

I. FELLIGI, *Statistics Canada* (to B. Bailar): I would like
to comment on the small area estimation problem particularly with
respect to the undercount. I think that it is necessary in any
research project of this kind to weigh two considerations. The
first is the payoff - how badly do you need the results - and the
second is the chance of attaining it. You want very much to have
small area underenumeration estimates so the payoff is very high.
However, I'm quite pessimistic about the likelihood of getting
them even with the kind of data we have in Canada, where the
data base is fairly rich in terms of having a sample of people
who were missed in the Census.

Yesterday, I showed that there were significant differences
between Provinces in the estimates of underenumeration. The
underenumeration rates range from about half of one percent up to
nearly 4% for the individual provinces. In order for a synthetic
estimator to reproduce these kinds of differences, the population
should have the following characteristics. First, there have to
be particular subgroups of the population that have unusually high

underenumeration rates, much higher than the differences that
exist between the provinces. Secondly, the distribution of the
independent variable has to be sufficiently different between the
provinces. For example, the largest underenumeration vote is for
young males. This is not taken into account in synthetic estimation procedures if the age distribution in every province is very
similar because in this case when you create your synthetic
estimates you simply reproduce the overall average.

Another problem is that those groups that have a particularly high underenumeration rate represent a very small subgroup
of the population and therefore don't affect the distribution
when you account for them in the synthetic estimate. Also, the
distribution by age among the states or provinces is very similar.
I've been looking at this problem for some time and I am rather
pessimistic about the possibility of a solution.

B. BAILAR, *U.S. Bureau of the Census:* We have this differentiation in the U.S. as well. Although our overall undercount
estimate in 1970 was 2.5%, it was 1.9% for the white population
and 7.7% for the black. We see a similar differentiation by age
groups as well.

Part of our problem is that we don't know why all of this
undercount is occurring. We do have some hypotheses explaining
why certain people in the cities may want to avoid being counted,
but we don't know if these same reasons apply to people in the
rural South, for example. When you make gross assumptions, on
top of not knowing very much, it leaves you with a sort of uneasy
feeling.

M.P. SINGH, *Statistics Canada* (to B. Bailar): I have a
two part question related to rotation patterns. First of all,
I'd like to know how you will formalize the approach by which you
decide upon a particular rotation pattern. Will this involve some
field experimentation? Secondly, you mentioned that the bias of
the composite estimator is increased using a rotation pattern.

I would like to know if attempts were made to adjust the composite estimator for this bias introduced by the rotation pattern.

B. BAILAR, *U.S. Bureau of the Census:* Right now we are ignoring the bias, and are trying to find that rotation pattern which will give us minimum variance for both our estimates of the month to month change and the national annual estimates for the U.S. We know how to solve this problem, but once we try to adjust for the bias we have no answers.

There are many things we can do to minimize the effects of the rotation group bias in the composite estimator (and we have done some of them), but each of these is really an attempt to make this estimate more consistent, not necessarily increasing its accuracy. We are then left with the question of choosing between accuracy and consistency. A few years ago I looked at what happens to the composite estimators for estimates of level, and it was clear that the bias was swamping the mean square error estimates. So, although we can select a rotation pattern which minimizes the variance of the estimates, the bias remains a tremendous problem which is as yet unsolved.

IV SUPERPOPULATION MODELS

SURVEY DESIGN UNDER SUPERPOPULATION MODELS

Wayne A. Fuller and Cary T. Isaki

Iowa State University

Sample designs and predictors that minimize the approximate anticipated variance are developed. Anticipated variance is the variance of the predictor computed with respect to the sampling design and the superpopulation model. The designer's information is expressed in terms of the parameters of a regression superpopulation model from which the finite population is a (conceptual) random sample. The limiting distribution of the estimator is presented.

1. INTRODUCTION

The problem of incorporating prior information into the construction of a design-estimator pair for the mean of a finite population will be considered. Let $U = \{u_i : i = 1, 2, \ldots, N\}$ denote the N units of the finite population. Let s denote a subset of the units of the population and let S be the set of all subsets, s. Let $d(s)$ (or d) be an estimator constructed from the units of s. Let $p(s)$ (or p) denote a sampling design which assigns probabilities to the elements s of S. Let A denote the information about U available at the design stage, let P denote a set of designs, and let D denote a set of estimators. We assume that the prior information can be quantified by treating the finite population as if it were a sample from an infinite population for which we know the first two moments.

Superpopulation models have been used by Cochran (1939, 1946), Deming and Stephan (1941), Madow and Madow (1944), Yates (1949), Godambe (1955), Hájek (1959), Rao, Hartley, and Cochran (1962), Brewer (1963), Godambe and Joshi (1965), Hanurav (1966), Isaki (1970), T.J. Rao (1971), Fuller (1975), Cassel, Särndal, and Wretman (1976), Brewer (1979), Ramachandran (1979), and Särndal (1980). The superpopulation approach has also been used by Royall (1970), Royall and Herson (1973), and Scott, Brewer and Ho (1978). Cassel, Särndal, and Wretman (1977, Ch.4) contains a discussion of superpopulation models. Robinson and Tsui (1979), Brewer (1979), Rosén (1972), Godambe (1969), Hájek (1964), and Madow (1945) have considered consistency for finite populations.

We shall develop sample designs and predictors for regression-type models. Estimators are constructed that are design consistent, but not necessarily design unbiased. A design-predictor pair is presented such that the predictor is, conditionally on the sample elements, best linear unbiased under the model. The limiting design behavior of the estimator is obtained. Some comparisons of this estimator with other estimators in the literature are made.

2.   MODEL AND DESIGN PROPERTIES

At the design stage we assume the finite population to be a sample from an infinite population for which information on the first two moments is available. We wish the moment information to be introduced into the design and estimation procedures, but we will use the population of samples created by randomization for inference. Because randomization furnishes a model independent basis for obtaining information on the accuracy of the estimators from the sample, we restrict consideration to the class of random designs.

A *predictor* is a function of the sample used to estimate a function of the finite population. A predictor d is conditionally model unbiased for $\overline{Y}$ if, given the sample s

$$E\{(d - \overline{Y}) | s\} = 0 . \qquad (2.1)$$

where E is the expectation with respect to the superpopulation and the conditioning is with respect to the elements of s , not the y-characteristics. This notation is somewhat redundant. The predictor d is a *model unbiased predictor* of $\overline{Y}$ if (2.1) holds for all s in S for which p(s) > 0 . A predictor d is *unbiased for* $\overline{Y}$ *with respect to the design* p (design unbiased or p-unbiased) if

$$E\{d\} = \overline{Y} , \qquad (2.2)$$

for all $(y_1, y_2, \ldots, y_N)$ contained in N-dimensional Euclidean space, where E denotes the expectation with respect to the design,

$$E\{d\} = \Sigma_s d(s) p(s)$$

and Σ_s denotes the summation over all samples s . A predictor d is the *best model unbiased predictor* of $\overline{Y}$ in the class $\mathcal{D}_a$ if (2.1) holds for all s in S for which p(s) > 0 and if $V(d - \overline{Y} | s)$ is less than or equal to $V(d_i - \overline{Y} | s)$ for all d_i in the class $\mathcal{D}_a$ and all s in S for which p(s) > 0 , where

$$V(d - \overline{Y} | s) = E\{[(d - \overline{Y}) - E(d - \overline{Y} | s)]^2 | s\} .$$

Cassel, Särndal and Wretman (1977, Ch.4) discuss model predictors, design unbiasedness, and the operators E and E .

The *anticipated variance* of $d - \overline{Y}$ is the variance of the random variable $d - \overline{Y}$ where the squares are averaged over the design and over samples from the superpopulation model,

$$AV\{d - \overline{Y}\} = E\{E[(d - \overline{Y})^2]\} - [E\{E(d - \overline{Y})\}]^2 .$$

For a superpopulation model with finite moments and a design whose probabilities are independent of the y-values; we have

$$\text{AV}\{d - \overline{Y}\} = E[V(d - \overline{Y}|s)] + V[E(d - \overline{Y}|s)] , \qquad (2.3)$$

where $V\{\cdot\}$ denotes the design variance.

We define a sequence of estimators $\{d_t\}$ constructed for a sequence of populations to be *design consistent* for $\overline{Y}_t$, where $\overline{Y}_t$ is the mean of the tth population, if, given $\varepsilon > 0$,

$$\lim_{t \to \infty} \text{Prob}\{|d_t - \overline{Y}_t| > \varepsilon\} = 0 ,$$

the probability being that created by the sample design. To investigate limiting properties of d_t the sequences must be well defined. We define the sequence of populations in terms of a sequence of elements $\{u_j\}$. Let (y_j, z_j, ω_j), where $\omega_j > 0$ for all j , be a vector of characteristics associated with the jth element. Let $\{U_t\}$ denote a sequence of finite populations of size $\{N_t\}$, where $0 < N_1 < N_2 < N_3 \dots$, where U_1 is composed of the first N_1 elements of $\{u_j\}$, $U_2 \supset U_1$ is composed of the first N_2 elements of $\{u_j\}$, etc. Let a sequence of samples $\{s_t\}$ of size $\{n_t\}$ be created from the sequence of populations by a sequence of designs such that s_1 is composed of n_1 distinct elements selected from U_1 , s_2 is composed of n_2 distinct elements selected from U_2 , etc., where $n_1 < n_2 < n_3 \dots$, and $n_t < N_t$ for all t . The sequence of populations is nested, but the sequence of samples is not. Let the probability that the ith element is included in the tth sample be given by

$$P\{u_i \in s_t\} = \pi_{i(t)} ,$$

where $0 < \pi_{i(t)} < 1$ and

$$\pi_{i(t)} = n_t \left(\sum_{j=1}^{N_t} \omega_j \right)^{-1} \omega_i \qquad (2.4)$$

We shall often assume

$$0 < \lambda_2 < \pi_{i(t)} < \lambda_1 < 1 \tag{2.5}$$

for all t . The joint probabilities of selection for the tth population are denoted by $\pi_{ij(t)}$ for $i,j \leq N_t$.

Lemmas 1, 2 and 3 give sufficient conditions for the Horvitz-Thompson estimator to be consistent.

Lemma 1. Let the sequence of populations $\{U_t\}$ and samples $\{s_t\}$ be as described. Let the sequence $\{y_i, \omega_i\}$ be a fixed sequence. Let the joint probabilities of inclusion satisfy

$$\pi_{i(t)} \pi_{j(t)} - \pi_{ij(t)} \leq \alpha_t \pi_{i(t)} \pi_{j(t)}$$

where

$$\lim_{t \to \infty} \alpha_t = 0 .$$

Let

$$d_t = \sum_{i \in s_t} (N_t \pi_{i(t)})^{-1} y_i , \tag{2.6}$$

$$N_t^{-2} n_t \sum_{j=1}^{N_t} \pi_{j(t)} (\pi_{j(t)}^{-1} Y_j - n_t^{-1} Y_t)^2 < M < \infty \tag{2.7}$$

for all t . Then d_t is a design consistent estimator of $\bar{Y}_t$. Furthermore

$$d_t - \bar{Y}_t = O_p(\alpha_t^{\frac{1}{2}}) .$$

Proof. We have

$$V\{d_t - \bar{Y}_t\} = (2N_t^2)^{-1} \sum_{i \neq j=1}^{N_t} (\pi_{i(t)} \pi_{j(t)} - \pi_{ij(t)}) (\pi_{i(t)}^{-1} Y_i - \pi_{j(t)}^{-1} Y_j)^2$$

$$\leq (2N_t^2)^{-1} \sum_{i \neq j=1}^{N_t} \alpha_t (\pi_{i(t)}^{-1} Y_i - \pi_{j(t)}^{-1} Y_j)^2 \pi_{i(t)} \pi_{j(t)}$$

$$= N_t^{-2} n_t \alpha_t \sum_{i=1}^{N_t} \pi_{i(t)} (\pi_{i(t)}^{-1} Y_i - n_t^{-1} Y_t)^2 . \qquad \square$$

For simple random sampling

$$\pi_{i(t)} \pi_{j(t)} - \pi_{ij(t)} = [n_t (N_t - 1)]^{-1} [N_t - n_t] (N_t^{-1} n_t)^2$$

and the quantity $\alpha_t = [n_t (N_t - 1)]^{-1} [N_t - n_t]$ satisfies the condition of Lemma 1.

Lemma 2. Let the sequence of populations $\{U_t\}$ and samples $\{s_t\}$ be as described. Let $|\pi_{j(t)}^{-1} Y_j| < M < \infty$ for all j and t. Let

$$g_{ij(t)} = \pi_{i(t)} \pi_{j(t)} - \pi_{ij(t)} \quad \text{if } \pi_{i(t)} \pi_{j(t)} - \pi_{ij(t)} \geq 0 \quad (2.8)$$

$$= 0 \qquad\qquad\qquad \text{otherwise.}$$

Assume

$$N_t^{-2} \sum_{i \neq j=1}^{N_t} g_{ij(t)} = O(n_t^{-2\delta}) ,$$

where $\delta > 0$ is a fixed number. Then $d_t - \overline{Y}_t = O_p(n^{-\delta})$, where d_t is defined in (2.6).

Proof. We have

$$v\{d_t - \overline{Y}_t\} \leq (2N_t^2)^{-1} \sum_{i \neq j=1}^{N_t} g_{ij(t)} (2M)^2$$

and the result follows. $\square$

The condition on the $\pi_{ij(t)}$ of Lemma 2 would be satisfied by a sequence of designs that assigned a fixed and bounded number of elements to an increasing number of strata. For example, consider the design with $\pi_{i(t)} = N_t^{-1} n_t = K^{-1}$ and let the sequence of elements be divided into groups of K adjacent elements. Let the design be the one-per-stratum design that selects one element at random in each of the n_t strata of the tth population. Then

$$\pi_{ij(t)} = 0 \qquad \text{if i and j are in the same stratum}$$

$$= \pi_{i(t)} \pi_{j(t)} = K^{-2} \quad \text{otherwise.}$$

It follows that

$$N_t^{-2} \sum_{\substack{i \neq j=1}}^{N_t} g_{ij(t)} = N_t^{-1}(K-1)K^{-2} = O(n_t^{-1})$$

and the condition of the lemma is satisfied. For this type of sequence of stratified designs the condition on the $\{y_i\}$ can be relaxed, as can be seen in Lemmas 3 and 4.

Lemma 3. Let the sequence of populations $\{\mathcal{U}_t\}$ and samples $\{s_t\}$ be as described. Let d_t be defined by (2.6) and let $g_{ij}(t)$ be defined by (2.8). Assume

$$\text{(i)} \qquad N_t^{-2} \sum_{\substack{i \neq j=1}}^{N_t} g_{ij(t)}^r = O(n_t^{-2\delta r})$$

$$\text{(ii)} \qquad N_t^{-1} \sum_{i=1}^{N_t} (\pi_{i(t)}^{-1} Y_i - n_t^{-1}Y_t)^{2k} < M < \infty$$

for all t, where $r, k > 1$ and $r^{-1} + k^{-1} = 1$. Then d_t is a design consistent estimator of $\overline{Y}_t$. Furthermore $d_t - \overline{Y}_t = O_p(n_t^{-\delta})$.

Proof. We have, by Hölder's Inequality,

$$V\{d_t - \overline{Y}_t\} \leq (2N_t^2)^{-1} \sum_{\substack{i \neq j=1}}^{N_t} g_{ij(t)} (\pi_{i(t)}^{-1} Y_i - \pi_{j(t)}^{-1}Y_j)^2$$

$$\leq \left[(2N_t^2)^{-1} \sum_{i \neq j=1}^{N_t} g_{ij}^r(t) \right]^{\frac{1}{r}} \left[(2N_t^2)^{-1} \sum_{i \neq j=1}^{N_t} (\pi_{i(t)}^{-1} Y_i - \pi_{j(t)}^{-1} Y_j)^{2k} \right]^{\frac{1}{k}}$$

$$\leq \left[N_t^{-2} \sum_{i \neq j=1}^{N_t} g_{ij}^r(t) \right]^{\frac{1}{r}} \left[4^k N_t^{-1} \sum_{i=1}^{N_t} (\pi_{i(t)}^{-1} Y_i - n_t^{-1} Y_t)^{2k} \right]^{\frac{1}{k}} . \quad \square$$

Note that for simple random sampling

$$(2N_t^2)^{-1} \sum_{i \neq j=1}^{N_t} g_{ij}^r(t) = O(n_t^{-r}) .$$

Lemma 4 demonstrates that it is possible to construct a sequence of designs such that the design variance of the Horvitz-Thompson estimator decreases at the rate n_t^{-1}.

Lemma 4. Let the sequences of populations $\{U_t\}$ and samples $\{s_t\}$ be as described. Let the fixed sequence $\{y_i\}$ satisfy

$$\lim_{t \to \infty} N_t^{-1} \sum_{i=1}^{N_t} y_i = \lim_{t \to \infty} \bar{Y}_t = \bar{Y} , \qquad (2.9)$$

$$\lim_{t \to \infty} N_t^{-1} \sum_{i=1}^{N_t} (y_i - \bar{Y}_t)^2 = \lim_{t \to \infty} \sigma_{y(t)}^2 = \sigma_y^2 , \qquad (2.10)$$

and let the sequence $\{\omega_i\}$ be such that inclusion probabilities $\pi_{i(t)}$ defined by (2.4) satisfy (2.5). Then there exists a sequence of designs with inclusion probabilities $\pi_{i(t)}$ such that

$$E\{(d_t - \bar{Y}_t)^2\} = O(n_t^{-1}) ,$$

where d_t is defined in (2.6).

Proof. A proof is given by Isaki and Fuller (1981). A sequence of samples is constructed by arranging the population in natural order and forming the cumulative sum of the probabilities $T_{i(t)} = \Sigma_{j=1}^{i} \pi_{j(t)}$. The population is divided into, for example, $\frac{1}{2} n_t$ (or $\frac{1}{2}(n_t + 1)$ for odd n_t) strata by placing a boundary at the points $T_{i(t)} = 2, 4, \ldots, n_t-2, (n_t-1$ for odd $n_t)$. Two elements are selected in each stratum using a two-per-stratum method such as that of Durbin (1967) combined with the method given in Fuller (1970, p.217).   □

We next give a central limit theorem for the estimator design pair of Lemma 4.

Lemma 5.   Let the assumptions of Lemma 4 hold and let

$$\lim_{t \to \infty} N_t^{-1} \sum_{i=1}^{N_t} |y_i - \overline{Y}_t|^{2+\delta} = B \qquad (2.11)$$

for some $\delta > 0$. Let the sequence of samples $\{s_t\}$ be constructed by the method described in the proof of Lemma 4.   Let d_t be defined by (2.6) and let

$$\lim_{t \to \infty} n_t V(d_t - \overline{Y}_t) = \lim_{t \to \infty} n_t V_t = A .$$

Then

$$V_t^{-\frac{1}{2}}(d_t - \overline{Y}_t) \xrightarrow{L} N(0,1) .$$

Proof.   Let $Y_{[j]}$ be the weighted sum of the population of y-elements in stratum j . (For an element completely in the stratum the weight in the sum for that element is one.   For an element crossing the boundary the weight in the sum is equal to the fraction of the inclusion probability that is in the stratum under consideration.)   Let

$$z_{[j]t} = \sum_{\substack{i \in s_t \\ i \in [j]}} \pi_{i(t)}^{-1} y_i - Y_{[j]}, \quad j = 1,2,\ldots,\alpha_t$$

where the sum is over the elements selected in the jth stratum for
the tth population and α_t is the number of strata. The random
variables $z_{[j]t}$ have zero means and form a triangular array such
that the elements in the tth row are m-dependent for all t. See
Hoeffding and Robbins (1948) and Fuller (1976, p.245). That is,
by the method of selection, elements in the sequence more than m
units apart are independent, where m is the smallest integer
greater than $\lambda_1(1-\lambda_1)^{-1}$.

Note that

$$N_t^{-1} \sum_{j=1}^{\alpha_t} z_{[j]t} = d_t - \bar{Y}_t .$$

Following the development in Fuller (1976, p.246) set

$$W_{it} = \sum_{r=1}^{k-m} z_{[(i-1)k+r]t} \qquad i = 1,2,\ldots,p ,$$

where k is the largest integer less than n_t^{α} for some
$0 < \alpha < \frac{1}{2}(1+\delta)^{-1}\delta$ and p is the largest integer less than
$k^{-1}n_t$. Let

$$S_{pt} = (N_t^2 V_t)^{-\frac{1}{2}} \sum_{i=1}^{p} W_{ij} .$$

Then

$$\operatorname*{plim}_{t \to \infty} (N_t^2 V_t)^{-\frac{1}{2}} \sum_{j=1}^{\alpha_t} z_{[j]t} - S_{pt} = 0 .$$

By assumption (2.11)

$$E\left\{ n_t^{-1} \sum_{j=1}^{\alpha_t} |z_{[j]}|^{2+\delta} \right\} = O(1) .$$

Using

$$E\left\{ \left| \sum_{r=1}^{k-m} z_{[(i-1)k+r]} \right|^{2+\delta} \right\} \leq k^{1+\delta} \sum_{r=1}^{k-m} E\left\{ \left| z_{[(i-1)k+r]} \right|^{2+\delta} \right\},$$

we have

$$\lim_{p \to \infty} \frac{\sum_{i=1}^{p} E\{ | (N_t^2 v_t)^{-\frac{1}{2}} w_{it} |^{2+\delta} \}}{\left| \sum_{i=1}^{p} E\{ [(N_t^2 v_t)^{-\frac{1}{2}} w_{it}]^2 \} \right|^{1+\frac{1}{2}\delta}}$$

$$= \lim_{p \to \infty} N_t^{(2+\delta)} V_t^{-\frac{1}{2}(2+\delta)} \sum_{i=1}^{p} \sum_{r=1}^{k-m} k^{1+\delta} E\{ | z_{[(i-1)k+r]} |^{2+\delta} \}$$

$$\leq \lim_{p \to \infty} (n_t V_t)^{-\frac{1}{2}(2+\delta)} \lambda^{2+\delta} n_t^{-\frac{1}{2}\delta} k^{1+\delta} n_t^{-1} \sum_{i=1}^{p} \sum_{r=1}^{k} E\{ | z_{[(i-1)k+r]} |^{2+\delta} \}$$

$$= 0 .$$

The result follows by the Liapounov central limit theorem.   □

3. MODEL AND RESULTS

We assume that the finite population of N elements was generated by the superpopulation model

$$y_i = z_i \beta + \varepsilon_i , \quad i=1,2,\ldots,N \tag{3.1}$$

$$E\{\varepsilon_i\} = 0 , \tag{3.2}$$

$$E\{\varepsilon_i \varepsilon_j\} = \gamma_{ii} \sigma^2 \qquad i = j \tag{3.3}$$

$$= \rho \gamma_{ii} \gamma_{jj} \sigma^2 \quad i \neq j, \ -(N-1)^{-1} < \rho < 1$$

where z_i is a q-dimensional row vector, β is a q-dimensional column vector, $\bar{Z}_t$ and γ_{ii} , $i = 1,2,\ldots,N$ are known, and β , ρ and σ^2 are unknown. Some or all of the elements of z_i may be known only for the sample elements.

Isaki and Fuller (1981) have given the following two linear

model results for the model (3.1), (3.2), and (3.3).

        __Lemma 6.__ Let model (3.1) hold. Assume (a) the vector $(\gamma_{11}^{\frac{1}{2}}, \gamma_{22}^{\frac{1}{2}}, \ldots, \gamma_{nn}^{\frac{1}{2}})'$ is in the column space of $Z_n' = (z_1', z_2', \ldots,$ $z_n')$, or (b) $\rho \equiv 0$. Let a sample of n elements be given. Then the best model unbiased linear predictor of $\bar{Y}$, conditional on $(z_1, z_2, \ldots, z_n)$ is

$$\bar{y}_p = f\,\bar{y} + (1-f)\bar{Z}_{N-n}\,\hat{\beta}\ , \tag{3.4}$$

where $f = N^{-1}n$, $\bar{z} = n^{-1}\sum_{i=1}^{n} z_i$,

$$\bar{Z} = (1-f)\bar{Z}_{N-n} + f\bar{z}\ ,$$

$$\hat{\beta} = (Z_n'\,\Gamma_n^{-1}\,Z_n)^{-1}\,Z_n'\,\Gamma_n^{-1}\,y_n\ , \tag{3.5}$$

$$\Gamma_n = \mathrm{diag}(\gamma_{11}, \gamma_{22}, \ldots, \gamma_{nn})\ ,$$

$$y_n' = (y_1, y_2, \ldots, y_n)\ ,$$

and, for convenience, the sample elements are assigned the first n subscripts.

        __Lemma 7.__ Let the assumptions of Result 1 hold. Assume in addition that the vector $(\gamma_{11}, \gamma_{22}, \ldots, \gamma_{nn})$ is in the column space of Z_n. Then the best model unbiased linear predictor of $\bar{Y}$ is

$$y_p = \bar{Z}\,\hat{\beta}\ . \tag{3.6}$$

        We now turn to an investigation of the design variance of regression estimators. We first consider the regression estimator constructed with fixed coefficients. The estimator we investigate is the design unbiased estimator

$$\bar{y}_c = \bar{\bar{y}} - (\bar{\bar{x}} - \bar{X})C\ , \tag{3.7}$$

where $\underset{\sim}{x}_i$ is a vector of auxiliary variables whose population mean $\bar{\underset{\sim}{X}}$ is known, $\underset{\sim}{C}$ is a vector of constants,

$$\bar{\bar{y}} = \sum_{i \in s} (N\pi_i)^{-1} y_i ,$$

$$\bar{\bar{\underset{\sim}{x}}} = \sum_{i \in s} (N\pi_i)^{-1} \underset{\sim}{x}_i .$$

Isaki (1970, p.40) studied this estimator for a single x-variable. The estimator has been considered by a number of authors including Basu (1971) and Cassel, Särndal and Wretman (1976).

In sampling without replacement the $\underset{\sim}{C}$ that minimizes the design variance is a function of the joint inclusion probabilities.

Lemma 8. Consider samples of fixed size n selected without replacement, with inclusion probabilities π_i and joint inclusion probabilities π_{ij} . Let the estimator be that defined in (3.7). Then the minimum design variance of $\bar{y}_c$ is obtained for $\underset{\sim}{C}$, where $\underset{\sim}{B}' = (B_1, \underset{\sim}{C}')$,

$$\underset{\sim}{B} = \underset{\sim}{G}^{-1} \underset{\sim}{H} , \tag{3.8}$$

$$\underset{\sim}{G} = \sum_{i \neq j=1}^{N} (\pi_i \pi_j - \pi_{ij}) (\pi_i^{-1} \underset{\sim}{z}_i - \pi_j^{-1} \underset{\sim}{z}_j)' (\pi_i^{-1} \underset{\sim}{z}_i - \pi_j^{-1} \underset{\sim}{z}_j) ,$$

$$\tag{3.9}$$

$$\underset{\sim}{H} = \sum_{i \neq j=1}^{N} (\pi_i \pi_j - \pi_{ij}) (\pi_i^{-1} \underset{\sim}{z}_i - \pi_j^{-1} \underset{\sim}{z}_j)' (\pi_i^{-1} y_i - \pi_j^{-1} y_j) ,$$

$$\underset{\sim}{z}_i = (\pi_i, \underset{\sim}{x}_i) .$$

If π_i is an element of $\underset{\sim}{x}_i$ then $\underset{\sim}{z}_i = \underset{\sim}{x}_i$.

Proof: The variance of the estimator is given by

$$V\{\overline{y}_c\} = V\{ \sum_{i \in s} (N\pi_i)^{-1}(y_i - \underset{\sim i}{x}\underset{\sim}{C}) - (\overline{Y} - \overline{\underset{\sim}{X}}\ \underset{\sim}{C})\} ,$$

$$= V\{ \sum_{i \in s} (N\pi_i)^{-1}[y_i - N\pi_i(\overline{Y} - \overline{\underset{\sim}{X}}\ \underset{\sim}{C}) - \underset{\sim i}{x}\underset{\sim}{C}]\} ,$$

$$= V\{ \sum_{i \in s} (N\pi_i)^{-1} e_i\} ,$$

where $e_i = y_i - \pi_i N(\overline{Y} - \overline{\underset{\sim}{X}}\ \underset{\sim}{C}) - \underset{\sim i}{x}\ \underset{\sim}{C}$. Using the Yates-Grundy
form of the variance, the variance is minimized by the $\underset{\sim}{C}$ of
(3.8).   □

If we assume that the design and variables are such that an
approximation of the type developed by Hartley and Rao (1962) is
appropriate, we obtain the approximate expressions[1]

$$\underset{\sim a}{G} = (N-n+1)^{-1}(N-n)N(N-1)^{-1} \sum_{i=1}^{N} \pi_i^{-1}[1-n^{-1}(n-1)\pi_i]\underset{\sim i}{z}'\underset{\sim i}{z} , \quad (3.10)$$

$$\underset{\sim a}{H} = (N-n+1)^{-1}(N-n)N(N-1)^{-1} \sum_{i=1}^{N} \pi_i^{-1}[1-n^{-1}(n-1)\pi_i]\underset{\sim i}{z}'y_i .$$

Design unbiased estimators of $\underset{\sim a}{G}$ and $\underset{\sim a}{H}$ are

$$\underset{\sim a}{\hat{G}} = \sum_{i \in s} (N-n+1)^{-1}(N-n)(N-1)^{-1}\pi_i^{-2}[1 - n^{-1}(n-1)\pi_i]\underset{\sim i}{z}'\underset{\sim i}{z} , \quad (3.11)$$

$$\underset{\sim a}{\hat{H}} = \sum_{i \in s} (N-n+1)^{-1}(N-n)(N-1)^{-1}\pi_i^{-2}[1 - n^{-1}(n-1)\pi_i]\underset{\sim i}{z}'y_i .$$

---

[1]   Expressions (3.10) are obtained from Hartley-Rao's equation
(5.17) by multiplying by $(N-n+1)^{-1}(N-1)^{-1}(N-n)N$.

It follows that an estimator of $\underset{\sim}{B}$ is

$$\hat{\underset{\sim}{B}}_a = \hat{G}_a^{-1} \hat{\underset{\sim}{H}}_a .$$ (3.12)

If one removes the factor $[1 - n^{-1}(n-1)\pi_i]$ in (3.11), one obtains the estimator

$$\bar{y}_\ell = \bar{\bar{y}} + \sum_{j=2}^{q} \hat{B}_j (\bar{z}_j - \bar{\bar{z}}_j)$$ (3.13)

where $\underset{\sim}{z}_i = (\pi_i, z_{2i}, z_{3i}, \ldots, z_{qi})$, $\hat{\underset{\sim}{B}}' = (B_1, B_2, \ldots, B_q)$,

$$\hat{\underset{\sim}{B}} = (\underset{\sim}{z}'_n \pi_n^{-2} \underset{\sim}{z}_n)^{-1} \underset{\sim}{z}'_n \pi_n^{-2} \underset{\sim}{y}_n ,$$

$$\pi_n = \text{diag}(\pi_1, \pi_2, \ldots, \pi_n) ,$$

$$\bar{\bar{z}}_j = \sum_{i \in s} (N \pi_i)^{-1} z_{ji} .$$

We note in passing that the approximation (3.10) is not appropriate for the design of Lemma 4.

A second regression estimator considered by Isaki (1970) is the location invariant estimator

$$\bar{\bar{y}}_{C1} = \bar{\underset{\sim}{x}} \underset{\sim}{C} + \left| \sum_{i \in s} (N \pi_i)^{-1} \right|^{-1} (\bar{y} - \bar{\underset{\sim}{x}} \underset{\sim}{C}) .$$ (3.14)

By the usual approximation for the variance of a ratio, we obtain

$$v\{\bar{\bar{y}}_{C1}\} \doteq v \left\{ \sum_{i \in s} (N \pi_i)^{-1} [y_i - \underset{\sim}{x}_i \underset{\sim}{C} - (\bar{Y} - \bar{\underset{\sim}{X}} \underset{\sim}{C})] \right\} .$$ (3.15)

Expression (3.15) is minimized for $\underset{\sim}{C}$ defined by (3.8) with

$$\underset{\sim}{z}_i = (1, \underset{\sim}{x}_i) .$$

These results demonstrate that the regression predictor has

considerable appeal from both the design and model points of view. Theorem 1, which has been proven by Isaki and Fuller (1981) gives the limiting design properties of the estimator.

   <u>Theorem 1.</u> Let a sequence of samples and populations be as described in Section 2. Let $\{y_i,\ \omega_i,\ z_i\}$ be a fixed sequence. Let the inclusion probabilities $\pi_{i(t)}$ satisfy (2.5). Let $\pi_{i(t)}$ be the first element of $\underset{\sim}{z}_{it}$. Let

$$\overline{y}_{\ell t} = \overline{\underset{\sim}{Z}}_t\ \hat{\underset{\sim}{B}}_t\ ,$$

$$= \overline{\overline{y}}_t + \sum_{j=2}^{q} \hat{B}_{jt}(\overline{Z}_{jt} - \overline{\overline{z}}_{jt})\ , \tag{3.16}$$

where

$$\hat{\underset{\sim}{B}}_t = (\underset{\sim}{z}'_{nt}\ \underset{\sim}{\pi}^{-2}_{nt}\ \underset{\sim}{z}_{nt})^{-1}\ \underset{\sim}{z}'_{nt}\ \underset{\sim}{\pi}^{-2}_{nt}\ \underset{\sim}{y}_{nt}\ , \tag{3.17}$$

$$\underset{\sim}{y}'_{nt} = (y_1,\ y_2,\ \dots,\ y_{nt})\ ,$$

$$\underset{\sim}{z}'_{nt} = (\underset{\sim}{z}'_{1t},\ \underset{\sim}{z}'_{2t},\ \dots,\ \underset{\sim}{z}'_{nt})\ ,$$

$$\underset{\sim}{\pi}_{nt} = \text{diag}(\pi_{1(t)},\ \pi_{2(t)},\dots,\ \pi_{n(t)})\ .$$

Let

$$\lim_{t \to \infty} n_t\ \text{Cov}\{\overline{\underset{\sim}{w}}_t\} = \underset{\sim}{\Phi}_{ww}\ , \qquad |\underset{\sim}{\Phi}_{ww}| > 0\ ,$$

$$\lim_{t \to \infty} N^{-1}_t \underset{\sim}{W}'_{nt}\ \underset{\sim}{\pi}^{-1}_{Nt}\ \underset{\sim}{W}_{Nt} = G, \qquad |\underset{\sim}{G}| > 0\ ,$$

$$\lim_{t \to \infty} n_t\ \text{Cov}\{\overline{\underset{\sim}{v}}_t\} = \underset{\sim}{\Phi}_{vv}\ ,$$

where

$$\underset{\sim}{W}'_{Nt} = (\underset{\sim}{w}'_{1t},\ \underset{\sim}{w}'_{2t},\dots,\ \underset{\sim}{w}'_{Nt})\ ,$$

$$\underset{\sim}{w}_{it} = (w_{1it}, w_{2it}, \ldots, w_{q+1,it}) = (y_i, \underset{\sim}{z}_{it}) ,$$

$$\underset{\sim}{v}_{it} = \pi_{i(t)}^{-1} (w_{1it}^2, w_{1it}w_{2it}, \ldots, w_{2it}^2, w_{2it}w_{3it}, \ldots, w_{q+1,it}^2) ,$$

$$\underset{\sim}{\bar{\bar{w}}}_t = \sum_{i \in s_t} (N_t \pi_{i(t)})^{-1} \underset{\sim}{w}_{it} ,$$

$$\underset{\sim}{\bar{W}}_t = N_t^{-1} \sum_{i=1}^{N_t} \underset{\sim}{w}_{it} ,$$

$$\underset{\sim}{\bar{\bar{v}}}_t = \sum_{i \in s_t} (N_t \pi_{i(t)})^{-1} \underset{\sim}{v}_{it} ,$$

$$\text{Cov} \{\underset{\sim}{\bar{\bar{w}}}_t\} = E\{(\underset{\sim}{\bar{\bar{w}}}_t - \underset{\sim}{\bar{W}}_t)' (\underset{\sim}{\bar{\bar{w}}}_t - \underset{\sim}{\bar{W}}_t)\} .$$

Then

$$\bar{y}_{\ell t} - \bar{Y}_t = \sum_{i \in s_t} (N_t \pi_{i(t)})^{-1} e_{it} + O_p(n_t^{-1}) , \qquad (3.18)$$

where

$$e_{it} = y_i - \underset{\sim}{z}_{it} \underset{\sim}{B}_t ,$$

$$\underset{\sim}{B}_t = (\underset{\sim}{Z}_{Nt}' \underset{\sim}{\pi}_{Nt}^{-1} \underset{\sim}{Z}_{Nt})^{-1} \underset{\sim}{Z}_{Nt}' \underset{\sim}{\pi}_{Nt}^{-1} \underset{\sim}{y}_{Nt} , \qquad (3.19)$$

$$\underset{\sim}{y}_{Nt}' = (y_1, y_2, \ldots, y_{Nt}) ,$$

and the order in probability statement is with respect to the
design.

Theorem 2, proven by Isaki and Fuller (1981), gives the
properties of the regression predictor under the superpopulation
model.

Theorem 2. Let model (3.1)-(3.3) hold and let

$$\pi_{i(t)} = n_t \left(\sum_{j=1}^{N_t} \gamma_{jj}^{\frac{1}{2}}\right)^{-1} \gamma_{ii}^{\frac{1}{2}} , \quad 0 < \lambda_2 < \pi_{i(t)} < \lambda_1 < 1 .$$

Let $\underset{\sim}{z}_i = (\gamma_{ii}^{\frac{1}{2}}, z_{2i}, \ldots, z_{qi})$, let $\{\underset{\sim}{z}_i\}$ be fixed and assume

$$E\,E\left\{||\hat{\underset{\sim}{B}}_t - \underset{\sim}{\beta}||^4\right\} = O(n_t^{-2})$$

$$E\,E\left\{||\bar{\bar{\underset{\sim}{z}}}_t - \bar{\underset{\sim}{z}}\,||^4\right\} = O(n_t^{-2}) \ .$$

Then

$$AV(\bar{y}_{\ell t} - \bar{Y}_t) = (1 - \rho_t)\sigma^2 N_t^{-2}\left[n_t^{-1}\left(\sum_{i=1}^{N_t}\gamma_{ii}^{\frac{1}{2}}\right)^2 - \sum_{i=1}^{N_t}\gamma_{ii}\right] + O(n_t^{-3/2}) \ .$$

Theorem 3, proven by Isaki and Fuller (1981), demonstrates that, in the limit, it is not possible to find a design-predictor pair superior to the pair of Theorem 2. The limit of the lower bound of the anticipated variance for design unbiased predictors obtained by Godambe and Joshi (1965) is the lower bound for the limit of the anticipated variance of consistent predictors.

Theorem 3. Let model (3.1) hold with $\gamma_{ii}^{\frac{1}{2}}$ and γ_{ii} elements of $\underset{\sim}{z}_i$. Let the sequence of samples and populations be that described in Section 2. Let

$$0 < \lambda_2 < n_t\left(\sum_{j=1}^{N_t}\gamma_{jj}^{\frac{1}{2}}\right)^{-1}\gamma_{ii}^{\frac{1}{2}} < \lambda_1 < 1$$

for all t . Let

$$\lim_{t \to \infty} N_t^{-1}\sum_{i=1}^{N_t}\underset{\sim}{w}_i = \lim_{t \to \infty}\bar{\underset{\sim}{W}}_t = \bar{\underset{\sim}{W}} \quad a.s.,$$

$$\lim_{t \to \infty} N_t^{-1}\sum_{i=1}^{N_t}(\underset{\sim}{w}_i - \bar{\underset{\sim}{W}}_t)'(\underset{\sim}{w}_i - \bar{\underset{\sim}{W}}_t) = \underset{\sim}{\Sigma}_{ww} \ , \quad a.s.,$$

where $\underset{\sim}{w}_i = (y_i, \underset{\sim}{z}_i)$. Let

$$\lim_{t \to \infty} n_t\,E\left\{\bar{\underset{\sim}{Z}}_t(\underset{\sim}{Z}_{nt}'\,\underset{\sim}{\Gamma}_n^{-1}\underset{\sim}{Z}_{nt})^{-1}\bar{\underset{\sim}{Z}}_t'\right\} = \underset{\sim}{\bar{Z}}\,\underset{\sim}{L}^{-1}\,\bar{\underset{\sim}{Z}}' \ ,$$

$$\lim_{t \to \infty} n_t^{-1} \, \underset{\sim}{Z}'_{Nt} \, \underset{\sim}{\pi}_{Nt} \, \underset{\sim}{\Gamma}^{-1} \, \underset{\sim}{Z}_{Nt} = \lim_{t \to \infty} \underset{\sim}{L}_t = \underset{\sim}{L}$$

for designs in P_c , where P_c is the class of fixed sample size nonreplacement designs admitting design consistent estimators. Let $\mathcal{D}_\ell$ be the class of linear predictors of the form

$$d_{\ell t} = \sum_{i \in s_t} \alpha_{is}(t) \, y_i \; ,$$

where

$$E\{d_{\ell t} - \overline{Y}_t | s_t\} = 0 \; ,$$

and the weights $\alpha_{is(t)}$ are permitted to be functions of the elements of the sample. Let $\overline{y}_{\ell \gamma t}$ denote the predictor (3.16) constructed for the design with inclusion probabilities proportional to $\gamma_{ii}^{\frac{1}{2}}$. Then

$$\lim_{t \to \infty} n_t [\mathrm{AV}(\overline{y}_{\ell \gamma t} - \overline{Y}_t) - \mathrm{AV}(d - \overline{Y}_t)] \leq 0$$

for all $d \subset \mathcal{D}_\ell$ and $p \subset P_c$.

4. DISCUSSION

Our discussion has touched upon several criteria for estimators and designs constructed in the presence of auxiliary information. The primary criteria mentioned were:

(1) Design consistency

(2) Model efficiency

(3) Design efficiency

(4) Model unbiasedness

(5) Minimum anticipated variance

(6) Design unbiasedness

Properties of estimators given less consideration include location

invariance and scale invariance. The requirement of design
unbiasedness was judged to be too restrictive and was replaced by
the requirement of design consistency. The anticipated variance
is one way of pooling measures of design and model efficiency. To
minimize the anticipated variance, to the level of approximation
employed, the inclusion probabilities should be proportional to
$\gamma_{ii}^{\frac{1}{2}}$, the square roots of the model variances. We assume π_i is
proportional to $\gamma_{ii}^{\frac{1}{2}}$ in the remainder of the discussion.

   For simple models such as the ratio model

$$y_i = \beta x_i + \varepsilon_i \, ,$$ (4.1)

$$E\{\varepsilon_i\} = 0 \, ,$$

$$E\{\varepsilon_i \varepsilon_j\} = \gamma_{ii} \, \sigma^2 \qquad i = j$$

$$= 0 \qquad\qquad \text{otherwise,}$$

there is a conflict between the criterion of design consistency
and the criterion of model efficiency. The competition between
these two criteria explains, in part, the number of estimators
that have been suggested for this model. Estimators that have
been suggested include:

   (i) Ratio estimator

$$\bar{y}_R = \left[\sum_{i \in s} (N\pi_i)^{-1} x_i\right]^{-1} \left[\sum_{i \in s} (N\pi_i)^{-1} y_i\right] \bar{X} \, .$$ (4.2)

   (ii) Brewer (1980)

$$\bar{y}_B = N^{-1}\left\{\sum_{i \in s} y_i + \left[\sum_{i \in s} (\pi_i^{-1} - 1) x_i\right]^{-1} \left[\sum_{i \in s} (\pi_i^{-1} - 1) y_i\right]\right.$$ (4.3)

$$\left.\left[N\bar{X} - \sum_{i \in s} x_i\right]\right\} \, .$$

(iii) Cassel, Sarndal and Wretman (1976)

$$\bar{y}_{CSW} = \bar{X}\,\hat{\beta} + \sum_{i \in s} (N\pi_i)^{-1}(y_i - x_i\hat{\beta}) \,, \qquad (4.4)$$

where

$$\hat{\beta} = \left| \sum_{i \in s} \pi_i^{-2} x_i^2 \right|^{-1} \sum_{i \in s} \pi_i^{-2} x_i y_i \;.$$

Sarndal (1980b) calls the estimator (4.4) a generalized regression estimator with the best linear unbiased (BLU) weights.

We have not considered model (4.1) directly in the discussion of this paper, but one might use the regression estimator (3.7) with $z_i = (\pi_i,\, x_i)$ as an estimator for such a model. The estimator of B might be constructed with weights π_i^{-2} or $\pi_i^{-2}[1 - n^{-1}(n-1)\pi_i]$. The estimator (3.14) constructed with $z_i = (1,\, x_i)$ and weights π_i^{-2} or $\pi_i^{-2}[1 - n^{-1}(n-1)\pi_i]$ is also a possibility. Both the ratio and regression estimators are model unbiased under model (4.1) and all estimators have the same large sample anticipated variance. The regression estimators are also unbiased under the model

$$y_i = \beta_1\pi_i + \beta_2 x_i + \varepsilon_i \;.$$

However, the large sample design variance is not the same for all estimators. To the extent that the joint inclusion probabilities are such that (3.10) is a good approximation for (3.9) the regression estimators will have smaller large sample design variances than the ratio estimators. This is analogous to the simple random sampling result that the regression estimator has smaller large sample design variance than the ratio estimator.

The model variance

$$E\{(\hat{\bar{Y}} - \bar{Y})^2 | s\} \,,$$

where $\hat{\bar{Y}}$ is the estimated mean, is given in Table 1 for the
estimators $\bar{y}_R$, $\bar{y}_B$, $\bar{y}_{CSW}$, $\bar{y}_\ell$ with $\underset{\sim}{z}_i = (\pi_i, x_i)$ and weights
π_i^{-2}, and $\bar{y}_{Cl}$ with $\underset{\sim}{z}_i = (1, x_i)$ and estimated $\underset{\sim}{C}$ with weights
π_i^{-2}, for five samples. All model variances are given as ratios
to the model variance of the best linear unbiased predictor (BLUP).
The model variances were computed assuming the elemental variances
were proportional to π_i^2. The samples are given in Table 2.
Sample 5 is equal to sample 3 with $x_i = \pi_i^2$. Table 1 demonstrates
that none of the design consistent estimators can be judged
uniformly superior or inferior under the model. For a sample
where the portion of x_i orthogonal to π_i has a mean of large
absolute value, the regression estimator $\bar{y}_\ell$ may have a considerably larger conditional model variance than the ratio estimators.
This is because two restrictions

$$\sum_{i \in s} w_i \pi_i = N^{-1} n$$

$$\sum_{i \in s} w_i x_i = \bar{X}$$

Table 1.   Model Relative Efficiency of Alternative Estimators

Estimator	Sample 1	Sample 2	Sample 3	Sample 4	Sample 5
$\bar{y}_R$	1.08	1.02	1.11	1.01	1.32
$\bar{y}_B$	1.08	1.00	1.11	1.02	1.36
$\bar{y}_{CSW}$	1.05	1.05	1.07	1.05	1.25
$\bar{y}_\ell$	1.02	6.97	1.04	2.52	1.03
$\bar{y}_{Cl}$	1.04	5.01	1.16	6.76	1.04
BLUP	1.00	1.00	1.00	1.00	1.00

Table 2.   Example Samples

Sample element	Sample 1 π_1	x_i	Sample 2 π_i	x_i	Sample 3 π_i	x_i	Sample 4 π_i	x_i
1	0.736	6.6	0.759	7.5	0.521	2.8	0.547	7.3
2	0.713	0.4	0.736	6.6	0.468	5.9	0.468	5.9
3	0.644	4.6	0.689	9.2	0.443	3.4	0.391	6.1
4	0.621	6.6	0.644	4.6	0.209	4.3	0.365	5.9
5	0.597	3.3	0.621	6.6	0.183	1.3	0.338	5.4
6	0.575	4.4	0.575	4.4	0.157	2.1	0.131	4.4
Population mean	0.667	4.9	0.667	4.9	0.333	3.94	0.333	3.94
Population mean π_i^2	0.4480		0.4480		0.1124		0.1124	

are imposed upon the weights of the linear estimator. On the other hand, if the y_i values do contain a term in π_i , the regression estimator can be superior to the ratio estimators because the regression estimator contains no bias term in the π_i. For example, in sample 2, the ratio estimator of the average of the π_i is 0.510 when the true value is 0.667. The other ratio-type estimators, $\bar{y}_B$, $\bar{y}_{CSW}$ and BLUP had similar biases. If the y_i were selected from a population satisfying the model

$$y_i = \beta_i \pi_i + \beta_2 x_i + \varepsilon_i \ ,$$

where the ε_i are independent $(0, \pi_i^2)$ random variables, the model mean square error of the regression estimator $\bar{y}_\ell$ will be less than that of the ratio estimator for $|\beta_1| > 2.02$.

In sample 4 the ratio-type estimators have a large bias if the model contains an intercept term. The ratio estimator estimates the constant one to be 0.729. Thus if the y_i were selected from a population satisfying the model

$$y_i = \beta_0 + \beta_2 x_i + \varepsilon_i \; ,$$

where the ε_i are independent $(0, \pi_i^2)$ random variables, the
model mean square error of the regression estimator $\bar{y}_{C1}$ will be
less than that of the ratio estimator for $|\beta_0| > 0.653$. In this
sample the estimator $\bar{y}_\ell$ with $z = (\pi, x)$ has an intercept bias
of -0.398, larger than the ratio intercept bias of -0.271.

These few small examples suggest that if one believes the
ratio model, but requires design consistency, the simple ratio
estimator $\bar{y}_R$ is a good choice. It is interesting, that in these
samples, Brewer's attempt to introduce prediction concepts into
the estimator did not lead to a uniform improvement in model
variance relative to the ratio estimator. If one is concerned
about an intercept in the model, as one must be in a large survey
where totals for subpopulations are being constructed, the
regression estimator $\bar{y}_\ell$ with $\underset{\sim}{z}_i = (1, \pi_i, x_i)$ or the regression estimator (3.14) with $\underset{\sim}{z}_i = (1, x_i)$ are reasonable estimators
from both model and design points of view.

The competition among the criteria generally decreases as
we move to regression models where the $\underset{\sim}{x}$-vector contains $\gamma_{ii}^{\frac{1}{2}}$.
For example, if the model is

$$y_i = \beta_1 \gamma_{ii}^{\frac{1}{2}} + \beta_2 \gamma_{ii} + \beta_3 x_i + \varepsilon_i$$

where the covariance structure is that defined in (4.1) and the
selection probabilities are proportional to $\gamma_{ii}^{\frac{1}{2}}$, the best linear
unbiased model predictor differs from the best design estimator of
the form (3.7) with estimated coefficients only to the extent that
the joint inclusion probabilities differ from a multiple of $\pi_i \pi_j$.

The addition of an intercept term to the model and the use
of weights π_i^{-2}, leads to an estimator that is the best linear

unbiased model predictor and is also location invariant.

ACKNOWLEDGEMENTS

This research was partly supported by Joint Statistical
Agreement 80-6 with the U.S. Bureau of the Census. We thank
Dr. Shriram Biyani, Mr. James Drew, Mr. Sastry Pantula,
Prof. Peter M. Robinson, and Prof. Carl Särndal for comments, and
Mr. Tin-Chiu Chua and Mr. Hsien-Ming Hung for computing the example.

REFERENCES

Basu, D. (1971). An essay on the logical foundations of survey
        sampling, Part One. In Godambe, V.P. and Sprott, D.A.(Eds.)
        Foundations of Statistical Inference. Toronto: Holt,
        Rinehart and Winston.

Brewer, K.R.W. (1963). Ratio estimation and finite populations:
        some results deductible from the assumption of an under-
        lying stochastic process. *Australian Journal of Statistics*
        5, 93-105.

Brewer, K.R.W. (1979). A class of robust sampling designs for
        large-scale surveys. *Journal of the American Statistical
        Association* 74, 911-915.

Cassel, C., Sarndal, C. and Wretman, J.H. (1976). Some results on
        generalized difference estimation and generalized regression
        estimation for finite populations. *Biometrika* 63, 615-620.

Cassel, C., Sarndal, C. and Wretman, J.H. (1977). *Foundations of
        Inference in Survey Sampling*. New York: John Wiley and
        Sons.

Cochran, W.G. (1939). The use of the analysis of variance in
        enumeration by sampling. *Journal of the American Statist-
        ical Association* 34, 492-510.

Cochran, W.G. (1946). Relative accuracy of systematic and
        Stratified random samples for a certain class of populations.
        Annals of Mathematical Statistics 17, 164-177.

Cochran, W.G. (1977). *Sampling Techniques*. New York: John Wiley
        and Sons.

Deming, W.E. and Stephan, F. (1941). On the interpretation of censuses as samples. *Journal of the American Statistical Association* 36, 45-49.

Durbin, J. (1967). Design of multistage surveys for the estimation of sampling errors. *Applied Statistics* 16, 152-164.

Fuller, W.A. (1970). Sampling with random stratum boundaries. *Journal of the Royal Statistical Society, Series B* 32, 203-226.

Fuller, W.A. (1971). A procedure for selecting nonreplacement unequal probability samples. Unpublished manuscript, Iowa State University.

Fuller, W.A. (1975). Regression analysis for Sample Survey. *Sankhyā, Series C* 37, 117-132.

Fuller, W.A. (1976). *Introduction to Statistical Time Series.* New York: John Wiley and Sons.

Godambe, V.P. (1955). A unified theory of sampling from finite populations. *Journal of the Royal Statistical Society, Series B* 17, 269-278.

Godambe, V.P. (1969). A law of large numbers for sampling finite populations with different inclusion probabilities for different individuals (abstract). *Annals of Mathematical Statistics* 40, 2218-2219.

Godambe, V.P., and Joshi, V.M. (1965). Admissibility and Bayes estimation in sampling finite populations, 1. *Annals of Mathematical Statistics* 36, 1707-1722.

Hájek, J. (1959). Optimum strategy and other problems in probability sampling. *Casopis Pro Pestovani Matematiky* 84, 387-423.

Hájek, J. (1964). Asymptotic theory of rejective sampling with varying probabilities from a finite population. *Annals of Mathematical Statistics* 35, 1491-1523.

Hanurav, T.V. (1966). Some aspects of unified sampling theory. *Sankhyā, Series A* 28, 175-204.

Hanurav, T.V. (1967). Optimum utilization of auxiliary information: πPS sampling of two units from a stratum. *Journal of the Royal Statistical Society, Series B* 29, 374-391.

Hanurav, T.V. (1968). Hyper-admissibility and optimum estimator
    for sampling finite populations. *Annals of Mathematical
    Statistics* 39, 621-642.

Hartley, H.O. and Rao, J.N.K. (1962). Sampling with unequal
    probabilities and without replacement. *Annals of Mathe-
    matical Statistics* 33, 350-374.

Hoeffding, W. and Robbins, H. (1948). The central limit theorem
    for dependent random variables. *Duke Math. J.* 15, 773-780.

Horvitz, D.G. and Thompson, D.J. (1952). A generalization of
    sampling without replacement from a finite universe.
    Journal of the American Statistical Association 47,663-685.

Isaki, C.T. (1970). Survey designs utilizing prior information.
    Unpublished Ph.D. thesis, Iowa State University, Ames,Iowa.

Isaki, C.T. and Fuller, W.A. (1981). Survey design under the
    regression superpopulation model. *Journal of the American
    Statistical Association* (To appear).

Koop, J.C. (1963). On the axioms of sample formation and their
    bearing on the construction of linear estimators in sampling
    theory for finite universe. *Metrika* 7, 81-114.

Madow, W.G. (1945). On the limiting distributions of estimates
    based on samples from finite universes. *Annals of Mathe-
    matical Statistics* 16, 535-545.

Madow, W.G. and Madow, L.H. (1944). On the theory of systematic
    sampling. *Annals of Mathematical Statistics* 15, 1-24.

Midzuno, H. (1952). On the sampling system with probability
    proportionate to sum of sizes. *Journal of the Institute of
    Statistical Mathematics* 2, 99-108.

Ramachandran, G. (1979). Some results on optimal estimation for
    finite populations. Paper presented at the 42nd session
    of the International Statistical Institute, Manila.

Rao, J.N.K., Hartley, H.O. and Cochran, W.G. (1962). On a simple
    procedure of unequal probability sampling without replace-
    ment. *Journal of the Royal Statistical Society, Series B*
    24, 482-491.

Rao, T.J. (1971). πPS sampling designs and the Horvitz-Thompson
    estimator. *Journal of the American Statistical Association*
    66, 872-875.

Robinson, P.M. and Tsui, K.W. (1979). On Brewer's asymptotic analysis in robust sampling designs for large scale surveys. Technical Report No. 79-43, University of British Columbia.

Rosen, B. (1972). Asymptotic theory for successive sampling with varying probabilities without replacement, I. *Annals of Mathematical Statistics* 43, 373-397.

Roy, J. and Chakravarti, I.M. (1960). Estimating the mean of a finite population. *Annals of Mathematical Statistics* 31, 392-398.

Royall, R.M. (1970). On finite population sampling theory under certain linear regression models. *Biometrika* 57, 377-387.

Royall, R.M. and Herson, J. (1973). Robust estimation in finite populations 1. *Journal of the American Statistical Association* 68, 880-889.

Sampford, M.R. (1967). On sampling without replacement with unequal probabilities of selection. *Biometrika* 54,499-513.

Särndal, C.E. (1980a). A two-way classification of regression estimation strategies in probability sampling. *Canadian Journal of Statistics* (to appear).

Särndal, C.E. (1980b). On π-inverse weighting versus BLU weighting in probability sampling. To appear.

Scott, A.J., Brewer, K.R.W. and Ho, E.W.H. (1978). Finite population sampling and robust estimation. *Journal of the American Statistical Association* 73, 359-361.

Yates, F. (1949). *Sampling Methods for Censuses and Surveys.* London: Griffin.

Zyskind, G. (1967). On canonical forms, non-negative covariance matrices and best and simple least squares linear estimators in linear models. *Annals of Mathematical Statistics* 38, 1092-1109.

ESTIMATION FOR NONRESPONSE SITUATIONS: TO WHAT EXTENT MUST WE RELY ON MODELS?

Carl Erik Särndal

Université de Montréal

Tak-Kee Hui

The University of British Columbia

The paper presents a Monte Carlo study illustrating a method of estimation in the presence of nonresponse proposed earlier by Cassel, Särndal and Wretman (1979). The method first derives estimates of unknown parameters of a response mechanism model, then uses these to obtain estimated individual response probabilities. These latter serve in turn for obtaining estimates of unknown regression parameters, which finally are needed in the estimation formula for the population mean or total. The Monte Carlo experiments confirm the hypothesis that either "correct" response mechanism modelling or "correct" regression modelling is essential for producing design unbiased population estimates. If at least one of the two models holds roughly, the results show that virtually unbiased estimates are obtained by the method, despite nonresponse, even for relatively modest sample sizes.

1.   INTRODUCTION

When data are missing because of nonresponse, unbiased estimation of the population mean becomes difficult since usually little is known about the processes that generate nonresponse. If one knew, for example, that individual response probabilities are roughly equal within strata, or that they increase, say, as a known function of size of household, then virtually unbiased estimates could easily be obtained. This paper emphasizes that model building - regression modelling and/or response mechanism modelling - is a necessary requirement for unbiased estimation with nonresponse. Although models are used, the approach in this paper is close in spirit to classical randomization theory, which means that bias and variance are defined by measures involving the chosen randomization distribution.

Cassel, Särndal and Wretman (1979) proposed a class of estimators for nonresponse data such that the form of each estimator in the class is determined in part by a regression model, in part by a response model. By the latter we mean a model for how individual response probabilities depend on strata or other groupings, or on auxiliary variable values, etc. Properties of the method are that if the regression is correctly modelled, then the resulting estimator is essentially unbiased, regardless of the response modelling. If the regression is incorrectly model-led, the estimator is still essentially unbiased, provided the response was correctly modelled; but the variance of the estimate is increased.

The method can be applied under any sampling design, and its step by step mechanics can be described as follows:

1. The unknown parameters of the response model are estimated.
2. The estimated parameters are used in estimating individual response probabilities.
3. These latter are used in the estimation of the unknown regression parameters.
4. The estimated regression slopes are used in forming an estimate of the unknown population mean.

In this paper, a Monte Carlo simulation is used to study properties of the method in sampling from artificially generated populations of size $N = 10,000$.

2.   A GENERAL ESTIMATION PROCEDURE FOR NONRESPONSE

In the debate on the foundations of inference in survey
sampling, a number of approaches to estimation of the population
mean have emerged, as discussed in Brewer and Särndal (1979).
In the case of complete response, a theoretically attractive but
in practice unlikely situation, a major dichotomy distinguishes
approaches based on classical randomization theory from those that
appeal more directly to statistical models.

Some statisticians who vigorously defend randomization
theory methods in the full response case admit that when non-
sampling errors such as nonresponse are present, some kind of
appeal to models is unavoidable. However, Hansen, Madow and
Tepping (1979) point out that, with some data missing, "no method
exists for guaranteeing by any statistical model that large biases
will not result from the application of the model. Only the
collection of almost all the data in the selected sample can
provide reasonable assurance of low bias due to missing data."

On the other hand, statisticians who reject randomization
theory in favor of more uncompromising use of models point to the
ineptness of randomization theory when it comes to dealing with
anything beyond simple sampling errors. Smith (1979) points out
that "the randomization approach can not offer a framework for
inference for missing values because these values are not controlled by the statistician's randomization, they occur according to
some unknown mechanism. So the only statistical approach to the
treatment of missing values is a model building approach ..."

We concur that randomization theory without models is insufficient when there is nonresponse, but maintain that the nonresponse estimator conjured by a randomization theorist may look
considerably different in form than the model theorist's suggestion. A typical attempt in the latter spirit (Schaible, 1979),
is to decompose the mean to be estimated, $\sum_1^N Y_k/N$, for example, as

$$\Sigma_r Y_k/N + \Sigma_{s-r} Y_k/N + \Sigma_{U-s} Y_k/N$$

where the three sums are over respondents, nonrespondents and non-sampled units, respectively, and then to attempt model dependent estimation of the latter two sums.

By contrast, the Cassel, Särndal and Wretman (1979) method, despite its appeal to models, is essentially a randomization theory approach, one aspect of which is inflation of the value associated with a responding unit k by an inflation factor $1/\Pi_k q_k$ to preserve (approximate) unbiasedness, where Π_k and q_k denote the inclusion probability and the response probability, respectively, of unit k. Since q_k, unlike Π_k, is beyond the control of the statistician, estimation of q_k is usually required. Some estimators involving inflation factors interpretable as estimated q-probabilities are currently in use, see for example Platek and Gray (1979); these factors are usually in the form of the inverse of the response rate in a "balancing area" or "weighting class".

A more general exploration of the methods involving estimated q-probabilities is obtained as follows. Consider the regression model ξ such that $Y_1, \ldots, Y_N$ are independent and

$$E_\xi(Y_k) = \sum_{j=0}^{p} \beta_j x_{kj} = \beta' x_k = \mu_k$$

$$(2.1)$$

$$V_\xi(Y_k) = \sigma_k^2$$

where $\beta = (\beta_0, \beta_1, \ldots, \beta_p)'$ is a (p+1)-vector of unknown coefficients, x_k is a known (p+1)-vector with $x_{ko} = 1$ for all k, and the σ_k^2 are unknown but usually of the form $\sigma_k^2 = \sigma^2 v(x_k)$ with σ^2 unknown and the function $v(\cdot)$ known.

A probability sample s is selected by a given sampling design. A subset r responds; s-r is the set of nonrespondents.

Let Σ_1^N , Σ_s , Σ_r , etc. denote summation over $k = 1,\ldots,N$,
$k \in s$, $k \in r$, etc.

Let

$$q_k = q(\underset{\sim}{x}_k,\ \underset{\sim}{\theta}) \qquad\qquad (2.2)$$

where $\underset{\sim}{\theta}$ is unknown, and where we make an explicit assumption
about the form of the function $q(\cdot,\cdot)$; this is our response
model. We estimate $\underset{\sim}{\theta}$ by $\underset{\sim}{\hat{\theta}}$ from the available data and then
calculate an estimate of q_k as

$$\hat{q}_k = q(\underset{\sim}{x}_k,\ \underset{\sim}{\hat{\theta}}) \qquad\qquad (2.3)$$

Cassel, Särndal and Wretman (1979) discuss the following
estimator of $\overline{Y} = N^{-1}\Sigma_1^N Y_k$,

$$T = \Sigma_r \frac{Y_k}{N\,\Pi_k\,\hat{q}_k} + \sum_{j=0}^{p} \hat{\beta}_j (\overline{x}_{.j} - \Sigma_r \frac{x_{kj}}{N\,\Pi_k\,\hat{q}_k}) \qquad (2.4)$$

where $\overline{x}_{.j} = \Sigma_1^N x_{kj}/N$ and, assuming the set r counts n_r
respondents,

$$\underset{\sim}{\hat{\beta}} = (\hat{\beta}_0,\ \hat{\beta}_1,\ldots,\ \hat{\beta}_p)' = (\underset{\sim}{X}'_r \underset{\sim}{V}_r^{-1} \underset{\sim}{\hat{Q}}_r^{-1} \underset{\sim}{\Pi}_r^{-1} \underset{\sim}{X}_r)^{-1} \underset{\sim}{X}'_r \underset{\sim}{V}_r^{-1} \underset{\sim}{\hat{Q}}_r^{-1} \underset{\sim}{\Pi}_r^{-1} \underset{\sim}{Y}_r,$$

where $\underset{\sim}{V}_r$, $\underset{\sim}{\hat{Q}}_r$ and $\underset{\sim}{\Pi}_r$ are $n_r \times n_r$ diagonal matrices with
diagonal elements, respectively, σ_k^2, $\hat{q}_k$ and Π_k for $k \in r$,
$\underset{\sim}{X}'_r$ is a $(p+1) \times n_r$ matrix with $\underset{\sim}{x}_k$, $k \in r$, as columns, and the
column vector $\underset{\sim}{Y}_r$ has Y_k for $k \in r$ as elements.

For an intuitive grasp of (2.4) pretend first that $\hat{q}_k = 1$
for all k. Then (2.4) is the "generalized regression estimator"
(Särndal,1979) which is approximately design unbiased for full
response. The data point k is then inflated by the factor
$1/\Pi_k$. However, when there is nonresponse, we need to further
inflate each of the $n_r \leq n$ responding data points; for point
k , we use the inflation factor $1/\Pi_k\hat{q}_k$. The effect of insertion
of $\hat{q}_k$, estimated according to (2.3), is to restore approximate

unbiasedness, provided the response model (2.2) is correct. The
first term of T has expected value approximately equal to $\overline{Y}$
and the expected value of the rest of T is roughly zero. If
the variance structure of the regression model (2.1) satisfies

$$\sigma_k^2 = \sum_{j=0}^{p} c_j x_{kj}$$

for all k and for arbitrary constants $c_0, c_1, \ldots, c_p$, then
(2.4) has the intuitively pleasing form

$$T = \Sigma_1^N \hat{Y}_k / N$$

where $\hat{Y}_k = \hat{\beta}_0 + \hat{\beta}_1 x_{k1} + \ldots + \hat{\beta}_p x_{kp}$ serves as a predicted value
of Y_k. For all cases considered in this paper, T has this
simple form, which implies that the possible occurrence of a unit
k with a very large inflation factor $1/\Pi_k \hat{q}_k$ will not throw the
value of T out of proportion. In other words, T is not
subject to the standard criticism raised against the Horvitz-
Thompson estimator.

3. REGRESSION MODELS AND RESPONSE MODELS FOR THE MONTE CARLO STUDY

The assumed regression model and the assumed response model
both contribute to determining the form of (2.4). Therefore, the
bias and variance characteristics of T depend on the correctness
of these models. Four cases can arise: the assumed regression
model is either true or false, and the same holds for the assumed
response model.

We shall apply (2.4) in a simple case involving a population
divided into two strata, U_1 of size N_1 and U_2 of size N_2;
$N_1 + N_2 = N$.

We shall consider populations for which the *true* regression
model is in terms of two *separate regressions* through the origin,
one for each stratum. Whenever this model is involved, the

auxiliary information is assumed to consist in $x_k = (x_k, \delta_k)'$,
known for $k = 1,...,N$, where $\delta_k = 1$ if $k \in U_1$ and $\delta_k = 0$ if
$k \in U_2$, and $x_1,..., x_k,..., x_N$ are positive values of a correlated variable, for example, the values of the variable of
interest at the previous survey. That is, stratum membership as
well as x-value is known for all N population units. The separate model is:

$$\mu_k = \beta_1 x_{k1} + \beta_2 x_{k2}$$

$$\sigma_k^2 = \sigma_1^2 x_{k1} + \sigma_2^2 x_{k2}$$

where $x_{k1} = \delta_k x_k$, $x_{k2} = (1-\delta_k)x_k$, and $\beta_1, \beta_2, \sigma_1^2, \sigma_2^2$ are
unknown.

Consider the alternative of a *common regression* through the
origin. When invoked, this model assumes that x_k , but not
necessarily stratum membership, is known for $k = 1,...,N$. The
common model is:

$$\mu_k = \beta x_k$$

$$\sigma_k^2 = \sigma^2 x_k$$

where β , σ^2 are unknown. Since the separate model is taken as
true, the common model is a *false* assumption for the populations
to be studied.

We consider only simple random sampling without replacement
(srs). An srs of size n is drawn from the unstratified population.
By poststratification we can if necessary identify the subset s_h
of size n_h of units belonging to stratum $h(h = 1,2)$; $n_1 + n_2 = n$;
$s_1 \cup s_2 = s$. Out of s , a subset r of size n_r responds. By
poststratification, the subset r_h of size n_{rh} can be attributed to stratum $h(h = 1,2)$; $n_{r1} + n_{r2} = n_r$; $r_1 \cup r_2 = r$.

A simple but usually grossly inadequate "solution" to the
nonresponse problem is to disregard it, saying in essence that

nonresponse occurs unsystematicaly with the same probability for every unit. Expressed as a model,

$$q_k = \theta \quad \text{for all} \quad k \tag{q0}$$

where θ is unknown $(0 < \theta < 1)$.

We consider three situations, each specified by a true response model, and one or more false alternatives. We shall study bias and variance of T under true and false response modelling, in each of the three situations.

Situation 1. The true response model specifies "constant response probabilities within strata", that is,

$$q_k = \theta_1^{\delta_k} \theta_2^{1-\delta_k} \tag{q1}$$

where $\underset{\sim}{\theta} = (\theta_1, \theta_2)$ is unknown $(0 < \theta_h < 1 ; h = 1,2)$. We shall compare the properties of T under (q1) with those under the false alternative (q0).

Situation 2. The true response model is

$$q_k = e^{-\theta x_k} \tag{q2}$$

where $\theta > 0$ is unknown. We shall compare properties of T under (q2) with those under two false alternatives, (q0) and (q1).

Situation 3. The true response model is

$$q_k = e^{-\theta_1 x_{k1} - \theta_2 x_{k2}} \tag{q3}$$

where $x_{k1} = \delta_k x_k$; $x_{k2} = (1 - \delta_k) x_k$; $\underset{\sim}{\theta} = (\theta_1, \theta_2)$ is unknown $(\theta_h > 0; h = 1,2)$. We shall compare results on T under (q3) with those under three false response models, (q0), (q1) and (q2).

We shall use the maximum likelihood estimates (MLE's) of θ and $\underset{\sim}{\theta} = (\theta_1, \theta_2)$. The likelihood function is

$$L(\underset{\sim}{\theta}) = \prod_{k \in r} q_k \prod_{k \in s-r} (1 - q_k)$$

where q_k depends on $\underset{\sim}{\theta}$ through whichever of the response models $(q0) - (q3)$ happens to have been assumed. The MLE's are, for model $(q0)$, $\hat{\theta} = n_r/n$, the overall response rate; for model $(q1)$, $\hat{\theta}_h = n_{rh}/n_h$ $(h = 1,2)$, the stratum response rate.

The MLE's $\hat{\theta}$ for model $(q2)$ and $(\hat{\theta}_1, \hat{\theta}_2)$ for model $(q3)$ have to be solved for by iterative methods. For model $(q2)$, the likelihood equation satisfied by $\hat{\theta}$ is

$$\Sigma_s x_k = \Sigma_{s-r} \{x_k/(1 - e^{-\hat{\theta}x_k})\} .$$

For model $(q3)$ the two equations satisfied by $\hat{\theta}_1$, $\hat{\theta}_2$ are

$$\Sigma_{s_h} x_k = \Sigma_{s_h-r_h} \{x_k/(1 - e^{-\hat{\theta}_h x_k})\} \quad (h = 1,2) .$$

The solutions to these equations will automatically meet the constraints $\theta > 0$ and $\theta_h > 0$ $(h = 1,2)$. Note that estimation of the separate θ's in models $(q1)$ and $(q3)$ requires that the sample can be poststratified whereas estimation of the common θ in $(q0)$ and $(q2)$ does not.

Another useful class of models for q_k , not considered here, is the logit type

$$\log\{q_k/(1 - q_k)\} = \alpha + \theta x_k .$$

For srs we have $\Pi_k = n/N$ for all k , so under the separate regression model and under response model (qI), the estimator (2.4) is given by

$$T_{SI} = \sum_{h=1}^{2} W_h \bar{x}_h \frac{\Sigma_{r_h} Y_k/\hat{q}_{kI}}{\Sigma_{r_h} x_k/\hat{q}_{kI}}$$

where $W_h = N_h/N$, $\bar{x}_h = \Sigma_{U_h} x_k/N_h$, and $\hat{q}_{kI}$ is the estimate of q_k obtained by replacing θ or (θ_1, θ_2) by their respective MLE's in the model formula (qI); $I = 0,1,2,3$. Note that T_{SI} requires knowledge of W_h and $\bar{x}_h$; in addition, poststratification of the sample is required.

For srs, under the common regression model and under response model (qI), (2.4) becomes

$$T_{CI} = \bar{x} \frac{\sum\limits_{h=1}^{2} \sum\limits_{r_h} Y_k/\hat{q}_{kI}}{\sum\limits_{h=1}^{2} \sum\limits_{r_h} x_k/\hat{q}_{kI}}$$

where $\bar{x} = \sum_1^N x_k/N$ and $\hat{q}_{kI}$ has the same meaning as in T_{SI}; $I = 0,1,2,3$. Poststratification of the srs sample is required for T_{C1} and T_{C3}.

Some of the estimators are particularly simple and in current use:

$$T_{S0} = T_{S1} = \sum\limits_{h=1}^{2} W_h \bar{x}_h \frac{\bar{Y}_{rh}}{\bar{x}_{rh}}$$

$$T_{CO} = \bar{x} \frac{\bar{Y}_r}{\bar{x}_r}$$

$$T_{C1} = \bar{x} \frac{\sum\limits_{h=1}^{2} n_h \bar{Y}_{rh}}{\sum\limits_{h=1}^{2} n_h \bar{x}_{rh}}$$

where $\bar{Y}_r = \sum_r Y_k/n_r$; $\bar{Y}_{rh} = \sum_{r_h} Y_k/n_{rh}$ and the x-means analogously defined.

The parameter estimations (β's and θ's) carried out to arrive at our eight estimators are summarized as follows:

Required Estimation	A Common β	Two Separate β's
A common θ	T_{CO}, T_{C2}	T_{S0}, T_{S2}
Two separate θ's	T_{C1}, T_{C3}	T_{S1}, T_{S3}

The eight estimators will be involved as follows in the three situations of the Monte Carlo study:

Situation	Response Model	Regression Model	
		False (Common)	True (Separate)
1	False: (q0)	T_{CO}	T_{SO}
	True: (q1)	T_{C1}	T_{S1}
2	False: (q0)	T_{CO}	T_{SO}
	False: (q1)	T_{C1}	T_{S1}
	True: (q2)	T_{C2}	T_{S2}
3	False: (q0)	T_{CO}	T_{SO}
	False: (q1)	T_{C1}	T_{S1}
	False: (q2)	T_{C2}	T_{S2}
	True: (q3)	T_{C3}	T_{S3}

4. THE SIMULATION PROCEDURE

Two artificial populations of size $N = 10,000$ were created as follows. First 5000 x_k-values for each of two strata were generated by a random number generator, x_k being gamma distributed with density $\{\Gamma(a)\}^{-1} b^a x^{a-1} e^{-bx}$, with $a = 2$, $b = 0.1$ for stratum 1, and $a = 2.5$, $b = 0.1$ for stratum 2. Conditional on x_k, an associated random number y_k was generated so that, in stratum h, y_k is gamma distributed with $a = \beta_h^2 x_k / \sigma_h^2$ and $b = \beta_h / \sigma_h^2$, so that, conforming to the separate regression model, y_k has expected value $\beta_h x_k$ and variance $\sigma_h^2 x_k$, given x_k. The values $(\beta_1, \beta_2) = (0.4, 0.6)$ and $(\sigma_1, \sigma_2) = (0.4, 0.6)$ were used for generating the population of Experiment 1. Experiment 2 used an artificial population with the same 10,000 x_k-values but with more distinct separation between the two regression lines, achieved by letting $(\beta_1, \beta_2) = (0.2, 0.8)$ and $(\sigma_1, \sigma_2) = (0.4, 0.6)$ in the generation of the y_k-values. The distribution of the 5000 y_k-values in each stratum looks approximately gamma shaped (although theoretically it is not an exact gamma). A summary of the two populations is given in Table 1.

Table 1. Description of Two Artificial Populations of Size N = 10,000 in Two Strata of Size 5000 Each.

Population/ Stratum		Experiment 1				Experiment 2			
		Mean	s.d.	sk.	c.v.	Mean	s.d.	sk.	c.v.
Entire	x	22.389	15.15	1.30	0.68	22.389	15.15	1.30	0.68
population	y	11.473	8.93	1.54	0.78	11.991	12.36	1.72	1.03
Stratum 1	x	20.073	14.17	1.37	0.71	20.073	14.17	1.37	0.71
	y	8.079	6.04	1.40	0.75	4.020	3.30	1.35	0.82
Stratum 2	x	24.723	15.74	1.24	0.64	24.723	15.74	1.24	0.64
	y	14.889	10.00	1.23	0.67	19.979	12.94	1.24	0.65

s.d. = standard deviation
sk. = skewness
c.v. = coefficient of variation.

The sampling procedure in Experiment 1 is as follows: A first srs sample s of size n = 100 is drawn by random number generation. For a unit $k \in s$, the response probability q_k is calculated according to the true response model for the situation, that is, model (qJ) for situation $J(J = 1,2,3)$, with values of θ, θ_1 and θ_2 shown in the left hand side of Table 2. These values were chosen such that the overall average response probability is $\bar{q} = N^{-1} \sum_1^N q_k = 0.8$ for each situation, and such that in Situations 1 and 3, the average response probability within a stratum is 0.9 for stratum 1 and 0.7 for stratum 2. A Bernoulli trial is then performed for unit $k \in s$ with probability q_k for "success" (response) and $1 - q_k$ for "failure" (nonresponse). Having completed such a Bernoulli trial for every unit $k \in s$, we have generated a random response set r , which can be broken down into poststratified response sets r_1 and r_2 $(r = r_1 \cup r_2)$, something that is required for the estimation of q_k under models

(q1) and (q3). Next, the MLE's (see Section 3) of θ or
$\underset{\sim}{\theta} = (\theta_1, \theta_2)$ are calculated for the models $q(I)$, $I = 0, 1, \ldots, J$,
considered in situation $J (J = 1, 2, 3)$. In other words, the true
model (qJ) and the false models (qI), $I = 0, 1, \ldots, J-1$, are
fitted in situation J. Then $\hat{q}_{kI}$, T_{SI} and T_{CI} are calculated
for $I = 0, 1, \ldots, J$. Each of the three situations is treated in
this way. The procedure is repeated for a total of 2500 samples
for each of sample sizes $n = 100$, 200 and 400. Finally, Bias =
$\sum_{i=1}^{K} (T_i - \overline{Y})/K$, MSE $= \sum_{i=1}^{K} (T_i - \overline{Y})^2/K$, and Var = MSE − Bias^2, are
calculated for each estimator T in each situation, where $K = 2500$.
Thus the sense of unbiasedness and variance that we use here is
one referring to repetitions of the double experiment (E_1, E_2),
where E_1 stands for "drawing of a sample s by srs from the
fixed size-N population" and E_2 stands for "a realization r
under the true response model, given s". This terminates
Experiment 1.

Experiment 2 was carried out exactly as Experiment 1 except
that it used the population described on the right hand side of
Table 1 (that is, more distinctly separated β's: $\beta_1 = 0.2$;
$\beta_2 = 0.8$) and the θ-values of the right hand side of Table 2
(that is, more distinct separation between average stratum
response rates: $\overline{q}_1 = 0.9$, $\overline{q}_2 = 0.5$, with $\overline{q} = 0.7$ as average
overall response rate).

5. RESULTS OF THE MONTE CARLO STUDY

In evaluating the Monte Carlo results of Tables 3-5, it
should be kept in mind that the T_S-estimators require more
auxiliary information (that is, knowledge of stratum x-means and
stratum sizes) for their computation than do the T_C-estimators.
In a given situation the two kinds of estimators are therefore
not direct substitutes for one another, so we could consider that
the choice of estimator consists in picking one from among the

Table 2. Values of θ and (θ_1, θ_2) in the True Response Model (qJ) for Situation $J = 1, 2, 3$.

Situation	Experiment 1	Experiment 2
1	$q_k = \begin{cases} 0.9; & k \in U_1 \\ 0.7; & k \in U_2 \end{cases}$	$q_k = \begin{cases} 0.9; & k \in U_1 \\ 0.5; & k \in U_2 \end{cases}$
	$\bar{q}_1 = 0.9; \quad \bar{q}_2 = 0.7$	$\bar{q}_1 = 0.9; \quad \bar{q}_2 = 0.5$
	$\bar{q} = 0.8$	$\bar{q} = 0.7$
2	$q_k = \exp\{-0.0104x_k\}$	$q_k = \exp\{-0.0172x_k\}$
	$\bar{q} = 0.8$	$\bar{q} = 0.7$
3	$q_k = \begin{cases} \exp\{-0.0054x_k\}; & k \in U_1 \\ \exp\{-0.0155x_k\}; & k \in U_2 \end{cases}$	$q_k = \begin{cases} \exp\{-0.0054x_k\}; & k \in U_1 \\ \exp\{-0.0324x_k\}; & k \in U_2 \end{cases}$
	$\bar{q}_1 = 0.9; \quad \bar{q}_2 = 0.7$	$\bar{q}_1 = 0.9; \quad \bar{q}_2 = 0.5$
	$\bar{q} = 0.8$	$\bar{q} = 0.7$

group of T_C-estimators or, if the additional information were available, one from among the group of T_S-estimators. As conjectured in Cassel, Särndal and Wretman (1979), there is, in every situation, little to choose between the T_S-estimators, but a great deal of difference among the T_C-estimators.

Some very clear patterns emerge:

1. Estimators based on a true regression model ($T_{S0} - T_{S3}$) have negligible bias, no matter whether they are based on a true or a false response model.

2. Estimators based on a false regression model ($T_{C0} - T_{C3}$) still have negligible bias if based on a true response model (that is, T_{C1} in Situation 1, T_{C2} in Situation 2 and T_{C3} in Situation 3). *Failure in regression modelling can be counteracted by correct response modelling.*

3. Estimators based on false regression *and* response models have a bias that is virtually unaffected by increases in sample size, and the bias tends to be larger as a result of either (a) increased separation in the stratum slopes β_1 and β_2, and (b) increased separation in the average stratum response rates $\bar{q}_1$ and $\bar{q}_2$. This latter fact was born out by a study of the effect of each of these factors separately; details are not reported here. The effect on bias of (a) and (b) together is seen by comparing results of Experiments 1 and 2.

4. The variance of any estimator decreases, as expected, in an approximate 1:2:4 fashion as the sample size increases in that same relation.

5. It follows from (3) and (4) that *the bias, unless removed by either true regression modelling or true response modelling, is likely to cause the estimate to be severely off target even though* n *is large and the variance neglible.*

6. The T_C-estimators have markedly higher variances than the T_S-estimators (by a factor of roughly 2 in Experiment 1; by a factor of 6 to 10 in Experiment 2). The T_C-estimators as a group have roughly comparable variances, as is true for the T_S-estimators, although within each group, estimators based on the more complex models (q2) and (q3) tend to have somewhat larger variances.

7. Since the goal of obtaining a virtually unbiased estimate can be achieved either by true regression modelling or by true response modelling, the variance aspect thus favors reaching this goal via true regression modelling.

8. Experiment 2 in Situation 3 offers a particularly potent example of the huge biases that can arise when both regression and response models are false. The true response is then in terms of response probabilities that decrease exponentially with x according to separate rates θ_h in the two strata. Fitting the

false, common θ models (q0) and (q2) gives estimators T_{C0} and T_{C2} with extremely large biases. Now, T_{C1}, although based on (q1) which is a false mathematical form, is derived from fitting two separate θ's, and while the bias is still substantial, it is much reduced compared to that of T_{C0} and T_{C2}. Finally, when the correct two parameter model (q3) is fitted we get T_{C3}, which is virtually unbiased.

9. In Situation 2, T_{C2} is essentially unbiased, confirming the theoretical argument. More surprisingly, so is T_{C1}, although based on a false response model. Moreover, the variance of T_{C1} is so much lower than that of T_{C2} that T_{C1} becomes the best choice in terms of MSE in Situation 2. In Situation 3, T_{C1} performs rather well in Experiment 1 but poorly in Experiment 2, in comparison to T_{C2}, which is favored by theoretical argument as well as by the Monte Carlo results. In striking contrast to the expected good performance of T_{C2} in Situation 2 stands the very poor performance of that estimator in Situation 3.

6. CONCLUDING REMARKS

We have given some evidence that the estimation method T given by (2.4) is successful, that is, essentially unbiased, for nonresponse data if at least one of the two models (regression model, response model) is correctly chosen. In practice, any regression or response model is likely to have some error, by misspecification of the mathematical form of the model, or by omission of important explanatory variables. In other words, the method of this paper is a strong tool, provided the model maker can make correct choices. Much work seems required especially in finding out more about the structure of response mechanisms and about the kinds of auxiliary variables that explain nonresponse. Inclusion of more explanatory variables into the regression and/or response models means extra parameters to estimate. It is suggested that rather too many than too few parameters should be estimated in both models, provided the sample size is large enough to overcome the instability introduced by these estimations. This point is illustrated by the rather extreme Experiment 2, Situation 3, where the bias of T_{C3} , although unimportant for every sample size compared to that of competing estimators, can not be described as truly negligible until n reaches 400 or more.

Table 3. Bias, Var and MSE for Situation 1.[*]

n	Esti-mator	Experiment 1			Experiment 2		
		Bias	Var	MSE	Bias	Var	MSE
100	T_{C0}	-0.293	0.162	0.248	-1.993	1.015	4.986
	T_{C1}	-0.013	0.155	0.156	-0.019	0.838	0.839
	T_{S1}	-0.004	0.085	0.085	-0.002	0.111	0.111
200	T_{C0}	-0.279	0.082	0.160	-1.990	0.511	4.473
	T_{C1}	0.000	0.079	0.079	-0.013	0.424	0.424
	T_{S1}	0.007	0.041	0.041	0.010	0.056	0.056
400	T_{C0}	-0.276	0.041	0.117	-1.972	0.256	4.145
	T_{C1}	0.002	0.039	0.039	-0.005	0.217	0.217
	T_{S1}	0.008	0.020	0.020	0.009	0.026	0.026

[*] True population mean: 11.473 (Experiment 1), 11.991 (Experiment 2).

Note: $T_{S0} \equiv T_{S1}$.

Table 4. Bias, Var and MSE in Situation 2.[*]

n	Esti-Mator	Experiment 1			Experiment 2		
		Bias	Var	MSE	Bias	Var	MSE
100	T_{C0}	-0.067	0.174	0.179	-0.228	1.072	1.124
	T_{C1}	-0.015	0.159	0.159	0.019	0.844	0.844
	T_{C2}	-0.007	0.198	0.198	0.001	1.533	1.533
	T_{S1}	-0.009	0.086	0.087	0.027	0.108	0.109
	T_{S2}	-0.000	0.089	0.089	0.006	0.120	0.120
200	T_{C0}	-0.064	0.086	0.090	-0.236	0.527	0.583
	T_{C1}	-0.012	0.081	0.081	0.014	0.420	0.420
	T_{C2}	-0.003	0.099	0.099	-0.012	0.791	0.791
	T_{S1}	-0.003	0.042	0.042	0.032	0.054	0.055
	T_{S2}	0.007	0.044	0.044	0.011	0.062	0.062
400	T_{C0}	-0.061	0.042	0.046	-0.235	0.256	0.311
	T_{C1}	-0.009	0.040	0.040	0.022	0.209	0.209
	T_{C2}	0.001	0.049	0.049	-0.006	0.381	0.381
	T_{S1}	-0.004	0.021	0.021	0.033	0.025	0.027
	T_{S2}	0.006	0.022	0.022	0.010	0.030	0.030

[*] True population mean: 11.473 (Experiment 1), 11.991 (Experiment 2). $T_{S0} \equiv T_{S1}$.

Table 5. Bias, Var and MSE for Situation 3.[*]

n	Esti-mator	Experiment 1			Experiment 2		
		Bias	Var	MSE	Bias	Var	MSE
100	T_{C0}	-0.402	0.164	0.326	-2.686	0.941	8.153
	T_{C1}	-0.123	0.164	0.179	-0.762	0.907	1.487
	T_{C2}	-0.376	0.190	0.331	-2.846	1.352	9.452
	T_{C3}	-0.021	0.186	0.187	-0.156	1.373	1.398
	T_{S1}	-0.116	0.095	0.095	0.041	0.149	0.151
	T_{S2}	-0.002	0.099	0.099	0.024	0.159	0.160
	T_{S3}	-0.002	0.104	0.104	0.020	0.180	0.181
200	T_{C0}	-0.398	0.086	0.244	-2.708	0.443	7.777
	T_{C1}	-0.118	0.085	0.098	-0.761	0.444	1.023
	T_{C2}	-0.369	0.100	0.236	-2.884	0.654	8.968
	T_{C3}	-0.010	0.095	0.095	-0.079	0.736	0.743
	T_{S1}	-0.008	0.048	0.048	0.042	0.073	0.074
	T_{S2}	0.002	0.050	0.050	0.021	0.078	0.079
	T_{S3}	0.002	0.052	0.052	0.016	0.094	0.094
400	T_{C0}	-0.393	0.044	0.198	-2.696	0.234	7.502
	T_{C1}	-0.113	0.042	0.055	-0.758	0.222	0.797
	T_{C2}	-0.365	0.051	0.184	-2.868	0.346	8.573
	T_{C3}	-0.003	0.046	0.046	-0.040	0.381	0.383
	T_{S1}	-0.004	0.024	0.024	0.036	0.035	0.037
	T_{S2}	0.007	0.024	0.025	0.013	0.039	0.039
	T_{S3}	0.006	0.025	0.025	0.005	0.050	0.050

[*] True population mean: 11.473 (Experiment 1), 11.991 (Experiment 2).
$T_{S0} \equiv T_{S1}$.

REFERENCES

Brewer, K. R. W. and Särndal, C. E. (1979). Six approaches to
    enumerative survey sampling. *Symposium on Incomplete Data:
    Preliminary Proceedings,* Washington, D.C., 499-507.

Cassel, C. M., Särndal, C. E. and Wretman, J. H. (1979). Some
    uses of statistical models in connection with the non-
    response problem. *Symposium on Incomplete Data: Preliminary
    proceedings,* Washington, D.C., 188-212.

Hansen, M. H., Madow, W.G. and Tepping, B. J. (1979). Discussion
    of paper by Brewer and Sarndal cited above.

Platek, R. and Gray, G. B. (1979). Methodology and application
    of adjustments for non-response. Invited paper, 42nd
    Session of the International Statistical Institute, Manila.

Särndal, C. E. (1979). On π-inverse weighting versus best linear
    unbiased weighting in probability sampling. *Biometrika* 67,
    639-650.

Schaible, W. L. (1979). Estimation of finite population totals
    from incomplete data: Prediction approach. *Symposium on
    Incomplete Data: Preliminary proceedings,* Washington, D.C.,
    170-187.

Smith, T. M. F. (1979). Discussion of paper by Brewer and Särndal
    cited above.

CHI-SQUARED TESTS FOR CONTINGENCY TABLES
WITH PROPORTIONS ESTIMATED FROM SURVEY DATA

A.J. Scott and J.N.K. Rao

University of Auckland and Carleton University

We investigate the effect of clustering and stratification on the standard chi-squared tests for homogeneity
and independence. Although the form of the asymptotic
distribution (a weighted sum of independent χ_1^2 random
variables with weights related to particular design
effects) is the same for both test statistics, the test
for homogeneity is more seriously affected in general
than that for independence. We suggest some simple
corrections to the usual chi-squared statistics and
investigate the performance of the modified tests
empirically and through the use of some simple models
for clustering.

1. INTRODUCTION

Over the last few years a great deal of attention has been
focussed on the problems that arise when standard chi-squared tests
based on the multinomial distribution are applied in more complex
situations such as with data from a stratified multi-stage survey.
(See Kish and Frankel (1974), Rao and Scott (1979), Fellegi (1980)
and Holt, Scott and Ewings (1980) for examples). It is now widely
recognized that stratification and clustering can have a substantial effect on the distribution of the chi-squared statistic and
that we can get some very misleading results if we fail to make
some adjustment for these factors.

In this paper, we look at the effects of stratification and
clustering on the analysis of two-way contingency tables, reviewing

and extending the results on chi-squared tests of homogeneity and
independence in Rao and Scott (1979) and Holt, Scott and Ewings
(1980). Test of homogeneity can arise in a survey context, for
example, when we want to compare the proportions of some categorical variable in different regions of a regionally stratified
survey, or when we want to compare proportions for different
national surveys (as in the World Fertility Study). An important
application involves the comparison of the results of two different
surveys, supposedly from the same population. Another example is
the problem of testing the agreement between interviewers based on
Mahalanobis' interpenetrating sub-samples using the same design.
Tests of independence arise, on the other hand, when we have two
categorical domain variables which cut across strata and clusters
in a single survey. In ordinary multinomial theory, of course,
the tests for homogeneity and independence turn out to be identical
because the multinomial has the nice property that, when we
condition on the marginal row totals, cell counts in different rows
become independent. This does not carry over to more general
sampling schemes and we shall see that the effect on the distribution of the chi-squared statistic can be very different in the two
situations considered.

2. TESTING HOMOGENEITY IN A TWO-WAY TABLE

2.1 Two Populations

Suppose we are interested in some categorical variable (with
c classes) and we have independent samples, of sizes n_1 and n_2,
from two populations. Let $\underset{\sim}{p}_i = (p_{i1}, \ldots, p_{i,c-1})^T$ represent the
vector of proportions in the first $c-1$ categories for the
variable in the ith population $(i = 1,2)$ and let $\underset{\sim}{\hat{p}}_i$ be the
corresponding vector of sample estimates. We assume that versions
of the Central Limit Theorem are available for both designs (see
Rao and Scott, 1979) and that $\sqrt{n_i}\,(\underset{\sim}{\hat{p}}_i - \underset{\sim}{p}_i)$ converges in distribution to $N_{c-1}(\underset{\sim}{0}, \underset{\sim}{V}_i)$ say as $n_i \to \infty$. For random sampling with

replacement, the distribution of $n_i \hat{p}_i$ is multinomial and $V_i = P_i = \text{diag}(p_i) - p_i p_i^T$. The hypothesis of interest is that the category proportions are the same for both populations, i.e., the null hypothesis is $H_0 : P_1 = P_2 = p$ where $p = (p_1, \ldots, p_{c-1})^T$. Ideally, we would have a consistent estimate, $\hat{V}_i$ say, of V_i available and could construct a test based on the Wald statistic

$$\chi^2_{WH} = (\hat{p}_1 - \hat{p}_2)^T \left[\frac{\hat{V}_1}{n_1} + \frac{\hat{V}_2}{n_2} \right]^{-1} (\hat{p}_1 - \hat{p}_2) \; ,$$

which is asymptotically distributed as a chi-squared random variable with $c-1$ degrees of freedom (χ^2_{c-1}) under H_0. This is the approach adopted by Koch, Freeman and Freeman (1975), for example.

In the real world, an estimate of V_i is often not available. This is especially true in secondary analyses of data that have been collected primarily for other purposes. In such cases, practitioners often resort to the ordinary Pearson statistic for testing homogeneity, using the observed cell counts n_{ij} in the case of a self-weighting design or, more generally, the estimated cell counts $\hat{n}_{ij} = n_i \hat{p}_{ij}$, i.e., they use the statistic

$$\chi^2_{PH} = \sum_{i=1}^{2} \sum_{j=1}^{c} \frac{(n_i \hat{p}_{ij} - n_i \hat{p}_{+j})^2}{n_i \hat{p}_{+j}} \; ,$$

where $\hat{p}_{+j} = (n_1 \hat{p}_{1j} + n_2 \hat{p}_{2j})/(n_1 + n_2)$ and $\hat{p}_{ic} = 1 - \hat{p}_{i1} - \cdots - \hat{p}_{i,c-1}$. Note that χ^2_{PH} can be written in the form

$$\chi^2_{PH} = \tilde{n}(\hat{p}_1 - \hat{p}_2)^T \hat{P}^{-1}(\hat{p}_1 - \hat{p}_2)$$

with $\tilde{n} = n_1 n_2/(n_1 + n_2)$, $\hat{P} = \text{diag}(\hat{p}_+) - \hat{p}_+ \hat{p}_+^T$ and $\hat{p}_+ = (\hat{p}_{+1}, \ldots, \hat{p}_{+,c-1})^T$, so that χ^2_{PH} is a modified form of the Wald statistic for the multinomial case since then $V_1 = V_2 = P$ under H_0. Since χ^2_{PH} is not asymptotically χ^2_{c-1} under H_0 for general sampling schemes, it is important to find out whether treating x^2_{PH} as x^2_{c-1} can be seriously misleading and, if so, whether it is possible to modify the statistic in a simple way to give better results.

The correct asymptotic distribution of x^2_{PH} under the above assumptions follows directly from standard results on quadratic forms (see for example Johnson and Kotz, 1970, p.150).

__Theorem 1.__ Under $H_0 : \underset{\sim}{p}_1 = \underset{\sim}{p}_2 = \underset{\sim}{p}$, $x^2_{PH} = \sum_{i=1}^{c-1} \delta_i z^2_i$,

where $z_1, \ldots, z_{c-1}$ are asymptotically independent $N(0,1)$ and $\delta_1, \ldots, \delta_{c-1}$ are the eigenvalues of $(n_2 D_1 + n_1 D_2)/(n_1 + n_2)$ with $\underset{\sim}{D}_i = \underset{\sim}{P}^{-1} \underset{\sim}{V}_i$ and* $\underset{\sim}{P} = \text{diag} (\underset{\sim}{p}) - \underset{\sim}{p} \underset{\sim}{p}^T$.

Following Rao and Scott (1977) we call $\underset{\sim}{D}_i = \underset{\sim}{P}^{-1} \underset{\sim}{V}_i$ the design effect matrix for the ith population by analogy with the ordinary definition of design effects for univariate random variables. It represents the inflation factor needed to transform $\underset{\sim}{P}$, the covariance matrix under multinomial sampling, to $\underset{\sim}{V}_i$, the true covariance matrix for the sampling scheme actually employed. For any given set of δ_i's, we can use the approximations given for linear combinations of χ^2 random variables in Solomon and Stephens (1975) to evaluate the correct percentage points of the distribution of x^2_{PH} and hence study the effect of using the percentage points of χ^2_{c-1} in their place. Note that in the case of testing interview differences using interpenetrating subsamples of the same design, $\underset{\sim}{V}_1 = \underset{\sim}{V}_2$ under H_0 and the δ_i's are the eigenvalues of $\underset{\sim}{D} = \underset{\sim}{P}^{-1} \underset{\sim}{V}.$

The results presented below are taken from a larger empirical study conducted by Holt, Scott and Ewings (1980). We consider four different sampling schemes for choosing a sample of 1000 households from a population of more than 13,000 households grouped into 345 clusters. (The population is actually the set of

---

* Strictly, $(n_2 D_1 + n_1 D_2)/(n_1 + n_2)$ should be replaced by $f_2 D_1 + f_1 D_2$, assuming that $n_i/(n_1 + n_2) \to f_i$ with $0 < f_i < 1$ as $n_i \to \infty$ $(i = 1,2)$.

sampled households from the 1971 General Household Survey (GHS) of the U.K. More details can be found in Holt, Scott and Ewings, 1980.) The four designs used were:

A : Two stage pps sampling with 50 primary sampling units (psu's) and 20 observations per psu.

B : Two stage pps sampling with 100 psu's and 10 observations per psu.

C : Simple random sampling with 1000 observations.

D : Proportionally allocated stratified random sampling using the 141 strata of the original GHS.

Sampling from the same population ensured that the hypothesis was true, and all population parameters are known exactly. Seven different variables were examined in the study. Table 1 shows the estimated (asymptotic) significance levels for all pairs of schemes (AA, AB, ..., DD) when the X^2_{PH} test is used at a nominal level of 0.05 (i.e., if H_0 is rejected when $X^2_{PH} > X^2_{c-1}(.05)$).

The main feature of the results is the severe distortion of significance level that occurs with either of the clustered designs A and B. Since neither design is at all unrealistic, the results give a strong warning that naive use of the ordinary chi-squared statistic with a multi-stage design can give very misleading results. The case CC in Table 1, of course, corresponds to the multinomial case and the estimated levels equal 0.05 (to two decimal places). The estimated levels for CD and DD do not exceed 0.05: in these cases $\delta_i \leq 1$ for all i so that X^2_{PH} test is conservative. The tabulated levels, however, are close to 0.05.

Is there any way of patching up the ordinary chi-squared test? If we could estimate $\delta_1, ..., \delta_{c-1}$, we could use the approximations of Solomon and Stephens (1975) to get asymptotically correct significance levels. However, complete knowledge of

Table 1. Estimated Significance Levels of X^2_{PH} Test for a Nominal Level of 0.05 for U.K. General Household Survey Data

Variable	No. of Categories	Designs*									
		AA	AB	AC	AD	BB	BC	BD	CC	CD	DD
Type of accommodation	4	.51	.43	.34	.32	.32	.19	.18	.05	.05	.04
Age of building	3	.48	.40	.30	.29	.29	.18	.17	.05	.04	.04
No. of bedrooms	5	.38	.32	.23	.22	.22	.13	.13	.05	.05	.04
Home Ownership	4	.49	.40	.30	.29	.29	.18	.16	.05	.04	.04
Bedroom Standard	5	.30	.23	.18	.17	.17	.10	.10	.05	.05	.05
No. of rooms	4	.39	.33	.23	.22	.22	.13	.12	.05	.05	.05
No. of cars	3	.30	.24	.18	.18	.17	.12	.11	.05	.05	.04

* AA, for example, means design A was used for both samples.

$\delta_1, \ldots, \delta_{c-1}$ is equivalent to knowledge of V_1 and V_2, in which case we would normally be better off using the Wald statistic, x^2_{WH}. In some situations, however, we may have estimates of the cell variances of $\hat{p}_{ij}$, $\hat{v}_{ijj}$ say, but not the covariance terms. A simple correction to x^2_{PH}, which requires only the knowledge of $\hat{v}_{ijj}$, can then be used as shown below. Now the average value $\bar{\delta}$ of $\delta_1, \ldots, \delta_{c-1}$ is given by

$$(c-1)\bar{\delta} = [n_2 tr(D_1) + n_1 tr(D_2)]/(n_1+n_2)$$

and it is possible to show (see Rao and Scott 1979) that $tr(D_i) = \sum_{j=1}^{c} (v_{ijj}/p_j)$. Thus $\hat{\delta}$ can be calculated from the variance terms above. Alternatively,

$$(c-1)\bar{\delta} = \tilde{n} \sum_{i=1}^{2} \sum_{j=1}^{c} (1 - p_j) d_{ij}/n_i .$$

where $d_{ij} = n_i Var(\hat{p}_{ij})/[p_j(1-p_j)]$ is the design effect for the (i,j)th cell. Thus $\bar{\delta}$ is a weighted average of the individual cell design effects. If we use the modified statistic $x^2_{MH} = x^2_{PH}/\hat{\delta}$ it will have the same asymptotic expected value $(c-1)$ as x^2_{c-1}, although its variance will be slightly smaller than that of x^2_{c-1}. Rao and Scott (1979) and Holt, Scott and Ewings (1980) have shown that treating x^2_{MH} as a x^2_{c-1} random variable under H_0 gives a very good approximation. For example, if we use this procedure with all the combinations of designs in Table 1, the estimated true significance level would never rise above 0.06. Given all the other problems with most survey data such as non-response, this is as close an approximation as we ever need in practice.

There is another possible approximation, which is asymptotically equivalent to x^2_{MH}, but which may have more intuitive appeal. Let $\bar{d}_i = tr(D_i)/(c-1)$ and define $n_i^* = n_i/\hat{d}_i$. We can regard n_i^* roughly as the sample size we would need to get the

same average accuracy across the c classes with simple random
sampling as with the design used. Then letting X_{EH}^2 be the value
we would get with the ordinary Pearson statistic when the cell
counts are replaced by $n_{ij}^* = n_i^* \hat{p}_{ij}$ (i.e., when the $\hat{p}_{ij}$'s are
weighted up by the equivalent, rather than the actual, sample
sizes), it is easy to see that X_{EH}^2 will have the same asymptotic
distribution as X_{MH}^2. Note that we are implicitly estimating the
common $\underset{\sim}{p}$ by $(n_1^* \hat{\underset{\sim}{p}}_1 + n_2^* \hat{\underset{\sim}{p}}_2)/(n_1^* + n_2^*)$ in this case which may be
preferable to $\hat{\underset{\sim}{p}}$ in small samples since the weights are inversely
proportional to the estimated variances of the $\hat{\underset{\sim}{p}}_i$'s.

2.2 Results for More Than Two Populations

Now consider the general situation in which we have
independent samples from r populations and want to check the
homogeneity of the category proportions across populations. Let
$\underset{\sim}{p}_i = (p_{i1},\ldots, p_{i,c-1})^T$ denote the vector of category proportions
for the ith population and let $\hat{\underset{\sim}{p}}_i$ be its sample estimate based
on a sample of size n_i $(i = 1,\ldots,r)$. As before, we assume that
$\sqrt{n_i} \, (\hat{\underset{\sim}{p}}_i - \underset{\sim}{p}_i)$ converges in distribution to $N_{c-1}(\underset{\sim}{0}, \underset{\sim}{V}_i)$. The
hypothesis of interest is

$$H_0 : \underset{\sim}{p}_i = \underset{\sim}{p} \; ; \quad i = 1,\ldots,r.$$

If consistent estimates $\hat{\underset{\sim}{V}}_1,\ldots,\hat{\underset{\sim}{V}}_r$ were available we could use a
Wald test based, for example, on the statistic $\hat{\underset{\sim}{q}}^T = [(\hat{\underset{\sim}{p}}_1 - \hat{\underset{\sim}{p}}_r)^T,\ldots,$
$(\hat{\underset{\sim}{p}}_{r-1} - \hat{\underset{\sim}{p}}_r)^T]$. The asymptotic covariance matrix of $\hat{\underset{\sim}{q}}$ is

$$\underset{\sim}{\Delta} = \left[\bigoplus_{i=1}^{r-1} (\underset{\sim}{V}_i / n_i) \right] + (\underset{\sim}{V}_r / n_r) \otimes \underset{\sim}{J}_{r-1}$$

where $\underset{\sim}{J}$ is a matrix of unit elements, $\oplus$ denotes the direct
sum and $\otimes$ the Kronecker product of matrices. Since $\hat{\underset{\sim}{q}}$ has mean
vector zero under H_0 it follows that

$$X_{WH}^2 = \hat{\underset{\sim}{q}}^T \, \hat{\underset{\sim}{\Delta}}^{-1} \, \hat{\underset{\sim}{q}}$$

has a limiting $\chi^2_{(r-1)(c-1)}$ distribution under H_0.

Again it is common practice, when no estimates $\hat{\underset{\sim}{V}}_1, \ldots, \hat{\underset{\sim}{V}}_r$ are available, to use the ordinary Pearson statistic with the estimated cell frequencies, that is

$$x^2_{PH} = \sum_{i=1}^{r} \sum_{j=1}^{c} \frac{(n_i \hat{P}_{ij} - n_i \hat{P}_{+j})^2}{n_i \hat{P}_{+j}} \, ,$$

where $\hat{P}_{+j} = \sum_{i=1}^{r} n_i \hat{P}_{ij}/n$ with $n = \sum_{i=1}^{r} n_i$. After some manipulation we can write x^2_{PH} in the form

$$x^2_{PH} = n \, \hat{\underset{\sim}{q}}^T (F \otimes \underset{\sim}{P}^{-1}) \hat{\underset{\sim}{q}}$$

where $\underset{\sim}{F} = \text{diag}(\underset{\sim}{f}) - \underset{\sim}{f} \, \underset{\sim}{f}^T$ with $\underset{\sim}{f} = (n_1/n, \ldots, n_{r-1}/n)^T$ and $\hat{\underset{\sim}{P}}$ is as defined in Section 2.1. Since $\underset{\sim}{\Delta}^{-1}$ reduces to $(P^{-1} \otimes F)$ when we have r multinomial samples and H_0 is true, we see that, as usual, x^2_{PH} is equivalent to a modified version of x^2_{WH} for the product-multinomial distribution. Actually the task of inverting $\underset{\sim}{\Delta}$ in general can be quite formidable if $(r-1)(c-1)$ is large and we may be happy to work with a reasonable approximation based on x^2_{PH} in place of x^2_{WH} even if estimates of the $\underset{\sim}{V}_i$'s are available.

As in the previous section, the correct asymptotic distribution of x^2_{PH} follows easily from standard results on quadratic forms.

<u>Theorem 2.</u> If H_0 is true, $x^2_{PH} = \sum_{i=1}^{(r-1)(c-1)} \delta_i z_i^2$ where $z_1, \ldots, z_{(r-1)(c-1)}$ are asymptotically independent $N(0,1)$ and the δ_i's are eigenvalues of $n(\underset{\sim}{F} \otimes \underset{\sim}{P}^{-1}) \underset{\sim}{\Delta}$.

Again, provided the δ_i's do not vary too much, we get a reasonable approximation to the null distribution by treating $x^2_{PH}/\hat{\bar{\delta}}$ as a $\chi^2_{(r-1)(c-1)}$ random variable where $\bar{\delta} = \sum_{i=1}^{(r-1)(c-1)} \delta_i/(r-1)(c-1)$. Notice that

$$(r-1)(c-1)\bar{\delta} = tr[n(\underset{\sim}{F} \otimes \underset{\sim}{P}^{-1})\underset{\sim}{\Delta}]$$

$$= \sum_{1}^{r} (1 - f_i) tr(\underset{\sim}{D}_i)$$

where $\underset{\sim}{D}_i = \underset{\sim}{P}^{-1}\underset{\sim}{V}_i$ is the design effect matrix for the ith population. Thus noting that $\bar{d}_i = tr(D_i)/(c-1)$, we have

$$\bar{\delta} = \sum_{i=1}^{r} (1 - f_i)\bar{d}_i/(r-1)$$

$$= \sum_{i=1}^{r} \sum_{j=1}^{c} (1-f_i)(1-p_j)d_{ij}/(r-1)(c-1)$$

which can be calculated directly from $Var(\hat{p}_{ij})$ ($i=1,...,r$;
$j=1,...,c$) since $n_i Var(\hat{p}_{ij}) = d_{ij}p_j(1-p_j)$.

Alternatively, we get the same asymptotic distribution by
substituting $n_i^* \hat{p}_{ij}$, where $n_i^* = n_i/\bar{d}_i$ is the effective sample
size, in place of the cell counts in X_{PH}^2. This is a simple
procedure, with considerable intuitive appeal, and results in
Holt, Scott and Ewings (1980) indicate that it will give very
accurate results in practice. We emphasize that using X_{PH}^2
without modification becomes steadily more misleading as r (and
hence the degrees of freedom $(r-1)(c-1)$) increases with $\bar{d}_i$ held
constant so that the unmodified chi-squared test is completely
worthless for large values for r .

So far, all of the results presented have been concerned with
hypothesis testing, which may not always be of great interest in
a large-scale survey where the huge sample sizes typically
involved almost guarantees statistical significance when there is
the slightest difference in the population proportions. However
the results apply equally to simultaneous confidence intervals.
For example, with $r = 2$ populations, following the standard
procedure for obtaining simultaneous confidence intervals of the
Schéffé type, we find that the probability is $(1-\alpha)$ that,
simultaneously for all coefficient vectors $\underset{\sim}{c}$, $\underset{\sim}{c}^T(\underset{\sim}{p}_1 - \underset{\sim}{p}_2)$ is

covered by the interval

$$\underset{\sim}{c}^T(\hat{\underset{\sim}{p}}_1 - \hat{\underset{\sim}{p}}_2) \pm K_\alpha[\underset{\sim}{c}^T(P_1/n_1 + P_2/n_2)\underset{\sim}{c}]^{\frac{1}{2}}$$

where K_α^2 is the upper α^{th} percentage point of the distribution

of $\sum_{i=1}^{c-1} \theta_i \chi_{1i}^2$ and the θ_i's are the eigenvalues of

$(P_1/n_1 + P_2/n_2)^{-1}(P_1D_1/n_1 + P_2D_2/n_2)$. As above, we get a reasonable approximation to K_α^2 by using $\bar{\theta}\,\chi_{c-1}^2(\alpha)$, where $\bar{\theta} = \sum_{i=1}^{c-1} \theta_i/(c-1)$.

3. TESTING INDEPENDENCE IN A TWO-WAY TABLE

3.1 General Theory

Now we turn to the situation in which we have estimates $\hat{p}_{ij}$ of the proportions for each cell in a two-way layout with r rows and c columns as in the previous section, but now the estimates are calculated from a single survey (of size n, say). In general the row and column categories may cut across strata and clusters so that the estimates in different rows (or in different columns) will not be independent. Let $\underset{\sim}{p} = (p_{11}, \ldots, p_{1c}, p_{21}, \ldots, p_{r,c-1})^T$ represent the vector of cell proportions for the population and let $\hat{\underset{\sim}{p}}$ represent the corresponding sample estimate. We also define the vectors of marginal row and column proportions $\underset{\sim}{p}_r = (p_{1+}, \ldots, p_{(r-1)+})^T$ where $p_{i+} = \sum_{j=1}^{c} p_{ij}$ and $\underset{\sim}{p}_c = (p_{+1}, \ldots, p_{+(c-1)})^T$ where $p_{+j} = \sum_{i=1}^{r} p_{ij}$. We shall assume that $\sqrt{n}(\hat{\underset{\sim}{p}} - \underset{\sim}{p})$ converges in distribution to $N_{rc-1}(\underset{\sim}{0},V)$. The hypothesis of interest is that of independence between rows and columns, i.e., the null hypothesis is

$$H_0 : p_{ij} = p_{i+}\,p_{+j} \quad (i = 1,\ldots,r;\ j = 1,\ldots,c).$$

Let $\hat{\underset{\sim}{h}} = (\hat{h}_{11}, \ldots, \hat{h}_{(r-1)(c-1)})^T$. In some situations we may be
able to get a direct estimate of the covariance matrix, $n \, V_{\underset{\sim}{h}}$,
of $\hat{\underset{\sim}{h}}$ using balanced repeated replication or by jackknifing, for
example, or we might use the Taylor series estimate $\hat{V}_{\underset{\sim}{h}} = \hat{\underset{\sim}{H}} \, \hat{\underset{\sim}{V}} \, \hat{\underset{\sim}{H}}^T$,
where $H = \partial h / \partial p$, if an estimate $\hat{\underset{\sim}{V}}$ of $\underset{\sim}{V}$ is available. In
such cases we can test H_0 using the Wald statistic

$$X^2_{WI} = n \, \hat{\underset{\sim}{h}}^T \, \hat{V}_{\underset{\sim}{h}}^{-1} \, \hat{\underset{\sim}{h}} \, ,$$

which is asymptotically $\chi^2_{(r-1)(c-1)}$ under H_0. For the reasons
we have outlined in previous sections, researchers often fall back
on the multinomial-based Pearson statistic

$$X^2_{PI} = \sum_{i=1}^{r} \sum_{j=1}^{c} (n\hat{p}_{ij} - n\hat{p}_{i+} \, \hat{p}_{+j})^2 / (n\hat{p}_{i+} \hat{p}_{+j})$$

when $\hat{V}_h$ is not available. This statistic can be written in the
form

$$= n \, \hat{\underset{\sim}{h}}^T (\hat{\underset{\sim}{P}}_r^{-1} \otimes \hat{\underset{\sim}{P}}_c^{-1}) \hat{\underset{\sim}{h}} \, ,$$

where $\hat{\underset{\sim}{P}}_r = \text{diag} \, (\hat{\underset{\sim}{p}}_r) - \hat{\underset{\sim}{p}}_r \hat{\underset{\sim}{p}}_r^T$, $\hat{\underset{\sim}{P}}_c = \text{diag} \, (\hat{\underset{\sim}{p}}_c) - \hat{\underset{\sim}{p}}_c \hat{\underset{\sim}{p}}_c^T$. Since
$H P H^T$ reduces to $P_r \otimes P_c$ when H_0 is true, we see that X^2_{PI}
is a modified form of X^2_{WI} for the multinomial. It follows in
the same way as in previous sections that the asymptotic distribution of X^2_{PI} under our more general assumptions has the form

$$X^2_{PI} = \sum_{i=1}^{(r-1)(c-1)} \delta_i \, z_i^2 \, ,$$

where the z_i's are asymptotically independent $N(0,1)$ and
$\delta_1 \geq \delta_2 \geq \cdots \geq \delta_{(r-1)(c-1)}$ are eigenvalues of $(\underset{\sim}{P}_r^{-1} \otimes \underset{\sim}{P}_c^{-1})(H V H^T)$.
As usual, treating $X^2_{PI} / \bar{\delta}$ as $\chi^2_{(r-1)(c-1)}$ under H_0
gives an adequate approximation in most practical situations, and
we note that

$$(r-1)(c-1)\bar{\delta} = \sum_{i=1}^{r} \sum_{j=1}^{c} Var(\hat{h}_{ij})/(p_{i+}p_{+j})$$

so that we can estimate $\bar{\delta}$ as long as we have the diagonal elements of $\hat{\underset{\sim}{V}}_{h}$. This sort of information is rare, however, and we will often only have information on the diagonal elements of $\underset{\sim}{V}$ at best. Rao and Scott (1979) show that $d_i \geq \delta_i \geq d_{i+r+c-2}$ where $d_1 \geq d_2 \geq \ldots \geq d_{rc-1}$ are the eigenvalues of $\underset{\sim}{P}^{-1}\underset{\sim}{V}$ so that $\bar{d}_u \geq \bar{\delta} \geq \bar{d}_\ell$ where $\bar{d}_u$ is the average of the $(r-1)(c-1)$ largest d_i's and $\bar{d}_\ell$ is the average of the $(r-1)(c-1)$ smallest d_i's. If the d_i's are not too variable then $\bar{\delta}$ might be expected to be close to $\bar{d} = \sum_{i=1}^{r} \sum_{j=1}^{c} V_{ij}/[(rc-1)p_{ij}]$, where $V_{ij} = Var(\hat{p}_{ij})$, and we can then estimate $\bar{d}$ using the estimates $\hat{V}_{ij}$. Fellegi (1980) and others have suggested modifying factors similar to $\bar{d}$. Even estimates of V_{ij} will be difficult to come by in a large two-way table, and the best we can hope for in many situations is to have some information on variances of the marginal row and column proportions. Ideally we would like an approximation for $\bar{\delta}$ in terms of the average design effect of the row proportions, $\bar{d}_r$, and the average design effect of the column proportions, $\bar{d}_c$. In the next section we look at a simple model for two-stage sampling to explore the relationship between $\bar{d}$, $\bar{\delta}$, $\bar{d}_r$ and $\bar{d}_c$.

3.2 Models for Two-Stage Sampling

Consider a simple model for two-stage sampling in which r psu's are selected at the first stage of sampling and a subsample of m_ℓ elements is selected from the M_ℓ units in the ℓth sampled psu $(\ell = 1,\ldots,r)$ at the second stage. Let $\underset{\sim}{m}_\ell = (m_{\ell 1},\ldots,m_{\ell,rc-1})^T$ be the vector of cell totals for the ℓth subsample and suppose that given the vector $\underset{\sim}{\pi}_\ell$, $\underset{\sim}{m}_\ell$ is multinomially distributed. In addition, we suppose that the $\underset{\sim}{\pi}_\ell$'s are drawn independently from a distribution with mean $\underset{\sim}{\pi}$ and covariance matrix $\underset{\sim}{V}_1$. This is a slight specialization of the model considered by Cohen (1976) and Altham (1976). Let

$$n = \sum_{\ell=1}^{r} m_\ell \quad \text{and} \quad \hat{p} = n/n \quad \text{where} \quad n = \sum_{\ell=1}^{r} m_\ell \ . \quad \text{It follows that} \quad \hat{p}$$

has mean vector π and covariance matrix

$$\text{Cov}(\hat{p}) = V/n = [\ P + (\overline{m}-1)V_1\ ]/n$$

where $P = \text{diag}(\pi) - \pi \pi^T$ as usual, $\overline{m} = \sum_{\ell=1}^{r} m_\ell^2 / \sum_{\ell=1}^{r} m_\ell$ and all

moments are with respect to the model. Consider the following simple special cases.

Example 1. Take $r = c = 2$ and suppose π_ℓ takes values $(1/2, 0, 0)^T$ and $(0, 1/2, 1/2)^T$ each with probability $1/2$. Then $\pi = (1/4, 1/4, 1/4)^T$,

$$P = \frac{1}{16} \begin{pmatrix} 3 & -1 & -1 \\ -1 & 3 & -1 \\ -1 & -1 & 3 \end{pmatrix} \quad \text{and} \quad V_1 = \frac{1}{10} \begin{pmatrix} 1 & -1 & -1 \\ -1 & 1 & 1 \\ -1 & 1 & 1 \end{pmatrix}.$$

It follows that $d_1 = \overline{m}$, $d_2 = d_3 = 1$, so $\overline{d} = 1 + (\overline{m}-1)/3$, while $\overline{\delta} = \overline{m}$ and $\overline{d}_r = \overline{d}_c = 1$. Thus $\overline{\delta}$ is larger than $\overline{d}$ here and very much larger than the marginal design effects.

Example 2. Again take $r = c = 2$ and suppose π_ℓ takes values $(1/2, 1/2, 0)^T$, $(0, 1/2, 0)^T$, $(0, 0, 1/2)^T$, and $(1/2, 0, 1/2)^T$ each with probability $1/4$. As above, $\pi = (1/4, 1/4, 1/4)^T$ but now

$$V_1 = \frac{1}{16} \begin{pmatrix} 1 & 0 & 0 \\ 0 & 1 & 0 \\ 0 & 0 & 1 \end{pmatrix}.$$

It follows that $d_1 = d_2 = 1 + (\overline{m}-1)/2$, $d_3 = 1$ so that $\overline{d} = 1 + (\overline{m}-1)/3$ as in the first example, but $\overline{\delta} = 1$ and $\overline{d}_r = \overline{d}_c = 1 + (\overline{m}-1)/2$. In this case $\overline{\delta}$ is much smaller than $\overline{d}$ which itself is smaller than either marginal design effect.

These examples show immediately that there is no hope of finding an approximation for $\overline{\delta}$ in terms of $\overline{d}$ or $\overline{d}_r$ and $\overline{d}_c$ which is universally applicable. To make further progress, we have to make more assumptions. Brier (1978) has considered the special case in which $m_\ell = m$ and π_ℓ is sampled from a Dirichlet distribution. In this case V_1 takes the very nice form aP ($0 \le a \le 1$) so that $d_i = 1 + (m-1)a$ for $i = 1,\ldots,k$. Thus if we are willing to assume that the π_ℓ's come from a Dirichlet distribution all design effects (and hence $\bar{d}$, $\bar{\delta}$, $\bar{d}_r$ and $\bar{d}_c$) have the value $1 + (m-1)a$ and $x^2_{PI}/[1 + (m-1)a]$ is asymptotically $\chi^2_{(r-1)(c-1)}$. Unfortunately, the implicit restriction that design effects for all cell totals and all linear combinations of cell totals have exactly the same value is a very strong one indeed, especially if the row and column variables are of quite different types. For example, average design effects for the socio-economic classifications reported by Kish, Groves and Krotki (1976) range from 4 to 8, while design effects for demographic variables range from 1.0 to 1.6.

Notice that the dependence hypothesis refers only to the overall vector π and places no restrictions on the psu probability vectors π_ℓ . We might consider a stronger hypothesis which postulates independence within each psu as well as overall independence, i.e.,

$$H_0^* : \pi_{ij}(\ell) = \pi_{i+}(\ell)\,\pi_{+j}(\ell);\ \ell = 1,\ldots,r$$

and $\quad \pi_{ij} = \pi_{i+}\,\pi_{+j};\ i = 1,\ldots,r-1;\ j = 1,\ldots,c-1$.

We note that this double requirement implies that $\pi_{i+}(\ell)$ and $\pi_{+j}(\ell)$ are uncorrelated for each i and j . If we make our hypothesis even stronger by requiring that $\pi_{i+}(\ell)$ and $\pi_{+j}(\ell)$ be independent we have the hypothesis

$$H_0^{**} : \pi_{ij}(\ell) = \pi_{i+}(\ell)\,\pi_{+j}(\ell);\ \ell = 1,\ldots,r$$

and $\pi_{i+}(\ell)$ is independent of $\pi_{+j}(\ell)$ ($i=1,\ldots,r-1$, $j=1,\ldots,c-1$).

We call this the hypothesis of *total independence*. Now $V_{\underset{\sim}{h}} =$ Cov $(\hat{\underset{\sim}{h}})$ can be expressed as

$$V_{\underset{\sim}{h}} = HPH^T + (\bar{m}-1) H \underset{\sim}{V}_1 \underset{\sim}{H}^T .$$

Moreover $\underset{\sim}{H} \underset{\sim}{V}_1 \underset{\sim}{H}^T$ is the covariance matrix of the vector with components

$$h^*_{ij}(\ell) = \pi_{ij}(\ell) - \pi_{i+}\pi_{+j}(\ell) - \pi_{i+}(\ell)\,\pi_{+j} + \pi_{i+}\,\pi_{+j}$$

$$= (\pi_{i+}(\ell) - \pi_{i+})(\pi_{+j}(\ell) - \pi_{+j})$$

if H_0^{**} is true. Since $\pi_{i+}(\ell)$ and $\pi_{+j}(\ell)$ are independent we have

$$\underset{\sim}{H} \underset{\sim}{V}_1 \underset{\sim}{H}^T = \mathrm{Cov}(\underset{\sim}{\pi}_r(\ell)) \otimes \mathrm{Cov}(\underset{\sim}{\pi}_c(\ell)) ,$$

where $\underset{\sim}{\pi}_r(\ell) = (\pi_{1+}(\ell),\dots,\pi_{(r-1)+}(\ell))^T$ and $\underset{\sim}{\pi}_c(\ell) = (\pi_{+1}(\ell),\dots,\pi_{+(c-1)}(\ell))^T$. Thus $V_{\underset{\sim}{h}}$ can be written in the form

$$V_{\underset{\sim}{h}} = (P_{\underset{\sim}{r}} \otimes P_{\underset{\sim}{c}}) + (\bar{m}-1)[\mathrm{Cov}(\underset{\sim}{\pi}_r(\ell)) \otimes \mathrm{Cov}(\underset{\sim}{\pi}_c(\ell))]$$

and $\delta_1, \delta_2, \dots, \delta_{(r-1)(c-1)}$, which are the eigenvalues of $(P_{\underset{\sim}{r}}^{-1} \otimes P_{\underset{\sim}{c}}^{-1})V_{\underset{\sim}{h}}$, can be expressed in the form $1 + (\bar{m}-1)\rho_i\gamma_j$ where $1 + (\bar{m}-1)\rho_i$ $(i = 1,\dots,r-1)$ are the generalized design effects for the row margins and $1 + (\bar{m}-1)\gamma_j$ $(j = 1,\dots,c-1)$ are the generalized design effects for the column margins. In particular

$$\bar{\delta} = 1 + (\bar{m}-1)\bar{\rho}\bar{\gamma}$$

where $\bar{d}_r = 1 + (\bar{m}-1)\bar{\rho}$, $\bar{d}_c = 1 + (\bar{m}-1)\bar{\gamma}$ and

$$\bar{\rho} = \sum_{i=1}^{r-1} \rho_i/(r-1) , \quad \bar{\gamma} = \sum_{j=1}^{c-1} \gamma_j/(c-1).$$

This has several very strong implications. First, $\bar{\delta}$ can be calculated directly from the marginal design effects when this strong hypothesis is true. Secondly, the value of $\bar{\delta}$ is going to be much smaller than either $\bar{d}_r$ and $\bar{d}_c$ under this hypothesis so that X^2_{PI} will need much less modification than in the homogeneity case. In particular if either row or column variable has an average design effect of 1 (i.e., if the marginal proportions are constant over psu's), then $\overline{\delta} = 1$ so that X^2_{PI} is asymptotically $\chi^2_{(r-1)(c-1)}$ without modification under the strong hypothesis of total independence.

At the very least, these results for different models warn us that the behaviour of X^2_{PI} is likely to be very different from that of X^2_{PH}. If the row design effects were large in the homogeneity case, then X^2_{PH} always gives misleading results if used without modification. In the independence case, however, a simple model for generating independence leads to a modifying factor that is much smaller than the average design effect of either marginal variable.

3.3 Conclusion

The results of the previous section indicate that no theoretical approximation for $\overline{\delta}$ in terms of $\overline{d}$ or $\overline{d}_r$ and $\overline{d}_c$ is possible without further assumptions, but the model incorporating the extra assumption of independence within psu's suggests that $\overline{\delta}$ might be expected to be smaller than $\overline{d}$ or $\min(\overline{d}_r, \overline{d}_c)$ even if within psu independence is only approximately true. Holt, Scott and Ewings (1980) have carried out an extensive empirical study, obtaining estimates of $\overline{\delta}$, $\overline{d}$, $\overline{d}_r$ and $\overline{d}_c$ for a wide variety of cross-classifications of variables measured in two large-scale U.K. surveys. In every case $\overline{\delta}$ turned out to be less than $\overline{d}$ or $\min(\overline{d}_r, \overline{d}_c)$. Similar results have been reported in other studies (see Kish and Frankel, 1974, for example) but more empirical work is needed in other fields and with surveys of different structures yet. One thing is clear: the tests for homogeneity and independence, which are identical in the multinomial case, are likely to behave very differently when the data is drawn by a more complex scheme.

Finally, we note that, even when suitable modifying factors can be calculated, an overall chi-squared test with cell estimates

averaged over strata should never be used blindly without careful
consideration of whether an overall hypothesis is appropriate.
Nathan (1975) gives examples of situations in which an overall
test ignoring the strata makes sense although in general collapsing over strata can lead to misleading inferences. A good
discussion of the dangers of collapsing indiscriminately is given
in Bishop, Fienberg and Holland (1975, p.47).

REFERENCES

Atham, P.A.E. (1976). Discrete variable analysis for individuals
    grouped into families. *Biometrika* 63, 263-269.

Bishop, Y.M.M., Fienberg, S.E., and Holland, P.W. (1975). *Discrete
    Multivariate Analysis: Theory and Practice*. Cambridge:
    MIT Press.

Brier, S.S. (1978). Discrete data models with random effects,
    Technical Report, University of Minnesota, School of
    Statistics.

Cohen, J.E. (1976). The distribution of the chi-squared statistic
    under cluster sampling from contingency tables, *Journal of
    the American Statistical Association* 71, 665-670.

Fellegi, I.P. (1980). Approximate tests of independence and
    goodness of fit based on stratified multistage samples,
    Journal of the American Statistical Association 75, 261-268.

Holt, D., Scott, A.J., and Ewings, P.O. (1980). Chi-squared tests
    with survey data. *Journal of the Royal Statistical Society,
    Ser. A* 143, 303-320.

Johnson, N.L., and Kotz, S. (1970). *Continuous Univariate
    Distributions*. Boston: Houghton Mifflin.

Kish, L., Groves, R.M., and Krotki, K.P. (1976). Sampling errors
    for fertility survey. Occasional Paper No. 17, London:
    World Fertility Survey.

Kish, L., and Frankel, M.R. (1974). Inference from complex
    samples. *Journal of the Royal Statistical Society Ser. B*,
    36, 1-37.

Koch, G.G., Freeman, D.H. Jr., and Freeman, J.L. (1975). Strategies
    in the multivariate analysis of data from complex surveys.

International Statistical Review 43, 59-78.

Nathan, G. (1975). Tests of independence in contingency tables from stratified proportional samples. *Sankhyā Ser. C,* 37, 77-87.

Rao, J.N.K., and Scott, A.J. (1979). Chi-squared tests for analysis of categorical data from complex surveys. *Proceedings of the Section on Survey Research Methods,* American Statistical Association, Washington, D.C., 58-66.

Solomon, H., and Stephens, M.A. (1977). Distribution of a sum of weighted chi-square variables. *Journal of the American Statistical Association* 72, 881-885.

REGRESSION ANALYSIS FOR COMPLEX SURVEYS

T. M. F. Smith

University of Southampton, U.K.

Possible approaches to aggregate regression
analysis when the data have been collected by means of
a complex survey are reviewed. The effect of survey
design is discussed and shown to be of importance for
all approaches. By introducing a design variable into
the problem results from missing value theory lead to
a maximum likelihood estimator which incorporates
design information in a model-based approach. The
various approaches are compared in a simulation study.

1.   INTRODUCTION

In this paper we examine the regression analysis of sample
survey data when the sample units have been selected by a complex
probability design. Formally we assume that there is a finite
population of N units which are assumed to have some structure,
for example, a known grouping into strata or a known grouping into
clusters, or both. The population structure can be used to devise
a complex sampling scheme which assigns a fixed probability, $p(s)$,
to every sample, s , of units. Any design other than simple
random sampling is a complex design and the resulting data
constitute a complex survey.

Sample surveys are multivariate and a major objective of
the analysis of survey data is to establish the pattern of relationships between sets of variables. When the variables are
quantitative, and knowledge of the subject matter suggests casual

relationships, then regression analysis may be an appropriate
method. Certainly it is a method which is widely employed in the
analysis of survey data.

When the population has a complex structure it is not
obvious what regression relationships should be examined. Should
regression lines be fitted within each group separately or should
an overall aggregate line be fitted? These are decisions for the
subject matter specialist as much as for the statistician, but if
there is doubt then lines should be fitted within groups and
various diagnostic tests carried out to see whether a single
aggregate line would suffice. The basic rule, "if in doubt plot
it out", applies just as much to survey data as it does to data in
the experimental sciences. A disaggregated analysis within strata
leads to no new problems since all strata are represented in the
sample. For clusters, however, disaggregation presents serious
problems for now only a subset of the clusters are represented in
the sample. How can inferences be made about the relationships
between the variables in the clusters not in the sample? Problems
of this type have recently been reviewed by Pfeffermann and
Nathan (1977).

In this paper we assume that the appropriate model for
study is an aggregate equation fitted to all the data rather than
separate equations fitted to subsets of the data. Within this
framework we can distinguish two types of inference; descriptive
inference and analytic inference. Descriptive inference may be
characterized by the property that the parameter of interest is a
known function of the values attached to the N units in the
finite population. Thus the object is to estimate a given property
of a given population at the time the sample was drawn. If all
the units were evaluated then in the absence of measurement errors
there would be no uncertainty in a descriptive inference. If the

parameter of interest cannot be expressed as a function of the
values attached to the N units then the inference is analytic,
the object being to estimate a parameter in another population
related in some way to the population being sampled. This other
population may often be a superpopulation proposed by the
statistician to represent the wider sphere of inference. Estimation of a finite population total is a descriptive inference while
estimation of a coefficient in an economic model is usually an
analytic inference.

There are two well developed theories for making descriptive inferences for samples drawn from finite populations. The
first theory is based on the distribution generated by random
sampling, the p-distribution, p(s) , and details can be found in
a standard text such as Cochran (1977). The second theory employs
stochastic models to represent the population structure and
inferences are based on the probability distribution specified in
the model, the so-called ξ-distribution. Details and references
can be found in Smith (1976) and Särndal (1978). In addition
there are hybrid theories that combine features of both approaches,
for example, Godambe and Thompson (1973). The recent book by
Cassel, Särndal and Wretman (1977) reviews all the above approaches.

One feature of inference with respect to the ξ-distribution
is that the survey design does not enter into the inference. This
result has been interpreted to mean that survey design does not
matter in the model based approach. This interpretation is incorrect since design always affects the efficiency of an inference
in terms of criteria such as mean square error. In addition
considerations of robustness and scientific validity will almost
always lead to randomisation being employed. Taking cost into
account both approaches to inference often lead to similar designs.
We do not consider further the choice of design.

The position with analytic inference is quite different.
Since the parameter to be estimated is not a function of the values

of the population units it is difficult to formulate a satisfac-
tory theory of inference employing only the p-distribution. An
extra assumption is required that relates the finite population
values to the parameter of analytic interest. Typically it is
assumed that the finite population values are a random sample
from an infinite superpopulation indexed by the parameter of
interest. Since survey populations are usually very large a
finite population estimator based on all N values will be "close
to" the unknown superpopulation parameter. This estimator then
constitutes a finite population parameter which can be estimated
using the p-distribution. However, the assumption of an underlying superpopulation is critical to interpretation and is itself
the specification of a ξ-distribution.

2. REGRESSION MODELS FOR FINITE POPULATIONS

It is possible to formulate regression models in either a
descriptive of analytic way and for inference to employ either
the p-distribution or the ξ-distribution. Hartley and Sielken
(1975) classify regression models according to these two criteria.
The classification is reproduced as Table 1.

Table 1. Sampling theories classified by sampling procedures and
          target parameters

Target Parameters	Sampling Procedure	
	Repeated sampling from a fixed finite population	Repeated two-step sampling from an infinite population
Parameters of finite population B or B$^+$	Case 1 Classical finite popu-lation sampling theory p-inference	Case 2 Super-population theory for finite population sampling. ξ-inference
Parameters of infinite super-population β	Infeasible	Case 3 Inference on infinite pop-ulation parameters from two-step sampling procedure ξ-inference

Let the values associated with the N units be represented
by the data matrix

$$
\underset{N\times r}{X} = \begin{pmatrix} X_{11}(1) & \cdots & X_{1q}(1) & X_{11}(2) & \cdots & X_{1p}(2) \\ \cdot & & & & & \\ \cdot & & & & & \\ \cdot & & & & & \\ X_{N1}(1) & \cdots & X_{Nq}(1) & X_{N1}(2) & \cdots & X_{Np}(2) \end{pmatrix}
$$

$$
= \begin{pmatrix} \underset{N\times q}{X_1} & \vdots & \underset{N\times p}{X_2} \end{pmatrix} , \tag{2.1}
$$

where $p + q = r$ and X_1 are dependent variables and X_2 are
independent variables.

Case 1. Descriptive p-inference

This has been studied by Kish and Frankel (1974), Jönrup
and Rennermalm (1976) and Shah, Holt and Folsom (1978). If
$q = 1$ and there is only one dependent variable then the parameters of interest are the numbers B_j , $j = 1,\ldots,p$, in the
finite population least squares vector

$$
B = \left(X_2^T X_2 \right)^{-1} X_2^T X_1 , \tag{2.2}
$$

where $B^T = (B_1,\ldots,B_p)$.

A fixed sample size design p(s) selects n units from N with
inclusion probabilities π_i , $i = 1,\ldots,N$. A p-unbiased estimator
of the population moment $T_{kj} = \sum_{i=1}^{N} X_{ik}X_{ij}$ is

$$
\hat{T}_{kj} = \sum_{i \in s} \frac{X_{ik} X_{ij}}{\pi_i} . \tag{2.3}
$$

Replacing each element in (2.2) by its estimator (2.3) gives

$$\hat{\underset{\sim}{B}} = \left(\widehat{X_{\sim 2}^T X_{\sim 2}}\right)^{-1} \left(\widehat{X_{\sim 2}^T X_{\sim 1}}\right). \qquad (2.4)$$

This estimator is called the p-weighted estimator and it is biased, although the bias will usually be small.

Kish and Frankel (1974) suggest three methods for estimating the sampling variance of $\hat{B}$ for repeated samples of size n using the design p(s). These are the Taylor series method, the jackknife method and balanced repeated replications. The Taylor series method can be applied to any design and has been used in this paper. Details are given in Appendix A.

Case 2. Descriptive ξ-inference.

This case has been studied by Fuller (1973) and Hartley and Sielken (1975). The finite population is assumed to be a random sample from an infinite population which has the regression structure

$$\underset{\sim}{X_1} = \underset{\sim}{X_2}\underset{\sim}{\beta} + \underset{\sim}{\varepsilon} \qquad (2.5)$$

where $\underset{\sim}{\varepsilon}$ has a probability distribution, ξ , with $E_\xi(\underset{\sim}{\varepsilon}) = \underset{\sim}{0}$ and $\text{Var}_\xi(\underset{\sim}{\varepsilon}) = \sigma^2 \underset{\sim}{V}$. A sample of n units is drawn using the sampling scheme p(s). The sampling scheme is assumed to be independent of both X_1 and X_2 and is ignored for making inference from the model (2.5).

The descriptive parameter of interest is defined to be the generalized least squares solution of (2.5), namely

$$\underset{\sim}{B}^+ = \left(X_{\sim 2}^T \underset{\sim}{V}^{-1} X_{\sim 2}\right)^{-1} X_{\sim 2}^T \underset{\sim}{V}^{-1} X_{\sim 1} . \qquad (2.6)$$

When the population is grouped into clusters the covariance matrix $\underset{\sim}{V}$ will include terms that represent the intra-cluster correlation coefficients. When $\underset{\sim}{V} = \underset{\sim}{I}$, $\underset{\sim}{B}^+$ reduces to $\underset{\sim}{B}$ in (2.2) and this

gives an interpretation to this choice of the finite population
parameter. When $\underset{\sim}{V}$ is unknown it must be estimated from the data
and some form of recursive procedure, such as that in Fuller and
Rao (1978), may be used.

If $\underset{\sim}{X}_{1s}$, $\underset{\sim}{X}_{2s}$ denote the sample values of $\underset{\sim}{X}_1$ and $\underset{\sim}{X}_2$
then the ordinary least squares estimator of $\underset{\sim}{B}$ or $\underset{\sim}{B}^+$ is

$$b = \left(X_{\sim 2s}^T \; X_{\sim 2s} \right)^{-1} X_{\sim 2s}^T \; X_{\sim 1s} \; . \qquad (2.7)$$

This is ξ-unbiased for $\underset{\sim}{B}$, $\underset{\sim}{B}^+$ or β. When $V = I$ it is an
efficient estimator of $\underset{\sim}{B} = \underset{\sim}{B}^+$. For descriptive inference the
variance of b is $\mathrm{Var}_\xi(\underset{\sim}{b}-\underset{\sim}{B})$, since both $\underset{\sim}{b}$ and $\underset{\sim}{B}$ are random
variables under the model (2.5), and this introduces finite
population correction terms in an interesting form. Details are
given in Fuller (1973) and Hartley and Sielken (1975).

Case 3. Analytic ξ-inference.

    This also assumes that the model (2.5) represents the data
but interest now centres on the superpopulation parameter β
rather than the finite population parameter $\underset{\sim}{B}$. This model is
implicit in much work in the social sciences in which the relation-
ship between $\underset{\sim}{X}_1$ and $\underset{\sim}{X}_2$ is of general scientific interest and
is not merely a description of the data at the time the survey was
carried out. If the sampling scheme $p(s)$ does not depend on $\underset{\sim}{X}_1$
and $\underset{\sim}{X}_2$ then the inferences follow from classical regression
theory.

3. THE USE OF COMPUTER PACKAGES

    Although programs exist for probability weighted estimation
of means and totals they are not yet generally available for
regression analysis. An exception is SUPERCARP (Hidiroglou,
Fuller and Hickman, (1979). Most survey analysts do not have

access to these specialist programs and so they tend to employ
standard regression programs in packages such as BMDP and SPSS.
These packages are very convenient for analyses based on ordinary
least squares. However, in both cases the variances given by the
programs are the conventional least squares variances and fail to
take into account population structure such as clustering or
stratification.

Weighted analyses are also available on SPSS and BMDP.
These correspond to using generalized least squares with the
covariance matrix $\underset{\sim}{V}$ replaced by a diagonal matrix $\underset{\sim}{W}$ of weights
to be attached to each unit. By choosing $w_i = \pi_i^{-1}$ probability
weighting is achieved for point estimation. Unfortunately the
variances in the programs are the generalized least squares
variances $(\underset{\sim}{X}^T \underset{\sim}{W}^{-1} \underset{\sim}{X}_2)^{-1} \hat{\sigma}^2$ and these are not the variances
associated with a p-inference. In addition the analysis of
variance in SPSS is based on a nominal sample size of $n^* = \sum_{i \in s} w_i$,
and this can play havoc with degrees of freedom. We understand
that this is being modified.

The general conclusion is that standard packages should be
used with great caution for the regression analysis of survey data.
Except where an ordinary least squares analysis is thought to be
appropriate, such as for a population with no known grouping into
strata or clusters, analyses based on standard regression packages
are not suitable for data from complex surveys. This conclusion
simply reinforces the conclusions in Kish and Frankel (1974).

4.   THE INFORMATION IN THE SAMPLING DESIGN, p(s)

One of the main conflicts between the model based approach
and the p-based approach to inference is in the use of the
information in the sample design to aid the inference. In a
p-based approach the only probabilities are those in the sample
design, p(s) , and so this is the only information available. In

the model based approach it is often assumed that the realized sample
s is independent of the observed values X_1 and X_2 and that the
design need not feature in the inference. Rubin (1976) discusses
the situations in which the design can be ignored.

Scott (1975) clarifies the position and explains the
dilemma. Let X_3 be a set of values known to the statistician
prior to the survey which summarize the prior information available
on each unit in the population. This information includes the
information on the grouping into clusters or strata. The survey
design uses this information and so should be expressed as
$p(s|X_3)$. The values $Y = (X_1 : X_2)$ to be observed are also
related to X_3 and have prior distribution $p(Y|X_3)$. Since s
refers only to the labels and X_3 contains all the information
including that on the labels which identifies the values in X_3
associated with the ith unit, we have

$$p(s, Y | X_3) = p(s | X_3)p(Y | X_3) , \qquad (4.1)$$

i.e. s and Y are conditionally independent given the values
of X_3. It is in this conditional sense that the sample units, s,
and the sample values Y , are independent. There is no extra
information in s not contained in X_3. However, if X_3 is not
known, then in general, unconditionally,

$$p(s,Y) \neq p(s) p(Y) , \qquad (4.2)$$

and the design is not independent of the values. It is rare for
X_3 to be made explicit and so it is often true that the design
$p(s)$ contains useful information about the values Y . The
question still remains as to how this information should be used?

Suppose we are in the position of knowing X_3 so that (4.1)
holds. This tells us that the design $p(s|X_3)$ contains no useful
additional information. However, it does not say that the particular selection of units in the observed sample s, and the values

$X_{\underset{\sim}{3s}}$ on which the design was based, should not affect the inference. The influence of selection on inference is well known and the first theoretical analysis was due to K. Pearson (1902). Pearson's results have been developed and rediscovered many times including the results on missing observations due to Anderson (1957). Holt, Smith and Winter (1980) give a full discussion and further references.

For sample surveys the main result derived from multivariate normal distributions is given in Appendix B. An equivalent result can be derived from wider linear model assumptions as in Lawley (1943). In this paper the empirical work concentrates on estimators of the slope of a simple linear regression and the required estimator adjusted for selection has the form

$$\hat{\beta} = \frac{\left\{ s_{12} + \dfrac{s_{13}\, s_{23}}{s_3^2}\left(\dfrac{\hat{\sigma}_3^2}{s_3^2} - 1\right) \right\}}{\left\{ s_2^2 + \dfrac{s_{23}^2}{s_3^2}\left(\dfrac{\hat{\sigma}_3^2}{s_3^2} - 1\right) \right\}} \quad , \qquad (4.3)$$

where we are considering the regression of scalar X_1 on scalar X_2 with scalar design variable X_3 , s_{ij} is the unweighted sample covariance between X_i and X_j , and $N\hat{\sigma}_3^2 = \sum\limits_{i=1}^{N} (X_{i3} - \overline{X}_3)^2$ is calculated from the full set of N values of X_3. If $s_{23} = 0$ or if $\hat{\sigma}_3^2/s_3^2 = 1$ the estimator reduces to the ordinary least squares estimator $b = s_{12}/s_2^2$. For many designs, and for large n , $s_3^2 \simeq \hat{\sigma}_3^2$ for most samples, in which case the ordinary least squares estimator will suffice. However, even for simple random sampling, a totally non-informative design, in some extreme samples $\hat{\sigma}_3^2/s_3^2 \neq 1$ and an adjustment for selection is worthwhile. The adjustment is due to the particular selection of units and it is not due to the choice of the design, $p(s|X_3)$, which does not feature in the inference. If $X_{\underset{\sim}{3}}$ is known sample selection

affects model based inference but sample design does not. If $X_{\sim 3}$ is
unknown the design should affect all forms of inference.

In a well designed survey stratification, clustering and
the use of variable selection probabilities reflect properties of
the population values and capture some of the features of the
design variable $X_{\sim 3}$. When $X_{\sim 3}$ is unknown the design, p(s) ,
may be used as a substitute and the use of inclusion probabilities
as weights is equivalent to the use of a type of ratio estimator
for the various population moments required in an aggregate
regression analysis. Obviously other approaches are possible
including the use of alternative substitute variables for $X_{\sim 3}$
which could be employed in estimators such as (4.3). This is a
problem which requires further study.

Clustering effects are qualitative rather than quantitative
and are best represented by introducing intra-cluster correlations
into the covariance matrix of the regression residuals, $\varepsilon_{\sim}$.
Fuller (1973), Campbell (1977) and Holt and Scott (1980) show
how this may be achieved. In one or two stage sampling the
variances of estimated regression parameters are increased in a
manner similar to that of the design effect employed by Kish and
Frankel (1974).

In order to examine the effects of a quantitative design
variable a simulation study was carried out which compared the
probability weighted estimator and its variance, the ordinary
least squares estimator and its least squares variance and the
maximum likelihood estimator and its normal theory variance in
terms of the coverage properties of confidence intervals. The
design variable was used to construct strata and to vary the
selection probabilities within strata.

5. THE SIMULATION STUDY

This section summarizes some of the results of a more

extensive study by Holt, Smith and Winter (1980). Real survey
data were used to obtain reasonable values for a population
covariance matrix, $\underset{\sim}{\Sigma}$, for design variables, independent variables
and dependent variables. The matrix, $\underset{\sim}{\Sigma}$, defines the population
values of the unknown aggregate regressions. A design variable,
X_3 , was selected and a fixed finite population of 10,000 values
of X_3 was obtained using a random normal generator. A design,
$p(s|X_3)$, was chosen and a sample of 1000 units was selected. For
each unit a pair of values (X_1,X_2) was generated from the
conditional normal distribution of $X_1,X_2|X_3$. For the fixed
values of X_3 this operation was repeated 1000 times by replic-
ating X_2 10 times and $X_1|X_2$ 100 times for each X_2. For each
of these 1000 samples β was estimated by the three methods.
Confidence intervals were computed for the 100 samples conditional
on the values of X_2 and X_3. These were then averaged over the
10 replications of X_2 to give average performance figures.

One set of data analysed were farms in a region in Canada.
The three variables used were as follows:

Design variable X_3 :   Total value of products sold in
                           previous year,

Independent variable X_2 :   Total acreage,

Dependent variable X_1 :   Total cropland acreage.

The values of X_3 were used to stratify the population into five
equal sized strata each of size 2,000. Six designs were used to
generate samples of size 1,000 as follows:

Design

D1.   Simple random sampling,

D2.   Proportionate allocation (200, 200, 200, 200, 200) ,

D3.   Increasing allocation (50, 150, 200, 250, 350) ,

D4.   Increasing allocation (50, 50, 100, 300, 500) ,

D5.   U-shaped allocation (300, 150, 100, 150, 300) ,

D6.   U-shaped allocation (450, 49, 2, 49, 450) .

If X_3 is positively correlated with X_2 then the U-shaped allocations should be relatively efficient for estimation of the regression of X_1 on X_2. In order to use the p-distribution the allocations had to be restricted to those which included at least two observations from each stratum.

Table 2 shows the average values of the estimators and of their standard errors for the three procedures for estimating the regression parameter β , given in Appendix C, namely,

(i)   ordinary least squares with least squares variance,

(ii) maximum likelihood with appropriate variance,

(iii) probability weighted with Taylor series variance.

Table 2. Average values and standard errors for the three
          procedures for the designs considered

Design	Ordinary least squares, b	Maximum likelihood, β	Probability weighted estimation, b^*
D1	.308	.308	.308
	.825(-2)	.826(-2)	.825(-2)
D2	.309	.307	.309
	.795(-2)	.794(-2)	.795(-2)
D3	.298	.308	.306
	.839(-2)	.849(-2)	.108(-1)
D4	.296	.307	.308
	.851(-2)	.863(-2)	.123(-1)
D5	.321	.307	.308
	.760(-2)	.770(-2)	.827(-2)
D6	.332	.307	.309
	.714(-2)	.752(-2)	.235(-1)

Note: (i)   Population value estimated = 0.3072.

       (ii) First entry is average value of estimator and second
             is average conditional standard error

       (iii) Standard error given with number of extra decimal
             places in brackets. Thus .825(-2) = .00825.

The bias of the least squares procedure for the unequally weighted designs D3 to D6 is immediately apparent. Little bias is apparent for the other two procedures but even a small bias can affect confidence intervals when the standard error is small and this is reflected in some of the results in Table 2. The standard errors are particularly interesting. For designs D1 and D2 all procedures are much the same. However, when variable selection probabilities are employed the standard error for the p-weighted procedure increases. In particular for the U-shaped allocation, which should be efficient for estimating a regression slope, the p-weighted standard error increases whereas the maximum likelihood standard error decreases in line with expectation. A possible explanation is that the probability weights are not reflecting the heteroscedasticity in the underlying population, so they are not reflecting the underlying population structure.

Table 3 shows the distribution of the frequencies with which confidence intervals computed for each procedure covered the true population value β . The expected frequencies are given at the top of the table.

As expected the bias of the ordinary least squares estimator dominates the intervals for unequal probability designs so that the coverage properties are badly distorted. For the equal probabilitiy designs D3 to D5 both maximum likelihood and p-weighting give adequate coverage properties. However, the intervals for the maximum likelihood procedure were shorter than those for p-weighting due to the smaller standard errors. For the extreme design D6 the coverage properties of the p-weighted procedure were poor whilst the maximum likelihood procedure still performs well.

One general conclusion is that the maximum likelihood procedure performs well, is more efficient than p-weighting in this rather favourable simulation study, and is worthy of further consideration by those applying regression analysis to survey data.

Table 3. Confidence interval coverage for the three procedures
          for the designs considered

Expected Frequencies		5	5	15	25	450	450	25	15	5	5
Design	Estimation procedure										
D1	b	7	3	19	32	500	394	28	9	4	4
	$\hat{\beta}$	8	4	18	31	465	419	30	13	4	8
	b^*	8	5	25	31	492	387	26	16	4	6
D2	b	11	5	17	30	517	394	14	11	1	0
	$\hat{\beta}$	9	6	9	23	433	476	19	15	8	2
	b^*	14	5	24	36	501	384	20	11	4	1
D3	b	0	0	2	1	142	618	79	68	42	48
	$\hat{\beta}$	5	8	16	26	464	431	27	16	3	4
	b^*	4	4	11	18	391	494	33	24	6	15
D4	b	0	0	0	2	98	577	97	90	49	87
	$\hat{\beta}$	11	2	16	22	455	443	25	21	2	3
	b^*	6	6	17	21	401	462	44	23	10	10
D5	b	185	69	117	130	458	41	0	0	0	0
	$\hat{\beta}$	11	7	11	32	432	454	26	18	3	6
	b^*	12	12	21	40	460	409	25	13	2	6
D6	b	777	83	74	31	34	1	0	0	0	0
	$\hat{\beta}$	9	5	18	24	422	461	32	18	5	6
	b^*	48	23	35	50	410	340	33	14	9	38

Another conclusion is that ordinary least squares analysis, such
as that based on standard computer packages, should only be
employed for equal probability designs when using stratified
samples. Since it is known that the ordinary least squares
variances are biased when the errors have a general covariance
matrix, V , which would be the case for clustered populations,

it is clear that for most surveys ordinary least squares procedures do not provide adequate inferences. The p-weighted procedure performs adequately except for the most extreme designs, but with some loss of efficiency.

A limited study was carried out by Holt, Smith and Winter (1980) to examine the properties of the maximum likelihood procedure when the values of X_3 were replaced by a substitute variable. The effect is to introduce a small bias which causes some distortion to the coverage properties of the intervals. For design D6 maximum likelihood is still superior but for the other designs the coverage properties of the p-weighted procedure are now slightly better than those of maximum likelihood although still relatively inefficient for some designs. However, p-weighting does not require evaluation of X_3 and this is a great advantage.

When this study was started it was hoped that it would provide the answer to the question whether to use p-weights or not in aggregate regression analysis for stratified samples. Ordinary least squares has been eliminated from consideration and the maximum likelihood procedure should now be seen as the alternative to probability weighting. However, the choice between these two procedures is not clear cut and the robustness of the probability weighted procedure in a study that was not inherently favourable to it is most noticeable. Further studies are needed before the question about the use of probability weights can be resolved.

APPENDIX A

Taylor series variance for a p-distribution

Let $\hat{\underline{Y}}$ be a $k \times 1$ vector of random variables with $E(\hat{\underline{Y}}) = \underline{Y}$. For a function $g(\hat{\underline{Y}})$ estimating $g(\underline{Y})$ if $Var(\hat{Y}_i)$ is $O(n^{-1})$ then

$$g(\hat{\underline{Y}}) \simeq g(\underline{Y}) + \sum_{i=1}^{k} (\hat{Y}_i - Y_i) \left. \frac{g(\hat{\underline{Y}})}{\hat{Y}_i} \right|_{\hat{Y}_i} = Y_i .$$

Thus $\quad E\{g(\hat{\underline{Y}}) - g(\underline{Y})\}^2 \simeq E\{ \sum_{i=1}^{k} (\hat{Y}_i - Y_i) \frac{\partial g(\underline{Y})}{\partial Y_i} \}^2$

$$= \text{Var} \left\{ \sum_{i=1}^{k} \hat{Y}_i \frac{\partial g(\underline{Y})}{\partial Y_i} \right\} .$$

For many problems in sample surveys $\hat{Y}_i = \sum_{j \in s} X_{ij}$, in which case

$$\text{Var}\{g(\hat{Y})\} \simeq \text{Var} \left\{ \sum_{i=1}^{k} \sum_{j \in s} X_{ij} \frac{\partial g}{\partial Y_i} \right\}$$

$$= \text{Var} \left\{ \sum_{j \in s} Z_j \right\} ,$$

where

$$Z_j = \sum_{i=1}^{k} X_{ij} \frac{\partial g}{\partial Y_i} .$$

Thus an approximate p-variance can be written down using standard results for the variance of a sum.

For the simple linear regression estimator (2.4),

$$b^* = \frac{\Sigma \pi_i^{-1} \Sigma X_{1i} X_{2i} \pi_i^{-1} - \Sigma X_{1i} \pi_i^{-1} \Sigma X_{2i} \pi_i^{-1}}{\Sigma \pi_i^{-1} \Sigma X_{2i}^2 \pi_i^{-1} - (\Sigma X_{2i} \pi_i^{-1})^2}$$

$$= \frac{\hat{Y}_1 \hat{Y}_2 - \hat{Y}_3 \hat{Y}_4}{\hat{Y}_1 \hat{Y}_5 - \hat{Y}_4^2} .$$

It then follows that

$$\sum_{i=1}^{5} \hat{Y}_i \frac{\partial g}{\partial Y_i} = \sum_{j \in s} \pi_j^{-1} W_j \ ,$$

where

$$W_j = \frac{(X_{2j} - \bar{X}_2)\{X_{1j} - \bar{X}_1 - B(X_{2j} - \bar{X}_2)\}}{\sum\limits_{j=1}^{N} (X_{2j} - \bar{X}_2)^2} \ .$$

$\sum\limits_{j \in s} \pi_j^{-1} W_j$ is just a Horvitz-Thompson estimator and its variance can be estimated using the Yates-Grundy estimator with W_j estimated by $\hat{W}_j$ by replacing unknown terms in W_j by p-estimators. For stratified random sampling this gives the result in Appendix C. One of the best descriptions of the Taylor series approach to variance estimation can be found in Rao (1975).

APPENDIX B

Maximum likelihood estimation

Let $z_i^T = (Y_i^T : X_i^T)$, $i = 1, \ldots, N$ be independent observations from a multivariate normal distribution with mean $\mu_1^T = (\mu_1^T : \mu_2^T)$ and covariance matrix $\Sigma = \begin{pmatrix} \Sigma_{11} & \Sigma_{12} \\ \Sigma_{21} & \Sigma_{22} \end{pmatrix}$. A sub-sample, s , of size n is drawn by any design $p(s|X)$ based on the known values X_i , $i = 1, \ldots, N$. The data comprise the labels i , the sample values Y_i , $i \in s$, and the known values X_i , $i = 1, \ldots, N$. Let $Y_s = \{Y_i; i \in s\}$, $X_s = \{X_i; i \in s\}$, where the label allows the appropriate X_i and Y_i to be associated. We then have

$$p(\text{data}) = p(i, Y_i, X_i, X_j; i \in s, j \notin s)$$
$$= p(Y_s|X,s)p(s|X)p(X)$$
$$= p(Y_s|X_s)p(s|X)p(X) \ . \qquad (1)$$

Let $\underset{\sim}{S} = \begin{pmatrix} \underset{\sim}{S}_{11} & \underset{\sim}{S}_{12} \\ \underset{\sim}{S}_{21} & \underset{\sim}{S}_{22} \end{pmatrix}$ be the usual sample sum of squares and products

matrix and $\overline{Y}_s$, $\overline{X}_s$ the sample means. The results follow those in Anderson (1957).

Theorem. The maximum likelihood estimator of $\underset{\sim}{\Sigma}_{11}$ for any design $p(s|\underset{\sim}{X})$ is

$$\hat{\underset{\sim}{\Sigma}}_{11} = \hat{\underset{\sim}{\Sigma}}_{1.2} + \hat{\underset{\sim}{B}}_{1.2} \hat{\underset{\sim}{\Sigma}}_{22} \hat{\underset{\sim}{B}}^T_{1.2},$$

where

$$\hat{\underset{\sim}{B}}_{1.2} = \underset{\sim}{S}_{12} \underset{\sim}{S}^{-1}_{22},$$

$$n \hat{\underset{\sim}{\Sigma}}_{1.2} = \underset{\sim}{S}_{11} - \hat{\underset{\sim}{B}}_{1.2} \underset{\sim}{S}_{22} \hat{\underset{\sim}{B}}^T_{1.2},$$

and

$$N \hat{\underset{\sim}{\Sigma}}_{22} = \sum_{i=1}^{N} (\underset{\sim}{X}_i - \overline{X})(\underset{\sim}{X}_i - \overline{X})^T.$$

Proof. From (1) the likelihood of $\underset{\sim}{\mu}$, $\underset{\sim}{\Sigma}$ can be written as

$$L(\underset{\sim}{\mu}, \underset{\sim}{\Sigma}) \propto p(\underset{\sim}{Y}_s | \underset{\sim}{X}_s) p(\underset{\sim}{X}) \quad \text{since } p(s|\underset{\sim}{X}) \text{ depends only on } \underset{\sim}{X}.$$

Now $p(\underset{\sim}{Y}_s | \underset{\sim}{X}_s) \propto \dfrac{1}{|\underset{\sim}{\Sigma}_{1.2}|^{n/2}} \exp\{-\dfrac{1}{2} \underset{i}{\sum}_s (\underset{\sim}{Y}_i - \underset{\sim}{\mu}_{1.2}(i))^T \cdot$

$$\underset{\sim}{\Sigma}^{-1}_{1.2}(\underset{\sim}{Y}_i - \underset{\sim}{\mu}_{1.2}(i))\},$$

where

$$\underset{\sim}{\mu}_{1.2}(i) = \underset{\sim}{\mu}_1 + \underset{\sim}{\Sigma}_{12} \underset{\sim}{\Sigma}^{-1}_{22}(\underset{\sim}{X}_i - \underset{\sim}{\mu}_2) = \underset{\sim}{\mu}_1 + \underset{\sim}{B}_{1.2}(\underset{\sim}{X}_i - \underset{\sim}{\mu}_2),$$

and

$$\underset{\sim}{\Sigma}_{1.2} = \underset{\sim}{\Sigma}_{11} - \underset{\sim}{\Sigma}_{12} \underset{\sim}{\Sigma}^{-1}_{22} \underset{\sim}{\Sigma}_{21};$$

and $p(\underset{\sim}{X}) \propto \dfrac{1}{|\underset{\sim}{\Sigma}_{22}|^{N/2}} \exp\{-\dfrac{1}{2}\sum_{i=1}^{N}(\underset{\sim}{X}_i - \underset{\sim}{\mu}_2)^T \underset{\sim}{\Sigma}^{-1}_{22}(\underset{\sim}{X}_i - \underset{\sim}{\mu}_2)\}.$

Reparameterising we find the maximum likelihood estimators in the conditional distribution are

$$\hat{\underset{\sim}{B}}_{1.2} = \sum_{i \in s} (\underset{\sim}{Y}_i - \underset{\sim}{Y}_s)(\underset{\sim}{X}_i - \underset{\sim}{X}_s)^T \Big\{ \sum_{i \in s} (\underset{\sim}{X}_i - \underset{\sim}{X}_s)(\underset{\sim}{X}_i - \underset{\sim}{X}_s)^T \Big\}^{-1}$$

$$= \underset{\sim}{S}_{12}\, \underset{\sim}{S}_{22}^{-1}$$

where $n\overline{X}_s = \sum\limits_{i \in s} \underset{\sim}{X}_i$ and $n\overline{Y}_s = \sum\limits_{i \in s} \underset{\sim}{Y}_i$,

$$n\,\hat{\underset{\sim}{\Sigma}}_{1.2} = \sum_{i \in s} (\underset{\sim}{Y}_i - \overline{\underset{\sim}{Y}}_s)(\underset{\sim}{Y}_i - \overline{\underset{\sim}{Y}}_s)^T - \hat{\underset{\sim}{B}}_{1.2}\, \underset{\sim}{S}_{22}\, \hat{\underset{\sim}{B}}_{1.2}^T$$

$$= \underset{\sim}{S}_{11} - \hat{\underset{\sim}{B}}_{1.2}\, \underset{\sim}{S}_{22}\, \hat{\underset{\sim}{B}}_{1.2}^T .$$

From the marginal distribution of $\underset{\sim}{X}$ we find

$$N\hat{\underset{\sim}{\mu}}_2 = \sum_{i=1}^{N} \underset{\sim}{X}_i ,$$

$$N\,\hat{\underset{\sim}{\Sigma}}_{22} = \sum_{i=1}^{N} (\underset{\sim}{X}_i - \overline{\underset{\sim}{X}})(\underset{\sim}{X}_i - \overline{\underset{\sim}{X}})^T .$$

Now $\underset{\sim}{\Sigma}_{1.2} = \underset{\sim}{\Sigma}_{11} - \underset{\sim}{\Sigma}_{12}\, \underset{\sim}{\Sigma}_{22}^{-1}\, \underset{\sim}{\Sigma}_{21}$

$$= \underset{\sim}{\Sigma}_{11} - \underset{\sim}{\Sigma}_{12}\, \underset{\sim}{\Sigma}_{22}^{-1}\, \underset{\sim}{\Sigma}_{22}\, \underset{\sim}{\Sigma}_{22}^{-1}\, \underset{\sim}{\Sigma}_{21}$$

$$= \underset{\sim}{\Sigma}_{11} - \underset{\sim}{B}_{1.2}\, \underset{\sim}{\Sigma}_{22}\, \underset{\sim}{B}_{1.2}^T .$$

Hence the maximum likelihood estimator of $\underset{\sim}{\Sigma}_{11}$ is

$$\hat{\underset{\sim}{\Sigma}}_{11} = \hat{\underset{\sim}{\Sigma}}_{1.2} + \hat{\underset{\sim}{B}}_{1.2}\, \hat{\underset{\sim}{\Sigma}}_{22}\, \hat{\underset{\sim}{B}}_{1.2}^T .$$

DeMets and Halperin (1977) examine the special case when $p = 2$, $q = 1$ and derive the maximum likelihood estimator of $\beta_{12} = \sigma_{12}/\sigma_{22}$, where $\underset{\sim}{\Sigma}_{11} = \begin{pmatrix} \sigma_{11} & \sigma_{12} \\ \sigma_{21} & \sigma_{22} \end{pmatrix}$. Their result is easily derived from the general result in the theorem.

The maximum likelihood estimator of the mean μ_1 is

$$\hat{\mu}_1 = \bar{Y}_s + S_{12} \, S_{22}^{-1} (\bar{X} - \bar{X}_s) \ .$$

For scalar variables Y_i and X_i this is just the well known regression estimator of a mean. This suggests that we can interpret the slope estimator (4.3) in a similar way as the regression estimator of a regression coefficient. This interpretation may appeal to those statisticians who distrust results derived from normal distribution assumptions.

APPENDIX C

The three interval estimation procedures

In the empirical study we use three procedures for estimating the simple linear regression coefficient,

$$\beta_1 = \sigma_{12}/\sigma_{22} \ ,$$

when there is a single design variable X_3. Let

$$ns_{jk} = \sum_{i \in s} (X_{ij} - \bar{x}_j)(X_{ik} - \bar{x}_k) \ ,$$

$$ns_j^2 = \sum_{i \in s} (X_{ij} - \bar{x}_j)^2 \ ,$$

$$n\hat{\sigma}_j^2 = \sum_{i=1}^{N} (X_{ij} - \bar{x}_j)^2 \ .$$

Procedure 1. Ordinary least squares.

Estimator: $\qquad b = s_{12}/s_2^2$

Estimated variance: $v(b) = \dfrac{s_1^2 - bs_{12}}{ns_2^2} \ .$

Procedure 2. Maximum likelihood.

$$\text{Estimator:} \quad \hat{\beta} = \frac{\left\{ s_{12} + \frac{s_{13} s_{23}}{s_3^2} \left(\frac{\hat{\sigma}_3^2}{s_3^2} - 1 \right) \right\}}{\left\{ s_2^2 + \frac{s_{23}^2}{s_3^2} \left(\frac{\hat{\sigma}_3^2}{s_3^2} - 1 \right) \right\}} .$$

$$\text{Estimated variance:} \quad v(\beta) = \frac{\hat{\sigma}_{1.23}^2 \left\{ s_2^2 + \frac{s_{23}^2}{s_3^2} \left(\frac{\hat{\sigma}_3^2}{s_3^2} - 1 \right)^2 + 2 \frac{s_{23}^2}{s_3^2} \left(\frac{\hat{\sigma}_3^2}{s_3^2} - 1 \right) \right\}}{n \left\{ s_2^2 + \frac{s_{23}^2}{s_3^2} \left(\frac{\hat{\sigma}_3^2}{s_3^2} - 1 \right) \right\}^2} ,$$

$$\text{where} \quad \hat{\sigma}_{1.23}^2 = s_1^2 - \frac{(s_{13}^2 s_2^2 + s_{12}^2 s_3^2 - 2 s_{12} s_{13} s_{23})}{(s_2^2 s_3^2 - s_{23}^2)} .$$

For the derivation of these expressions see Nathan and Holt (1980).

Procedure 3. p-weighted estimation (stratified sampling).

Whilst the general form of the probability weighted estimator is given in Appendix A the variance and an appropriate estimator will depend on the joint inclusion probabilities of elements. The results given here are appropriate for a stratified sample with the notation X_{ihk} referring to the kth variable and the ith individual in the hth stratum.

$$\text{Estimator:} \quad b^* = \frac{N \sum_{h} \sum_{i \epsilon s} W_h X_{ij1} X_{ih2} - \sum_{h} \sum_{i \epsilon s} W_h X_{ih1} \sum_{h} \sum_{i \epsilon s} W_h X_{ih2}}{N \sum_{h} \sum_{i \epsilon s} W_h X_{ih2}^2 - (\sum_{h} \sum_{i \epsilon s} W_h X_{ih2})^2}$$

$$\text{where} \quad W_h = N_h / n_h .$$

Estimated variance: $v(b^*) = \sum_h n_h^2 (\frac{1}{n_h} - \frac{1}{N_h}) \sum_{i \in s} (Z_{ih} - \bar{Z}_h)^2 / (n_h - 1)$,

where

$$Z_{ih} = \frac{W_h(X_{ih2} - \hat{\bar{X}}_2)\{X_{ih1} - \hat{\bar{X}} - b^*(X_{ih2} - \hat{\bar{X}}_2)\}}{\sum_h \sum_{i \in s} W_h X_{ih2} - N\hat{\bar{X}}_2^2}$$

and

$$N\hat{\bar{X}}_j = \sum_h \sum_{i \in s} W_h X_{ihj} = \sum_h N_h \bar{X}_{jh} \ ,$$

$$n\bar{Z}_h = \sum_{i \in s} Z_{ih} \ .$$

Strictly speaking a p-based approach to analytic inference is not possible. However, with the added assumption that the finite population is a random sample from the superpopulation with regression parameter β , then since $E_p(b^*) \simeq B$ and $B = \beta + O_p(N^{-\frac{1}{2}})$, b^* should be a reasonable estimator of β . Similarly if n/N is small $v(b^*)$ should be a reasonable estimator of $E_\xi E_p(b^* - \beta)^2 = E_\xi\{V_p(b^*)\} + E_\xi\{E_p(b^*) - \beta\}^2$. Since for large N the second term should be negligible.

For each procedure we assume n is large, n/N is small and that the conditions of the central limit theorem apply. In each case an interval estimate of β is given by

$$\beta^* \pm Z_{\alpha/2} \sqrt{v(\beta^*)}$$

where β^* is any estimator of β, $v(\beta^*)$ is an estimator of the variance of β^* and $Z_{\alpha/2}$ is the upper $100\alpha/2$ % point of a standard normal distribution.

ACKNOWLEDGEMENTS

This work was supported by Grant Number HR4973/1 from the Social Science Research Council. The author is grateful for advice from and discussions with Professor D. Holt, Dr. G. Nathan, Professor J.N.K. Rao, and Professor A.J. Scott.

Everybody who has worked in the area of sample surveys must acknowledge their tremendous debt to Bill Cochran. Although I met him only twice, having taught extensively from all his books, I have always felt that I knew him well. He will be greatly missed by the whole community of statisticians.

REFERENCES

Anderson, T.W. (1957). Maximum likelihood estimates for a multi-variate normal distribution when some observations are missing. *J. Amer. Statist. Ass.* 52, 200-203.

Campbell, C. (1977). Properties of ordinary and weighted least squares estimators of regression coefficients for two-stage samples. *Proceedings of the Social Statistic Section*. American Statistical Association, Washington,D.C. 800-805.

Cassel, C-M., Särndal, C-E. and Wretman, J.H. (1977). *Foundations of Inference in Survey Sampling*. John Wiley and Sons, New York.

Cochran, W.G. (1977). *Sampling Techniques* (3rd ed.) John Wiley & Sons, New York.

Demets, D. and Halperin, M. (1977). Estimation of a simple regression coefficient in samples arising from a sub-sampling procedure. *Biometrics* 33, 47-56.

Fuller, W.A. (1973). Regression analysis for sample surveys. First meeting of the International Association of Survey Statisticians, Vienna, August 1973.

Fuller, W.A. and Rao, J.N.K. (1978). Estimation for a linear regression model with unknown diagonal covariance matrix. *Ann. Statist.* 6, 1149-1158.

Godambe, V.P. and Thompson, M.E. (1973). Estimation in sampling theory with exchangeable prior distributions. *Ann. Statist.* 1, 1212-1221.

Hartley, H.O. and Sielken, R.L. (1975). A "super-population viewpoint" for finite population sampling. *Biometrics* 31, 411-422.

Hidiroglou, M.A., Fuller, W.A., and Hickman, R.D. (1979). SUPER CARP (5th edition) Survey Section, Statistical Laboratory, Iowa State University.

Holt, D. and Scott, A.J. (1981). Regression analysis using survey data. *The Statistician* (to appear).

Holt, D., Smith, T.M.F. and Winter, P.D. (1980). Regression analysis of data from complex surveys. *J. Roy. Statist. Soc. A.* 143, 474-487.

Jonrup, H. and Rennermalm, B. (1976). Regression analysis in samples from finite population. *Scand. J. Statist.* 3,33-37.

Kish, L. and Frankel, M.R. (1974). Inference from complex samples (with Discussion). *J. Roy. Statist. Soc. B* 1, 1-37.

Lawley, D.N. (1943). A note on Karl Pearson's selection formula. *Proc. Roy. Soc. Edin., Sect. A* 62, 28-30.

Nathan, G. and Holt, D. (1978) The effect of survey design on regression analysis. *J. Roy. Statist. Soc., B* 42, 377-386.

Pearson, K. (1902) On the influence of natural selection on the variability and correlation of organs. *Phil. Trans. Roy. Soc. A* 200, 1-66.

Pfeffermann, D. and Nathan, G. (1977). Regression analysis of data from complex samples. Third Meeting of the International Association of Survey Statisticians, New Delhi, December 1977.

Rao, J.N.K. (1975) Analytic studies of sample survey data. *Survey Methodology* 1 (Supplementary Issue). Statistics Canada, Ottawa.

Rubin, D.B. (1976). Inference and missing data. *Biometrika* 63, 581-592.

Särndal, C-E. (1978). Design-based and model-based inference in survey sampling. *Scand. J. Statist.* 5, 27-52.

Scott, A.J. (1975). Some comments on the problem of randomisation
    in surveys. Second Meeting of the International Association
    of Survey Statisticians, Warsaw, September 1975.

Shah, B.V., Holt, M.M. and Folsom, R.E. (1977). Inference about
    regression models from sample survey data. Third Meeting
    of the International Association of Statisticians, New
    Delhi, December, 1977.

Smith, T.M.F. (1976). The foundations of survey sampling: a
    review. *J. Roy. Statist. Soc. A* 139, 183-195.

SUPERPOPULATION MODELS: GENERAL DISCUSSION

D. BINDER, *Statistics Canada* (to W. Fuller): You've been assessing this particular estimator on the basis of anticipated variance, which of course depends on the model. Have you looked at how this estimator compares under perturbations of the model in the sense that the model itself isn't satisfied? What conditions do you need between the y's and the z's in order for this to still be a better estimator than what you would normally use in standard practice? I also noticed that you showed the inclusion probabilities for stratified sampling. How do these relate to the Neyman allocation inclusion probabilities in the case of fixed cost per stratum?

W. FULLER, *Iowa State University*: On the first question, this estimator will never be any worse than the simple Horvitz-Thompson estimator with respect to the design variance in large samples. In other words, those estimated coefficients do nearly minimize the design variance. When we're talking about design variance, there's just the usual sampling setup of n-tuples. Thus, if you look at this kind of estimator and form that linear combination, that's the best you can do. In that sense, just as regression estimation is a pretty robust technique, we know that when the relationship deviates a long way from linearity there is bias. We also know that in large samples that bias really isn't that bad, and then this does lead essentially to the Neyman allocation. In that case, if you have only zeros and ones, then you can in fact do a conditional maximization which is sort of what Neyman allocation is. You can fix the n's, which is equivalent to choosing a particular set of π_{ij}'s and they are a reasonable and perfectly legitimate set of π_{ij}'s.

A. CHAUDHURI, *Indian Statistical Institute* (to W. Fuller):
I have two questions. First, how do other strategies such as the
Rao-Hartley-Cochran, Murthy and Sen-Midzuno-Lahiri schemes compare
with the present approach? Secondly, what criteria should be used
as a measure of error of the estimates you propose?

W. FULLER, *Iowa State University*: Any unequal probability
sampling scheme that provides reasonable π_{ij}'s is all right. In
other words, although there are restrictions on the joint inclusion
probabilities, they are the kind that you would want. What is
being ruled out here is the selection of a number of adjacent x's.
The requirement of consistency is basically a reasonableness
requirement. As in simple random sampling the π_{ij}'s behave in a
way that leads to a decrease in variance. Thus any unequal
probability sampling scheme that comes up with reasonable π_{ij}'s
fits in here.

The estimation of error also goes through basically as you
would expect. The $\hat{B}$'s converge to the corresponding population
parameters so that if you just take the $\hat{B}$'s and use a Yates-
Grundy kind of variance estimator it will be a consistent estimator
of the design variance.

T. DALENIUS, *Brown University* (to W. Fuller): There is one
aspect that worries me a little about these approaches. If your
regression model is not realistic, attempts should be made by
someone to answer the question of its robustness. I think we
could try to use something analogous to the difference estimator
with an ordinary regression model where you do not make any
assumptions about linearity or anything else. Would that be
possible?

W. FULLER, *Iowa State University*: My feeling is that
estimating the B's is a more robust procedure than using the
difference estimator because you are basically letting the sample

tell you what the coefficients are. You could say that's an adaptive procedure, but the $\hat{B}$'s are the sample estimates of what's best. It's also easier to carry out the calculations this way as a matter of fact.

The robustness of the regression estimator depends on the sample size. Obviously, if your sample is very small and you have something highly non-linear, then you are going to have large biases and maybe large variances too. I think it's sort of like ratio estimation. In lots of practical applications, the bias for highly non-linear cases is not that bad.

Tore points out that if you use a difference estimator you are assuming nothing. That's correct. But the variance is also a function of those deviations that go with that difference and they can be much larger than the smallest deviations you can find. In the sense that I am talking about, you definitely have hope for having smaller things to square.

J.N.K. RAO, *Carleton University*: I think we should define what the model-based and model-dependent approaches are. Model-based means that you use the models to construct estimators that are design consistent or design unbiased. Model-dependent means that all the criteria such as consistency and unbiasedness relate only to the model, with the design being totally irrelevant at the inference stage.

The work that I've seen so far in both areas uses models which are mathematically convenient. I don't think anyone has really done any serious work in model building. One can see the complexities of this approach as soon as you get into multistage sampling. Whichever approach you use, either the model-based or the model-dependent approach, you're going to have many parameters, including correlation coefficients, which all have to be estimated. This may be quite complicated and the gain in efficiency, if any (over say the standard ratio estimates, which have the versatility of adapting to multistage designs), may be quite trivial. This would be worth investigating.

I think most of the research to date has been only for
unistage sampling. I have seen some research in two stage sampling
by Richard Royall and others which indicates the complexities.
Once you have too many parameters, you start making simplifying
assumptions, whether justified or not. Perhaps Wayne could
comment on the versatility of the model-based approach in multi-
stage designs.

W. FULLER, *Iowa State University*: In one sense the multi-
stage designs are completely covered in my work. In other words,
the π_{ij}'s again do not preclude multistage designs and so the
estimators that arise are the usual multistage kind of estimators.
Those $\hat{B}$'s are just what you would use as a straight jack-up kind
of regression estimator. Now if you try to put in more covariance
structure, then the estimators might become more complicated.
(Maybe you should put in this structure if you go to multistage
sampling). However, in one sense, what I talked about generalizes
immediately to multistage designs.

Whether or not a model person would be completely happy
with this model in the multistage situation is another question.
You could put a lot of structure in, a lot of covariances and so
forth, and it is true they are not here now. But I think, to turn
it around, you could construct a model that would look reasonable
and would give you the same estimator. The use of this straightforward jack-up kind of estimator in multistage sampling would
then be fairly reasonable.

K. EBERHARDT, *National Bureau of Standards* (to C. Sarndal):
Dividing by the q-probabilities doesn't strike me as intuitively
pleasing right off for reasons which I guess are in the direction
of Basu's example. If there are a lot of people with very small
q's, using this procedure with known q's doesn't appeal to me.
The fact that you estimate them sort of saves it for the following

reasons. If you knew all the q's and used them in the denominator
in your formulas, and you were unlucky in that all the people with
low probabilities of responding did in fact respond, then they
would be very heavily weighted. Did you take known q's in your
simulation just to see if that effect does show up?

C. SARNDAL, *University of British Columbia*: No we didn't.
We felt that assuming the q-probabilities known would be unrealistic. We made it more difficult for ourselves by assuming that
they were unknown and had to be estimated. Small q's are not a
serious problem though, because although the first term of the
estimator is inflated in the case where both π and $\hat{q}$ turn out
to be small, the first term cancels the last term for many models,
so that what just remains is $\Sigma \hat{\beta}_j \bar{x}_{.j}$ (which is not seriously
affected by a small inflation factor). Special cases of the
estimator have been proposed in the literature. I think our
estimator will work well in general, and small inflation factors
are not a problem.

J.N.K. RAO, *Carleton University*: I would like to make one
comment on the Basu example. I think that example has been exag-
gerated. In 1961 when I was in Iowa State writing my thesis, I
knew at that time that the Horvitz-Thompson estimator was not
supposed to be used indiscriminately. If you look at the Horvitz-
Thompson paper, the authors carefully make recommendations on when
to use it. If some people use this estimator in spite of these
recommendations, that is not the fault of the whole sampling
community. The point is that the Horvitz-Thompson estimator can
be rectified very simply by taking the ratios of the probabilities.
This has been pointed out by Hajék repeatedly. Basu's example,
however, is very amusing and it does warn us when to be careful
and in this sense the model-inspired approach saves us from making
silly mistakes. Otherwise, an intelligent practioner isn't likely
to make such mistakes.

D. KREWSKI, *Health and Welfare Canada* (to A. Scott):
Although this is a symposium on survey sampling, I would like to
mention a potential application of your results in the area of
biostatistics. In two generation carcinogenicity studies using
rats and mice, tumor incidence data is often available for a
control and a test group on a litter basis. The structure of this
data is that of a stratified cluster sample with treatments corresponding to strata and litters to clusters. Ignoring the
possibility of intralitter correlation and applying the usual
chi-square statistic for 2×2 tables may lead to highly inflated
false positive rates in this case. Your results indicate that it
may be possible to apply a simple adjustment factor to this chi-
square statistic which would allow for the presence of litter (or
design) effects. This approach would be much simpler than a
maximum likelihood or randomization analysis of these data on a
litter basis.

A. SCOTT, *University of Auckland*: I didn't really have
time to talk about all the relevant references, although I should
have said perhaps that there is a lot of work by Altham and Cohen
in the biostatistics literature which is really quite similar to
ours. We actually started building our own work from that. Since
most people have heard Jon or myself talk about that before, I
rushed through that.

H. CHAUDHRY, *Statistics Canada* (to A. Scott): Do you have
a consistent estimator of the covariance matrix of $f_{ij} = p_{ij} - p_{i.} p_{.j}$?
If so, suppose we calculate the ratio of the determinant of that
covariance matrix to that of the covariance matrix of f_{ij}
estimated under the assumption of multinomial sampling (perhaps
this ratio could be called a generalized design effect). If we
now adjust the Wald statistic calculated under the multinomial
assumption using this ratio, would it follow a chi-square distribution?

A. SCOTT, *University of Auckland*: I think it does, but the
point is that you usually don't have a consistent estimator of the
covariance matrix.

H. CHAUDHRY, *Statistics Canada*: Didn't you assume in your
talk that you had a consistent estimator of the covariance matrix?

A. SCOTT, *University of Auckland*: No, I said that if you
had, you would use the Wald statistic. We tried to develop some
simple modifications for those situations where such an estimate
was not available.

D. BINDER, *Statistics Canada* (to A. Scott): I would like
to congratulate you on an excellent paper. I have encountered
this topic in practice and can vouch for the fact that people are
concerned about this problem today. Clearly there are a lot of
unanswered questions, particularly with tests of independence. I
think this is something that people will be thinking about very
much in the near future.

My particular question concerns the next step. Suppose we
perform a test of significance and we come to the conclusion that
in fact it is not significant. In your particular case you look
at two hypotheses. The first is that p_1 equals p_2. What I
would think the practitioner would do in this case is to just take
the average or the weighted average of the estimates p_1 and p_2
as the pooled estimate of the common value of p (which is not
necessarily the most efficient thing to do compared to weighting
by the covariance structure). The second hypothesis is that of
independence. How do you think we should estimate the marginal
probabilities, now knowing that we don't have multinomial sampling?
Just taking the row sums of our estimates may not be the most
appropriate thing to do. Nowadays people think in terms of log-
linear models when they see contingency tables. Suppose for
example that we have a three-way table where only the three-way

interaction is zero but everything else isn't. We know in the
multinomial situation that the sufficient statistic consists of
the marginal two-way totals and we can base our estimates on that.
What do you suppose we should do in the case of complex surveys?

A. SCOTT, *University of Auckland*: I don't know. Obviously
it is a big problem and we have only made a start. We know what
needs doing and we have thought about it, but we haven't done
anything yet.

D. BINDER, *Statistics Canada*: I just mentioned this so
that perhaps some of the people in this audience might start
thinking about these problems. They are very real problems today
in applications.

W. FULLER, *Iowa State University* (to T.M.F. Smith): I
think you're correct if your experiment did favour $\hat{\beta}$, particularly
in the case of normality. For example, if your auxillary information had been just a zero-one stratification variable, then I
believe the conditional estimator would have larger variance. This
is because you have been pooling two within-stratum estimators
whereas the p-weighted estimators would also be adding in a between
stratum component.

T.M.F. SMITH, *University of Southampton*: We have done some
robustness studies on this where, instead of using the $X_{\sim 3}$ values,
we replace them by a single stratum indicator variable. Not a
zero-one indicator for each stratum but an indicator such as
1,2,3,4,5 because we knew the strata were organized by size. I
think the serious effect is on the bias of the maximum likelihood
estimator. Because the standard error is small, that bias shifts
the confidence intervals causing a definite detereoration in the
maximum likelihood approach.

D. BINDER, *Statistics Canada* (to T.M.F. Smith): I find it
interesting that you use this model but that you use auxiliary
information to try to approximate your probabilities of selection.
I think one of the reasons why your p-based design performed so
well is because it was in fact correlated with the auxiliary
variables. What I'd like to point out is that in many surveys
the original design was developed for one purpose with other
surveys later "hooking onto" these surveys. Since the objectives
of these latter surveys are different from the original ones,
p(s) does not really always reflect the sort of subjective notions
that you would like it to reflect in terms of having a good design.
If in that case you can find auxiliary data such as $\underset{\sim}{X}_3$ which
would really reflect what you would have liked to do rather than
what you did do, I think you could improve on your p-based
estimators.

T.M.F. SMITH, *University of Southhampton*: I agree that
there will be situations where one can generate the design and
then the p-estimator in such a way that it does badly. It's got
a certain robustness, but I think the main effect would be to
inflate the variance. To be fair to the simulation study, I would
consider only conditional inference if I was working within a
model-based framework and I wouldn't have to simulate. Here,
however, we've taken a set of conditional variances and then
averaged them over another set so that we've got a sort of semi-
conditional rather than totally conditional estimate. That is a
move towards favouring the p-based approach.

There is a loss of efficiency through using the p-based
approach relative to the strict model-based approach. The question
is whether in large samples that sort of inefficiency matters very
much. For example, if the design that you use is unrelated to the
regression you're analyzing, it is just like simple random sampling
and we know that that works fine. I suppose you might somehow have
an inverse relationship coming in between the design and the analysis

which might throw things off.

K. EBERHARDT, *National Bureau of Standards* (to T.M.F. Smith):
I think I saw an occasion where balanced samples reared their heads.
It was in the q-factor that adjusts ordinary least squares to
maximum likelihood. When q = 1 you have a balanced sample in
the first two moments of the X_3 variables.

T.M.F. SMITH, *University of Southhampton:* Yes, the adjustment factor depends on the second moment, balancing on the second
moment in this case. If you work not with the regression coefficients but with the mean, the adjusted estimator is the standard
regression estimator of the mean. All this maximum likelihood
estimator is, if you like, is a regression estimator of a regression coefficient. It is not really anything foreign to ordinary
p-based procedures. It's like having a regression equation where
you've got one variable which doesn't appear in the regression
equation but is related to X_1 and X_2 . All you are doing is
adjusting for this extra bit of knowledge.

J.N.K. RAO, *Carleton University* (to T.M.F. Smith): Your
model-based estimator $\hat{\beta}$ can be adjusted so as to be p-based like
Carl Sarndal's. That would protect you against the probabilities
being fouled up. It could be weighted up, giving an improvement
on b^* .

V VARIANCE ESTIMATION

ESTIMATION OF THE MEAN SQUARE ERROR
OF THE RATIO ESTIMATOR

Poduri S.R.S. Rao

University of Rochester

The conventional and weighted least squares estimators and a simplified version of the jackknife estimator
for the mean square error of the ratio estimator of a
finite population mean are considered. The biases and
the mean square errors of these estimators are compared
under a suitable linear regression model.

1. INTRODUCTION

Let $(\bar{X}, \bar{Y})$ denote the means of an auxiliary variable x
and the characteristic of interest y of a finite population of
size N. Denote the means of a sample of size n by $(\bar{x}, \bar{y})$ and
let $r = \bar{y}/\bar{x}$. The ratio estimator for $\bar{Y}$ is

$$\hat{\bar{Y}}_R = r\bar{X} \qquad (1.1)$$

with approximate variance

$$V(\hat{\bar{Y}}_R) = \frac{(1-f)}{n} (S_y^2 + R^2 S_x^2 - 2RS_{xy}), \qquad (1.2)$$

where $f = n/N$, $R = \bar{Y}/\bar{X}$, S_x^2 and S_y^2 are the population variances
of x and y and S_{xy} is the covariance between them. The
variance in (1.2) is smaller than $V(\bar{y}) = (1-f)S_y^2/n$ whenever
$\rho > (C_x/2C_y)$, where ρ is the correlation coefficient between x
and y and $C_x = (S_x/\bar{X})$ and $C_y = (S_y/\bar{Y})$ are the coefficients

of variation.

A regression model suitable for evaluating the ratio type
of estimator is

$$Y_i = \beta x_i + e_i , \quad (i = 1,\dots,N) \tag{1.3}$$

with $E(e_i|x_i) = 0$. Then,

$$E[\hat{\overline{Y}} - \overline{Y})|x_i] = 0 , \tag{1.4}$$

i.e., $\hat{\overline{Y}}_R$ is unbiased under the above model. If $V(e_i|x_i) = \delta x_i$
and $E(e_i e_j|x_i x_j) = 0$ $(i \neq j)$, $\hat{\overline{Y}}_R$ is the minimum variance
unbiased estimator. We consider the general case where
$V(e_i|x_i) = \delta x_i^t$. Now, the variance of $\hat{\overline{Y}}_R$ is

$$V(\hat{\overline{Y}}_R|x_i) = E[(\hat{\overline{Y}}_R - \overline{Y})^2|x_i] = E\left(\frac{\overline{e}}{\overline{x}}\overline{X} - \overline{E}\right)^2$$

$$= \left(\frac{\overline{X}}{n\overline{x}}\right)^2 \sum_i^n x_i^t + \frac{\sum_i^N x_i^t}{N^2} - 2\frac{\overline{X}}{\overline{x}}\frac{\sum_i^N x_i^t}{Nn}\delta$$

$$= (1-f)^2\left[\left(\frac{\overline{X}'}{n\overline{x}}\right)^2 \sum_i^n x_i^t + \frac{\sum_i^\ell x_i^t}{\ell^2}\right]\delta, \tag{1.5}$$

where $n\overline{e} = \sum_i^n e_i$, $N\overline{E} = \sum_i^N e_i$, $\ell = N-n$, and $\ell\overline{x}' = \sum_i^\ell x_i$.

With the model in (1.3), $E[(\overline{y} - \overline{Y})|x_i] = \beta(\overline{x} - \overline{X})$. Thus,
$\overline{y}$ is conditionally unbiased only for balanced samples $(\overline{x} = \overline{X})$.
Its MSE is

$$M(\overline{y}|x_i) = E[(\overline{y} - \overline{Y})^2|x_i] = \beta^2(\overline{x} - \overline{X})^2 + \left[\frac{\sum_i^n x_i^t}{n^2} + \frac{\sum_i^N x_i^t}{N^2} - 2\frac{\sum_i^n x_i^t}{Nn}\right]\delta. \tag{1.6}$$

From (1.5) and (1.6)

$$D = M(\overline{y}|x_i) - V(\hat{\overline{Y}}_R|x_i)$$

$$= \beta^2(\overline{x}-\overline{X})^2 + \frac{(1-f)}{n^2}\frac{(\overline{x}-\overline{X})(\overline{x}+\overline{X}')}{\overline{x}^2}(\sum_i x_i^t)\delta. \tag{1.7}$$

Thus, from (1.7), for any value of t, $D > 0$ for $\bar{x} > \bar{X}$ when $|\beta|$ is small. If $\bar{x}$ is smaller than $\bar{X}$, $D > 0$ for sufficiently large values of β.

The above results exhibiting the superiority of $\hat{Y}_R$ over $\bar{y}$ are only two reasons for the popularity of the former estimator. In the next section, six estimators for the variance of $\hat{Y}_R$ in (1.5) are considered. In Section 3, the biases (conditional on x_i) of these variance estimators are derived under the model in (1.3) for the general case where $V(e_i|x_i) = \delta x_i^t$, and they are compared for the important case when t lies between zero and two. Sections 4 and 5 contain comparisons of the expected biases and MSE's when x has the gamma distribution $f(x) = e^{-x}x^{h-1}/\Gamma(h)$; neither N nor n is assumed to be large. These comparisons are based on the results obtained by the author in Rao (1979).

2. THE ESTIMATORS FOR THE VARIANCE OF $\hat{Y}_R$

The conventional estimator is

$$\hat{V}_C = \frac{(1-f)}{n} s_d^2 , \qquad (2.1)$$

where $d_i = (y_i - rx_i)$ and $(n-1)s_d^2 = \sum_{}^{n} d_i^2$. The weighted least squares estimator based on the model in (1.3) with $t = 1$ is

$$\hat{V}_W = \frac{(1-f)}{n} \frac{\overline{XX}'}{\bar{x}} \hat{\delta} , \qquad (2.2)$$

where $(n-1)\hat{\delta} = \sum_{}^{n} (d_i^2/x_i)$; see Royall and Eberhardt (1978).

The jackknife estimator for $V(\hat{Y}_R)$, based on the procedure suggested by Quenouille (1956) and Tukey (1958), is

$$\hat{V}_J = \frac{(1-f)(g-1)}{g} \bar{x}^2 \sum_{}^{g} (r'_j - \bar{r}')^2 . \qquad (2.3)$$

In this procedure, the sample is divided into g groups of size $m = (n/g)$ each. Now, $(\bar{x}'_j, \bar{y}'_j)$ are the means of the $(n-m)$

observations $(m = n/g)$, $r'_j = (\bar{y}'_j/\bar{x}'_j)$ and $g\bar{r}' = \sum\limits_j^g r'_j$. The alternative to $\hat{V}_J$ suggested by Kish and Frankel (1970) is to replace $(r'_j - \bar{r}')$ by $(r'_j - r)$ to obtain

$$\hat{V}_{KF} = \frac{(1-f)(g-1)}{g} \bar{x}^2 \sum_j^g (r'_j - r)^2 . \qquad (2.4)$$

Replacing $(r'_j - r)$ in (2.4) by $(\bar{y}'_j - r\bar{x}'_j)\bar{x}$, we may consider another alternative to $\hat{V}_{KF}$ or $\hat{V}_J$ as

$$\hat{V}_R = \frac{(1-f)}{g} \frac{\bar{x}^2}{\bar{x}} \frac{\sum (y_j - rx_j)^2}{g-1} . \qquad (2.5)$$

When $g = n$, clearly,

$$\hat{V}_R = \frac{(1-f)}{n} \frac{\bar{x}^2}{\bar{x}} \frac{\sum\limits^n (y_i - rx_i)^2}{n-1}$$

$$= \frac{\bar{x}^2}{\bar{x}} \hat{V}_C . \qquad (2.6)$$

3. BIASES OF $\hat{V}_C$, $\hat{V}_W$ AND $\hat{V}_R$ CONDITIONAL ON x

From (1.5), the conditional variance of $\hat{\bar{Y}}_R$ is given by

$$V(\hat{\bar{Y}}_R | x_i) = \frac{(1-f)}{n} \left[\frac{\overline{XX}'}{\bar{x}^2} + \frac{f(\bar{x} - \bar{X}')}{\bar{x}} \right] \delta , \quad t = 0$$

$$V(\hat{\bar{Y}}_R | x_i) = \frac{(1-f)}{n} \frac{\overline{XX}'}{\bar{x}} \delta , \qquad t = 1$$

and

$$V(\hat{\bar{Y}}_R | x_i) = \left[\frac{\bar{x}^2 (nc^2 + 1)}{n} + \frac{\bar{x}^2 (NC^2 + 1)}{N} - 2 \frac{\overline{Xx}(nc^2 + 1)}{N} \right] \delta , t = 2 \quad (3.1)$$

where $n\hat{\sigma}^2 = \sum\limits^n (x_i - \bar{x})^2$, $N\sigma^2 = \sum\limits^N (x_i - \bar{X})^2$, $c^2 = (\hat{\sigma}^2/n\bar{x}^2)$ and $C^2 = (\sigma^2/N\bar{X}^2)$. We note that both c^2 and C^2 are smaller than unity.

As for the cases of $t = 0$ and $t = 1$, the above variance takes simple forms for special values of c^2. For instance, as c^2 approaches C^2, the variance in (3.1) for $t = 2$ becomes

$$V(\hat{Y}_R | x_i) = \left[2 \frac{(1-f)}{n} \bar{X}\bar{X}' (nc^2 + 1) - \frac{(1-f)}{n} x^2 \right] \delta . \quad (3.1a)$$

Further, if c^2 approaches zero, (3.1a) becomes

$$V(\hat{Y}_R | x_i) = \left[\frac{(1-f)}{n} \bar{X}\bar{X}' + \frac{\bar{X}(\bar{X} - \bar{x})}{N} \right] \delta . \quad (3.1b)$$

From (1.3) and (2.1),

$$E(\hat{V}_c | x_i) = \frac{(1-f)}{n(n-1)} \left[\sum x_i^t + \frac{(\sum x_i^2)(\sum x_i^t)}{(\sum x_i)^2} - 2 \frac{\sum x_i^{t+1}}{\sum x_i} \right] \delta . \quad (3.2)$$

From the results in Section 3.1 and equation (3.2), the bias of $\hat{V}_c$ for $t = 0$ and 1 can be written as

$$B(\hat{V}_c | x_i) = \frac{(1-f)}{n}(1 + c^2 - \frac{\bar{X}\bar{X}'}{\bar{x}^2}) \delta + (\frac{\bar{X} - \bar{x}}{\bar{x}}) \frac{\delta}{N}, \quad t = 0 \quad (3.2a)$$

and

$$B(\hat{V}_c | x_i) = \frac{(1-f)}{n\bar{x}} \left[(1 - c^2)\bar{x}^2 - \bar{X}\bar{X}' \right] \delta, \quad t = 1. \quad (3.2b)$$

For $t = 2$ it can be shown that

$$B(\hat{V}_c | x_i) \leq \frac{(1-f)}{n} (\bar{x}^2 - \bar{X}\bar{X}') \delta , \quad (3.2c)$$

provided $m_3 \geq 0$ and $c^2 = C^2 = 0$. The expectation of $\hat{V}$ is clearly obtained by multiplying the right side of (3.2) by $(\bar{X}/\bar{x})^2$. The magnitude of its bias relative to that of $\hat{V}_c$ clearly depends on the closeness of $\bar{x}$ to $\bar{X}$.

From (1.3) and (2.2)

$$E(\hat{V}_W | x_i) = \frac{(1-f)}{n(n-1)} \frac{\bar{X}\bar{X}'}{\bar{x}} (\sum_{i}^{n} x_i^{t-1} - \frac{\sum_{i}^{n} x_i^t}{n\bar{x}}) \delta . \quad (3.3)$$

From the results in Section 3.1 and Equation (3.3), when $t = 0$ it can be shown that

$$B(\hat{V}_W|x_i) > - \frac{(1-f)}{n} \frac{(\bar{x} - \bar{X}')}{\bar{x}} \delta .$$ (3.3a)

When $t = 1$ clearly

$$B(\hat{V}_W|x_i) = 0 .$$ (3.3b)

As c^2 approaches C^2, for $t = 2$, the bias becomes

$$B(\hat{V}_W|x_i) = - \frac{(1-f)}{n} \bar{X}\bar{X}'[(2n+1)C^2 + 1]\delta + \frac{(1-f)}{n} \bar{X}^2\delta .$$ (3.3c)

If $c^2 = 0$ in addition, for $t = 2$,

$$B(\hat{V}_W|x_i) = \frac{\bar{X}(\bar{x} - \bar{X})}{N} \delta .$$ (3.3d)

When $t = 0$ and $\bar{X} > \bar{x}$, from (3.2a),

$$B(\hat{V}_C|x_i) \leq \frac{(1-f)}{n} C^2\delta + \frac{\bar{X} - \bar{x}}{\bar{x}} \frac{\delta}{N}$$ (3.4)

and from (3.3a)

$$B(\hat{V}_W|x_i) > (\frac{\bar{X} - \bar{x}}{\bar{x}})\frac{\delta}{N} .$$ (3.5)

Thus the bias of $\hat{V}_C$ becomes smaller than that of $\hat{V}_W$ as $t = 0$ and $\bar{X} > \bar{x}$ and as c^2 approaches zero.

When $t = 1$, $\hat{V}_W$ is of course unbiased. From (3.2b), the bias of $\hat{V}_C$ becomes small as $\bar{x}^2(1 - c^2) - \bar{X}\bar{X}'$ becomes small. Thus, as $\bar{x}$ approaches $\bar{X}$ (i.e., $\bar{x}$ also approaches $\bar{X}'$) and c^2 approaches zero, the bias of $\hat{V}_C$ becomes small.

When $t = 2$, $c^2 = 0$ and $m_3 \geq 0$, from (3.2c) and (3.3d) the bias of $\hat{V}_C$ becomes smaller than that of $\hat{V}_W$ if $\bar{x} < \bar{X}$.

4. COMPARISONS OF THE EXPECTED BIASES

Let $\hat{V}$ denote any one of the estimators $\hat{V}_C$, $\hat{V}_W$ and $\hat{V}_R$. The expected (unconditional) biases $E[\hat{V} - V(\bar{Y}_R|x_i)]$ are derived by the author in Rao (1979) when x has the gamma distribution as described in Section 1. We have computed these biases for several values of N, n and h when $t = 0$, 1 and 2. The biases are presented in Table 1 for selected values of these parameters. The following conclusions are drawn from all the computed values.

(1) $\hat{V}_C$ underestimates V when $t = 0$, 1 and 2. $\hat{V}_W$ is, of course, unbiased when $t = 1$, but overestimates V when $t = 0$ and underestimates when $t = 2$. $\hat{V}_R$ overestimates when $t = 0$ and underestimates when $t = 1$ and 2.

(2) The classical estimator $\hat{V}_C$ has much smaller absolute bias than $\hat{V}_W$ when $t = 0$ and 2, although $\hat{V}_W$ is unbiased when $t = 1$.

(3) The newly constructed estimator $\hat{V}_R$ has much smaller absolute bias than $\hat{V}_C$ when $t = 0$ and 1, and slightly larger absolute bias than the former when $t = 2$. The absolute bias of $\hat{V}_R$ is considerably smaller than that of $\hat{V}_W$ when $t = 0$ and 2.

(4) As n increases, the absolute biases of $\hat{V}_C$ and $\hat{V}_R$ become smaller for $t = 0$, 1 and 2, and almost become insignificant when $n \geq 30$. The absolute bias of $\hat{V}_W$ also decreases as n increases, but its magnitude does not become negligible, unlike the biases of $\hat{V}_C$ and $\hat{V}_R$.

5. EXPECTED MSE'S

Cochran (1977) presents a comprehensive summary of the relative merits of $\hat{V}_C$ and $\hat{V}_J$. In a joint paper of the author, see Rao and Rao (1971), biases of these two estimators are compared. J.N.K. Rao and Kuzik (1974) examine the merits of these two estimators through a "semi-empirical" study. Krewski

Table 1. Unconditional Biases of $\hat{V}_C$, $\hat{V}_W$, and $\hat{V}_R$ for Selected Values of n, t and h and N = 100

n	Esti-mator	t = 0				t = 1				t = 2			
		h = 1	2	3	4	1	2	3	4	1	2	3	4
5	$\hat{V}_C$	-.1678	-.0542	-.0316	-.0221	-.0792	-.0768	-.0763	-.0762	-.0855	-.1698	-.2518	-.3328
	$\hat{V}_W$	*	.2636	.1177	.0744	0.0	0.0	0.0	0.0	-.1631	-.3524	-.5433	-.7347
	$\hat{V}_R$	.0585	.0221	.0137	.0099	-.0416	-.0403	-.0401	-.0400	-.1231	-.2455	-.3646	-.4823
10	$\hat{V}_C$	-.0246	-.0100	-.0062	-.0045	-.0182	-.0180	-.0180	-.0180	-.0284	-.0499	-.0703	-.0905
	$\hat{V}_W$	*	.1052	.0500	.0325	0.0	0.0	0.0	0.0	-.0843	-.1749	-.2657	-.3565
	$\hat{V}_R$	.0095	.0044	.0028	.0021	-.0100	-.0099	-.0099	-.0099	-.0398	-.0700	-.0988	-.1272
20	$\hat{V}_C$	-.0049	-.0018	-.0012	-.0008	-.0040	-.0040	-.0040	-.0049	-.0081	-.0132	-.0181	-.0230
	$\hat{V}_W$	*	.0433	.0211	.0139	0.0	0.0	0.0	0.0	-.0392	-.0796	-.1200	-.1604
	$\hat{V}_R$	.0017	.0008	.0005	.0004	-.0024	-.0024	-.0024	-.0024	-.0108	-.0176	-.0241	-.0306
30	$\hat{V}_C$	-.0013	-.0006	-.0003	-.0003	-.0016	-.0016	-.0016	-.0016	-.0036	-.0057	-.0078	-.0098
	$\hat{V}_W$	*	.0254	.0121	.0080	0.0	0.0	0.0	0.0	-.0233	-.0468	-.0704	-.0940
	$\hat{V}_R$	.0005	.0003	.0002	.0001	-.0010	-.0010	-.0010	-.0010	-.0046	-.0073	-.0099	-.0125

* Not defined for t = 0 and h = 1.

and Chakrabarty (1981) compared $\hat{V}_C$ and $\hat{V}_J$ when N is large. Royall and Eberhardt (1975) and Royall and Cumberland (1978) considered the case of $t = 1$ only and compared $\hat{V}_C$, $\hat{V}_W$ and $\hat{V}_J$ under the model (conditional on x_i) in (1.3) when either N or n is large or both of them are large.

The conditional MSE's are derived by the author in Rao (1979). However, comparisons among them involve complicated conditions in addition to simple balancing $(\bar{x} = \bar{X})$. The unconditional MSE's, $E[\hat{V} - \hat{V}(\hat{\bar{Y}}_R|x_i)]^2$ are also derived in Rao (1979), and in this Section we draw some conclusions based on them. For illustration, we present them in Table 2 for some selected cases. The MSE of $\hat{V}_R$ is not presented there since we have found it to have considerably larger MSE than $\hat{V}_C$ and $\hat{V}_W$. Two major conclusions that can be drawn are as follows.

(1)   The MSE of $\hat{V}_W$ is larger than that of $\hat{V}_C$ when $t = 0$ and 1. When $t = 2$, the former is smaller than the latter provided $n \geq 10$. We note that $\hat{V}_W$ is unbiased when $t = 1$.

(2)   As n increases, the MSE's of $\hat{V}_C$ and $\hat{V}_W$ decrease, and the difference between them becomes small.

We note that the conclusions in J.N.K. Rao and Kuzik (1974) are based on the average (over x) of $E\{\hat{V} - E[V(\hat{\bar{Y}}_R|x_i)]\}^2$ through a "semi-empirical study." The above conclusions of this study are based on the average of $E[\hat{V} - V(\hat{\bar{Y}}_R|x_i)]^2$ as mentioned above.

ACKNOWLEDGEMENTS

The author would like to thank Dr. Barbara Bailar for her encouragement during the course of this investigation. This research is partially supported by a Joint Statistical Agreement 1979-80 between the U.S. Bureau of the Census and the University of Rochester. The suggestions of the editors for presenting the material have been very helpful. Thanks to Ms. Lorraine Ziegenfuss for her efficient typing of the manuscript.

Table 2. Unconditional MSE's of $\hat{V}_C$ and $\hat{V}_W$ for Selected Values of n, t and h and N = 100

n	Estimator	t = 0				t = 1				t = 2			
		h = 1	2	3	4	1	2	3	4	1	2	3	4
5	$\hat{V}_C$	.8022	.0598	.0282	.0189	.0434	.0707	.1091	.1554	.0500	.4184	1.4835	3.7127
	$\hat{V}_W$	*	*	.2405	.1192	.0672	.1173	.3539	.5941	.0758	.7476	3.2200	9.4538
10	$\hat{V}_C$	.0125	.0040	.0027	.0022	.0049	.0114	.0202	.0311	.0180	.1237	.4099	.9950
	$\hat{V}_W$	*	*	.0190	.0104	.0048	.0161	.0342	.0590	.0144	.1092	.4213	1.1690
15	$\hat{V}_C$	.0021	.0009	.0007	.0006	.0014	.0034	.0061	.0095	.0069	.0430	.1371	.3262
	$\hat{V}_W$	*	*	.0048	.0026	.0011	.0040	.0086	.0150	.0051	.0342	.1233	.3301
20	$\hat{V}_C$	.0006	.0003	.0003	.0002	.0005	.0014	.0024	.0038	.0031	.0183	.0578	.1359
	$\hat{V}_W$	*	*	.0018	.0010	.0004	.0014	.0031	.0055	.0024	.0146	.0503	.1307
25	$\hat{V}_C$	.0003	.0001	.0001	.0001	.0003	.0006	.0011	.0018	.0016	.0092	.0282	.0656
	$\hat{V}_W$	*	*	.0008	.0005	.0002	.0006	.0014	.0024	.0013	.0074	.0245	.0621
30	$\hat{V}_C$	.0001	.0001	.0001	.0001	.0001	.0003	.0006	.0009	.0009	.0050	.0151	.0349
	$\hat{V}_W$	*	*	.0004	.0002	.0001	.0003	.0007	.0012	.0008	.0042	.0133	.0330

* Not defined for t = 0 and h = 1,2.

REFERENCES

Cochran, W.G. (1977). *Sampling Techniques*. 3rd ed. John Wiley & Sons, New York.

Kish, L. and Frankel, M.R. (1970). Balanced repeated replications for standard errors. *Journal of the American Statistical Association* 65, 1071-1094.

Krewski, D. and Chakrabarty, R.P. (1981). On the stability of the Jackknife variance estimator in ratio estimation. *Journal of Statistical Planning and Inference* (in press).

Quenouille, M.H. (1956). Note on bias in estimation. *Biometrika* 43, 353-360.

Rao, J.N.K. and Kuzik, R.A. (1974). Sampling errors in ratio estimation. *Sankhyā C* 36, 43-58.

Rao, P.S.R.S. and Rao, J.N.K. (1971). Small sample results for ratio estimators. *Biometrika* 58, 625-630.

Rao, P.S.R.S. (1979). Variance estimation for the ratio estimator of the mean of a finite population. Research report, U.S. Bureau of the Census and the University of Rochester.

Royall, R.M. and Eberhardt, K.R. (1975). Variance estimates for the ratio estimator. *Sankhyā C* 37, 43-52.

Royall, R.M. and Cumberland, W.G. (1978). Variance estimation in finite population sampling. *Journal of the American Statistical Association* 73, 351-358.

Tukey, J.W. (1958). Bias and confidence in not-quite-large samples. *Annals of Mathematical Statistics (Abstract)* 29, 614.

NON-NEGATIVE UNBIASED VARIANCE ESTIMATORS

Arijit Chaudhuri

Indian Statistical Institute

Results on necessary and sufficient conditions
for non-negativity of variance estimators are briefly
reviewed. Some new sufficient conditions are also
given.

1. INTRODUCTION

In making design-based inference concerning finite population parameters (such as totals, means and variances) the usual
practice is to obtain a point estimate along with an unbiased
estimate of its variance. The implication is thereby to help one
to form a confidence interval and an idea of a measure of error.
Naturally one should insist on non-negativity of the variance
estimators. But it is well-known that for several uni-stage
sampling strategies, particularly the popular ones involving the
Horvitz-Thompson (1952) estimator (HT & HTE, in brief) and the
ratio estimator based on sampling with probabilities proportional
to sample aggregates of size measures (the Lahiri-Midzuno-Sen,
1952, 1952, 1953, or LMS strategy), the variance estimators need
not necessarily be uniformly non-negative. A brief review of the
literature in this area is given in Section 2 with some new
results given in section 3. These latter results are in the
nature of providing necessary or, if possible, necessary and
sufficient conditions on sampling designs which would guarantee

uniform non-negativity of the variance estimators. However, these conditions usually relate to size measures which are of necessity 'fixed' in any particular situation and cannot be changed to suit one's requirements. Thus, a problem arises when the so-called necessary or necessary and sufficient conditions do not hold for the situation at hand and the particular variance estimate obtained for the sample data in fact turns out negative. As a way out of this dilemma, we suggest an admittedly inefficient solution: a collection of estimates along with estimates of their respective variances may be formed with the provision to choose one from a subset of it for which the variance estimate is positive for the sample actually surveyed.

2. A BRIEF REVIEW OF THE LITERATURE

In a recent article J.N.K. Rao (1979) gives the following general result concerning a simple derivation of the mean square error (MSE) of linear estimators in a unified way for a finite population total along with the necessary form for its (uniformly) non-negative quadratic unbiased estimators.

For a finite population U of N units with variate-values y_i $(i = 1, \ldots, N)$ a general linear estimator of $Y = \Sigma\, y_i$ based on a sample s selected with probability $p(s)$ is $\hat{Y} = \Sigma\, d_{is} y_i$ with $d_{is} = 0$ for $i \notin s$ and d_{is} independent of the y_i's. If $\hat{Y}$ is such that its mean square error $MSE(\hat{Y}) = E(\hat{Y} - Y)^2$ equals zero if $y_i = c w_i$ $(i = 1, \ldots, N)$ where $c (\neq 0)$ is an arbitrary constant and the w_i's are non-zero constants, then one can write

$$MSE(\hat{Y}) = -\sum_{i<j} \sum d_{ij} w_{ij} (z_i - z_j)^2 \,,$$

where $z_i = y_i / w_i$ and $d_{ij} = E(d_{is} - 1)(d_{js} - 1)$. (Here E denotes the expectation operator with respect to sampling design p.) Then it follows that a uniformly non-negative quadratic unbiased estimator of $MSE(\hat{Y})$ is

$$mse(\hat{Y}) = - \sum_{i<j} \sum d_{ij}(s)w_i w_j (z_i - z_j)^2 , \qquad (2.1)$$

where the $d_{ij}(s)$'s do not depend on y_i's, $d_{ij}(s) = 0$ if s
does not contain both i and j and

$$\sum_{s \ni i,j} p(s)d_{ij}(s) = d_{ij} , \quad i < j .$$

He also illustrates how specific selections from the necessary
form lead to well-known variance estimators for many uni-stage
sampling strategies and also yield new alternatives. For the
important special case of sampling only 2 units he further
specifies the form of the only possible available non-negative
unbiased quadratic estimator of the MSE. Extensions to multi-
stage sampling are also indicated. His investigation stresses
the role of the properties of the sampling design in yielding
non-negative estimators of MSE and de-emphasizes the need for
tedious manipulations involving the variate-values as done by
several earlier authors. However, the relative merits of his
alternative non-negative estimators, when available, are not yet
sufficiently well known.

For the ratio estimator based on the LMS scheme, four
different estimators have recently been proposed by T.J. Rao
(1972, 1977) and Chaudhuri (1976). For each of them, a sufficient
condition for non-negativity has been presented. Unfortunately,
none of them is of the necessary form given by J.N.K. Rao (1979).
Moreover, it may be verified (J.N.K. Rao and Vijayan, 1977;
Chaudhuri and Arnab, 1979) that these sufficient conditions
cannot hold either at all or except in the trivial case when all
size measures are equal. The same result has been observed with
respect to the estimators of variance due to Sharma (1970) and
Banduopadhyay, Chattopadhyay and Kundu (1977) (see Chaudhuri and
Arnab, 1979). Rao and Vijayan (1977) identified two distinct

estimators for the variance of the ratio estimator based on the
LMS sampling scheme, when the sample size n exceeds 2 , both
satisfying Rao's (1979) condition for the necessary form of non-
negative quadratic unbiased estimators. They also demonstrated
that for n = 2 , at most one non-negative unbiased estimator is
available. They also gave sufficient conditions for their non-
negativity, studied empirically the probability of their being
non-negative and investigated their stability.

     For the HTE based on any fixed-size design it follows from
Rao's (1979) general result that the estimator for the variance
V(HTE) due to Yates and Grundy (1953), to be called v_{YG}
satisfies his necessary form (2.1). Vijayan (1975) demonstrated
the necessary form of a uniformly non-negative unbiased quadratic
estimator for the variance of the HTE based on a fixed sample
size design and found the necessary and sufficient condition on
a design based on n = 2 to yield such an estimator. He further
showed that if a NNUQVE (non-negative unbiased quadratic variance
estimator) is to exist it must be v_{YG} . Vijayan also pointed
out that the HTE of V(HTE) , to be called v_{HT} , for any design
is capable of assuming negative values.

     Sufficient conditions for non-negativity of v_{YG} are
well-known and several sampling schemes satisfying them have been
identified   (see Sen, 1953; Raj, 1956; J.N.K. Rao, 1962,63,65;
Seth, 1966 and Prabhu-Ajgaonkar,1967, among others). A well-known
sufficient condition is

$$\pi_{ij} \leq \pi_i \pi_j \quad \forall\ i,j \tag{2.2}$$

where π_i , π_{ij} are the first and second order inclusion probabilities. Lanke (1974) showed that (2.2) is satisfied for every
rejective sampling scheme (where a sample of size n , taken by
PPSWR, is retained provided all units are distinct and discarded
otherwise). For successive sampling (where units are drawn one
by one using PPSWR until a stipulated number n of distinct units

is realized), this holds if $n = 2$ but may not hold if $n > 2$; still v_{YG} may be positive for such a scheme. In fact in general for $n > 2$, (2.2) is neither necessary for the non-negativity of v_{YG} nor for the existence of any other non-negative, unbiased variance estimator (NNUVE) for $V(HTE)$, as observed by Lanke (1974). He also showed that in case $n = 2$ and $\pi_{ij} > 0$, (2.2) is a necessary and sufficient condition for the existence of an NNUE (the estimator need not be necessarily quadratic as he observed).

Murthy's (1957) estimator, however, has a uniformly NNUVE (Pathak and Shukla, 1966) provided it is based on PPSWOR scheme; if one bases it on a different sampling scheme, a variance estimator may not necessarily be positive (noting, in particular that the Murthy estimator for the LMS scheme is just the ratio estimator). For estimating a finite population variance, Liu (1974) proposed several unbiased estimators and Chaudhuri (1978) added a few more and supplied conditions on their non-negativity.

3.   SOME RESULTS FOR UNI-STAGE SAMPLING

We will write y_i, x_i $(i = 1,\ldots,N)$ respectively for the variate values, s for a sample of size n from a population U of size N ,

$$y_s = \sum_{i \in s} y_i \ , \quad x_s = \sum_{i \in s} x_i \ , \quad Y = \sum y_i \ , \quad X = \sum x_i \ ,$$

$$P_i = x_i/X \ , \quad P_s = \sum_{i \in s} P_i \ , \quad M_r = \binom{N-r}{n-r}, \quad r = 0,1,2 \ ;$$

$$a_{ij} = (y_i/P_i - y_j/P_j)^2 P_i P_j$$

and $P(s) = (1/M_1)P_s$. Then, for the LMS scheme, $P(s)$ is the selection probability for s and the ratio estimator $t = X\, y_s/x_s$ (which is a special case of the general estimator $\hat{Y} = \sum d_{is} y_i$) is unbiased for Y with variance which we may write as

$$V(t) = \sum_{i<j} \sum a_{ij} (1 - \frac{1}{M_1} \sum_{s \ni i,j} \frac{1}{P_s}) \ .$$

For any fixed (not depending on s) constants α, β in [0,1], we have an unbiased estimator for $V(t)$ as

$$\hat{V}_1(t) = \sum_{i<j\in s} \sum a_{ij} \left[\{ \frac{\alpha}{\pi'_{ij}} + \frac{M_1(1-\alpha)}{M_2 P_s} \} - \{ \frac{M_1 \beta}{M_2 P_s} + \frac{1-\beta}{\pi'_{ij}} \} \frac{1}{M_1} \sum_{s \ni ij} \frac{1}{P_s} \right]$$

where $\pi'_{ij} = ab + ac(p_i + p_j)$, $i,j = 1,\ldots,N$, $(i \neq j)$ are the 2nd order inclusion probabilities for LMS scheme, with

$a = (n-1)/(N-1)$, $b = (n-2)/(N-2)$, $c = (N-n)/(N-2)$.

For the most natural choices $(\alpha,\beta) = (0,0)$ and $(\alpha,\beta) = (1,1)$ respectively, $\hat{V}_1(t)$ takes the forms

$$\hat{V}_{11}(t) = \sum_{i<j\in s} \sum a_{ij} (\frac{M_1}{M_2 P_s} - \frac{1}{M_1 \pi'_{ij}} \sum_{s \ni ij} \frac{1}{P_s})$$

and

$$\hat{V}_{12}(t) = \sum_{i<j\in s} \sum a_{ij} (\frac{1}{\pi'_{ij}} - \frac{1}{M_2 P_s} \sum_{s \ni ij} \frac{1}{P_s}) \ .$$

Sufficient non-negative conditions for these estimates are

$$x_s \leq \frac{1}{a} (\sum_{s \ni i,j} x_s) (\sum_{s \ni i,j} \frac{1}{x_s}) \quad \forall \ s \qquad\qquad (3.1)$$

and

$$x_s \geq \frac{1}{M_1 M_2} (\sum_{s \ni ij} x_s) (\sum_{s \ni ij} \frac{1}{x_s}) \quad \forall \ s \qquad\qquad (3.2)$$

respectively. For the two estimators for $V(t)$ proposed by J.N.K. Rao and Vijayan (1977), namely,

$$\hat{V}_2(t) = \sum_{i<j\in s} \sum a_{ij} \frac{1}{P_s} (\frac{1}{a} - \frac{1}{P_s})$$

and

$$\hat{V}_3(t) = \sum_{i<j\epsilon s} \sum \frac{a_{ij}}{\pi_{ij}'}(1 - \frac{1}{M_1}\sum_{s \ni i,j}\frac{1}{P_s}) ,$$

a common sufficient condition for non-negativity is

$$x_s \geq Xa \quad \forall \ s . \tag{3.3}$$

Apparently, (3.1) and (3.2) should obtain more frequently than (3.3) and they are seemingly less stringent. Again, in case (3.3) holds, another unbiased estimator for V(T) , namely

$$\hat{V}_4(t) = \frac{N-1}{n-1}\frac{1}{P_s}\sum_{i<j\epsilon s}\sum a_{ij}(1 - \frac{1}{M_1}\sum_{s \ni i,j}\frac{1}{P_s})$$

is also non-negative.

Regarding efficiency, it is not easy to discriminate among these five estimators but $\hat{V}_2(t)$ is the simplest to use. Incidentally, for the variance $V(\bar{e})$ of the HTE $\bar{e} = \sum_{i\epsilon s} y_i/\pi_i$, one may propose the estimator

$$\hat{V}(\bar{e}) = \sum_{i<j\epsilon s}(\frac{y_i}{\pi_i} - \frac{y_j}{\pi_j})^2\left[\pi_i\pi_j\{\frac{\alpha}{P(s)M_2} + \frac{1-\alpha}{\pi_{ij}}\} \right.$$

$$\left. - \pi_{ij}\{\frac{\beta}{P(s)M_2} + \frac{1-\beta}{\pi_{ij}}\}\right]$$

for fixed constants α,β in [0,1] with P(s) denoting the selection probability of a sample s of n units. For the natural choices $\alpha = 0$, $\beta = 1$, and $\alpha = 1$, $\beta = 0$ respectively a sufficient condition for non-negativity of $\hat{V}(\bar{e})$ is respectively

$$P(s) > (\pi_{ij}/\pi_i\pi_j)\pi_{ij}/M_2 \quad \forall \ s,(i,j)$$

and

$$P(s) < \pi_i\pi_j/M_2 \quad \forall \ s,(i,j) .$$

Clearly, the positivity of a variance estimator here depends on parameters of sampling designs. Hence, in the case of HTE, one may choose a design satisfying the requisite conditions. In the case of ratio estimator based on LMS scheme, however, one may not, because the size measures are fixed and beyond control on every particular situation. Thus, it is possible that for a sample drawn and surveyed each unbiased variance estimate tried out may turn out negative. In such a case one may proceed as follows.

Let s_n be a sample of n distinct units actually chosen with probability $p(s_n)$ and the data surveyed be $d_n = (s_n, Y_i : i \in s_n)$, $t = t(d_n)$ an unbiased estimate for Y, $V(t)$ its variance and v an unbiased estimate for $V(t)$ based on d_n.

Let a subsample s_{n-1} of $n-1$ distinct units be chosen from s_n with conditional probability

$$P_c(s_{n-1} \mid s_n) = [1/(n-1)](y_{s_{n-1}}/y_{s_n}) .$$

Then, $e = e(d_n) = (y_{s_n}/y_{s_{n-1}})t$ is unbiased for Y. Since

$$V(e) = VE(e \mid d_n) + EV(e \mid d_n) = V(t) + E(w) ,$$

where

$$w = t^2[y_{s_n}\{1/(n-1)\} \sum_{s_{n-1} \subset s_n} (1/y_{s_{n-1}}) - 1] > 0 ,$$

(since w is a conditional variance) e is less efficient than t but $z = v+w$ is an unbiased estimate for $V(e)$ and, w being necessarily positive, z may yet be positive even if v be negative. Repeating the process of sub-sampling from s_{n-1}, one may consider a sequence of such estimates (of decreasing efficiency) and their unbiased variance estimates. In practice, one may decide to choose one of them for which the variance estimate is positive for the data at hand. However, this involves a preliminary test for positivity from observed data and hence destroys unbiasedness in a strict sense.

In the particular case $t = X y_{s_n}/x_{s_n}$ for the LMS scheme, one may, following Lanke (1976), choose the undernoted additional alternative course. Let $P_c(s_{n-1}|s_n) = [1/(n-1)]P_{s_{n-1}}/P_{s_n}$; then for $\bar{e} = y_{s_{n-1}}/P_{s_{n-1}}$, we have $E(\bar{e}|d_n) = t$ and

$$V(\bar{e}) = V(t) + EV(\bar{e}|d_n)$$

$$= V(t) + \{1/[(n-1)P_{s_n}]\} \sum_{s_{n-1}\subset s_n} (y^2_{s_{n-1}}/P_{s_{n-1}})\}$$

$$- (y_{s_n}/P_{s_n})^2]$$

$$= V(t) + E(w)$$

say, where

$$w = (1/P_{s_n})[\sum_{j=1}^{n}\{1/(P_{s_n} - P_j)\}\{y_j-(y_{s_n}/P_{s_n})P_j\}^2] > 0$$

and $z = v+w$ is unbiased for $V(\bar{e})$ and w is positive so that z may also be positive even if v may be negative. More importantly, the marginal distribution of s_{n-1} is $q(s_{n-1})$ with

$$q(s_{n-1}) = \sum_{s_n \supset s_{n-1}} P_c(s_{n-1}|s_n)P(s_n) = [1/(n-1)] \sum_{s_n \supset s_{n-1}} (P_{s_{n-1}}/P_{s_n})$$

$$\cdot (P_{s_n}/M_1)$$

$$= P_{s_{n-1}} / \binom{N-1}{n-1} \ .$$

Thus, $q(s_{n-1})$ corresponds to Lahiri's version of LMS scheme (with sample size $n-1$ instead of n). Hence, $V(\bar{e})$ can be expressed in the same form as $V(t)$ because $\bar{e}$ is simply the

ratio estimator for LMS scheme for a sample of n-1 units.
$V(\bar{e})$ may be unbiasedly estimated directly by u (say) which may
be one of the estimators of the type v considered above but of
a size n-1 . With z, u is also a possible candidate for
selection in case it is positive.

We have so far considered only uni-stage sampling. With
multi-stage sampling additional difficulties concerning non-
negativity in variance estimation arise. Two well-known general
unbiased variance estimators for multi-stage sampling are due to
Raj (1966) and J.N.K. Rao (1975), of which the latter is more
generally applicable. Treatment of such general situations is
not easy and a further discussion on the topic is now postponed.

ACKNOWLEDGEMENT

The author gratefully acknowledges the fruitful discussion
he had with Professor J.N.K. Rao in preparing this draft following
a revision of an earlier version presented at the symposium.
This work was done while visiting Carleton University, April -
June, 1980 and supported by a grant from the Natural Sciences and
Engineering Research Council of Canada.

REFERENCES

Banduopadhyay, S., Chattopadhyay, A.K. and Kundu, S. (1977). On
    estimation of population total. *Sankhyā*, C39, 28-42.

Chaudhuri, A. (1976). A non-negativity criterion for a certain
    variance estimator. *Metrika*, 23, 201-205.

Chaudhuri, A. (1978). On estimating the variance of a finite
    population. *Metrika* 25, 65-76.

Chaudhuri, A. and Arnab, R. (1980). On unbiased variance estima-
    tion with various multi-stage sampling strategies. *Metrika*
    27 (in press).

Horvitz, D.G., and Thomson, D.J. (1952). A generalization of
    sampling without replacement from a finite universe.
    J. Amer. Statist. Ass. 47, 663-685.

Lahiri, D.B. (1951). A method of sample selection for providing
    unbiased ratio estimates. Bull. Int. Statist. Inst. 33,
    133-140.

Lanke, J. (1974). On non-negative variance estimators in survey
    sampling. Sankhyā C36 , 33-42.

Lanke, J. (1976). Another way to look at the jackknife with an
    application in survey sampling. Private communication.

Liu, T.P. (1974). A general unbiased estimator for the variance
    of a finite population. Sankhyā C36, 23-32.

Midzuno, H. (1952). On the sampling system with probability
    proportional to sum of sizes. Ann. Inst. Statist. Math. 3,
    99-107.

Murthy, M.N. (1957). Ordered and unordered estimators in sampling
    without resplacement. Sankhyā 18, 379-390.

Pathak, P.K. and Shukla, N.D. (1966). Non-negativity of a
    variance estimator. Sankhyā A28, 41-46.

Prabhu-Ajgaonkar, S.G. (1967). Unbiased estimators of the
    variances of the Narain, Horvitz and Thompson estimator.
    Sankhyā, A29, 55-60.

Raj, D. (1956). Some estimators in sampling with varying
    probabilities without replacement. J. Amer. Statist. Ass.
    51, 269-284.

Raj, D. (1966). Some remarks on a simple procedure of sampling
    without replacement. J. Amer. Statist. Ass. 61, 391-396.

Rao, J.N.K. (1962). On the estimate of the variance in unequal
    probability sampling. Ann. Inst. Statist. Math. 13, 57-60.

Rao, J.N.K. (1964). On two systems of probability sampling with-
    out replacement, Ann. Inst. Statist. Math. 15, 67-72.

Rao, J.N.K. (1965). On two simple schemes of unequal probability
    sampling without replacement, J. Ind. Statist. Ass. 3,
    173-180.

Rao, J.N.K. (1975). Unbiased variance estimation for multi-stage
    designs. Sankhyā C37, 133-139.

Rao, J.N.K. (1979). On deriving mean square errors and their non-negative unbiased estimators in finite population sampling. *J. Ind. Statist. Ass.* 17, 125-136.

Rao, J.N.K. and Vijayan, K. (1977). On estimating the variance in sampling with probability proportional to aggregate size. *J. Amer. Statist. Ass.* 72, 579-584.

Rao, T.J. (1972). On the variance of the ratio estimator for the Midzuno-Sen sampling scheme. *Metrika* 18, 209-215.

Rao, T.J. (1977). Estimating the variance of the ratio estimator for the Midzuno-Sen sampling scheme. *Metrika* 24, 203-208.

Sen, A.R. (1953). On the estimate of the variance in sampling with varying probabilities. *J. Ind. Soc. Agri. Statist.* 5, 119-127.

Seth, G.R. (1966). On estimators of variance of estimate of population total in varying probabilities. *J. Ind. Soc. Agri. Statist.* 18, 52-56.

Sharma, S.S. (1970). On an estimation in T_3-class of linear estimators in sampling with varying probabilities from a finite population. *Ann. Inst. Statist. Math.* 22, 495-500.

Vijayan, K. (1975). On estimating the variance in unequal probability sampling, *J. Amer. Statist. Ass.*, 70, 713-716.

Yates, F. and Grundy, P.M. (1953). Selection without replacement from within strata with probability proportional to size, *J. Roy. Statist. Soc.* B15, 253-261.

VARIANCE ESTIMATORS FOR A
SEQUENTIAL SAMPLE SELECTION PROCEDURE

James R. Chromy

Research Triangle Institute

The concept of probability minimum replacement
(PMR) sampling is defined. A sequential procedure
for PMR sampling is described along with associated
estimators and their variances. An unbiased estimator
of variance and some alternative biased estimators of
variance are proposed and discussed.

1. BACKGROUND

Large surveys routinely employ deeply stratified multi-stage sample designs with probability proportional to size (PPS) selection at the first stage. Hansen, Hurwitz & Madow (1953) discuss the desirability of equal stratum sizes for PPS sample designs with one or two units per stratum as well as the handling of large primary sampling units (units larger than the average stratum) as certainty strata or self-representing primary sampling units. It should be noted that when the sampling frame consists of unequal-sized units, it is rarely possible to exactly equalize the size of strata. Therefore, intended PPS sampling schemes succeed in being PPS only within strata.

The problem of equalizing stratum sizes is solved by Kish (1965) by using implicit strata or zones. The method can be applied to PPS selection most simply by utilizing systematic sampling with fractional intervals as suggested by Madow (1949)

and modified by Hartley & Rao (1962); it can also be applied with
more general PPS designs including both probability replacement
(PR) and probability nonreplacement (PNR) designs. When zones
are used with general PR or PNR designs, certain primary sampling
units straddle the stratum boundaries. These boundary-straddling
units may become selected from two different strata even when a
one unit per stratum PR design is used.

Sequential methods of probability sample selection are
discussed for equal probability selection designs by Fan, Muller
& Rezucha (1962). A sequential selection method applicable to
unequal probabilities of selection is given by Sunter (1977).
Sunter's method sometimes requires the sampler to change some
selection probabilities or to let the sample size vary.

The selection method proposed and discussed in this paper
has the following properties.

(1)   Exact PPS selection is achieved for arbitrary unequal
       size measures.

(2)   The number of selections n is fixed.

(3)   The selection scheme is probability minimum
       replacement (PMR).

The concept of PMR selection is defined in Section 2. The selection algorithm and associated unbiased estimation procedures are
discussed in Section 3. Some alternative variance estimation
procedures are discussed in Section 4.

2.   DEFINITIONS

The following notation is used in this paper:

N = number of sampling units in the sampling frame;

i = sampling unit label;

$S(i)$ = size measure associated with sampling unit i;

$S(+) = \sum_{i=1}^{N} S(i)$;

n = total sample size; and

n(i) = number of times sampling unit i is selected.

Given this notation, a *PPS sampling design* is defined as any
sample design for which the expected number of selections of
sampling unit i is proportional to unit i's size measure, i.e.,
En(i) = nS(i)/S(+). Probability nonreplacement (PNR) sample
designs are restricted in application to sampling frames with
nS(i)/S(+) < 1 for all i = 1,...,N. When this condition holds,
a *PNR sample design* is one for which Pr{n(i) = 1} = nS(i)/S(+)
and Pr{n(i) = 0} = 1 - Pr{n(i) = 1}. Probability minimum replacement (PMR) sample designs allow any positive size measures S(i).
The standardized size measure may be written in terms of two
components as nS(i)/S(+) = m(i) + f(i) , where m(i) is a non-
negative integer and $0 \le f(i) < 1$. Given this notation, a *PMR
sample design* is one for which Pr{n(i) = m(i) + 1} = f(i) and
Pr{n(i) = m(i)} = 1 - f(i). Note that PNR sample designs are a
special case of PMR designs with all m(i) equal to zero.

3. THE SELECTION ALGORITHM

3.1 The Algorithm for a Fixed Ordering and a Fixed Start

Some additional notation facilitates discussion of the
selection algorithm. The partial sum of standardized size
measures will be written as

$$\sum_{j=1}^{i} nS(j)/S(+) = I(i) + F(i) ,$$

where I(i) is a nonnegative integer and $0 \le F(i) < 1$. Further,
let M(i) and $\overline{M}(i)$ denote the events that n(i) = m(i) + 1
and m(i) respectively. Similarly, let C(i) and $\overline{C}(i)$ denote
the events that

$$\sum_{j=1}^{i} n(j) = I(i) + 1$$

and I(i), respectively. By convention, we will take I(0) and
F(0) to be zero and set Pr[C(0)] = 0 and $Pr[\overline{C}(0)] = 1$.

The algorithm is applied by determining $n(1),\ldots,n(N)$ sequentially. Equivalently, the algorithm may be applied to determine the partial sums

$$n(1), \ldots, \sum_{j=1}^{N} n(j) .$$

In the latter case, the events $C(i)$ or $\bar{C}(i)$ are selected sequentially according to the conditional probabilities given in Table 1. (In the former case, the events $M(i)$ or $\bar{M}(i)$ are selected through the use of Table 2.)

Table 1. Conditional sequential probabilities for the event $C(i)$

Case	Deterministic Conditions	$Pr\{C(i)\|\bar{C}(i-1)\}$	$Pr\{C(i)\|C(i-1)\}$
(1)	$F(i) = 0$	0	0
(2)	$F(i) > F(i-1) \geq 0$	$[F(i) - F(i-1)]/[1-F(i-1)]$	1
(3)	$F(i-1) \geq F(i) > 0$	0	$F(i)/F(i-1)$

Table 2. Conditional sequential probabilities for the event $M(i)$

Case	Deterministic Conditions	$Pr\{M(i)\|\bar{C}(i-1)\}$	$Pr\{M(i)\|C(i-1)\}$
(1a)	$F(i)=0,\ F(i-1)=0$	0	0
(1b)	$F(i)=0,\ F(i-1)>0$	1	0
(2)	$F(i) > F(i-1) \geq 0$	$[F(i)-F(i-1)]/[1-F(i-1)]$	0
(3a)	$F(i-1) > F(i) > 0$	1	$F(i)/F(i-1)$
(3b)	$F(i-1) = F(i) > 0$	0	0

3.2 Properties of the Basic Design

It can be shown that sample designs generated by the selection algorithm have fixed sample size n, are PPS, and are PMR. The following lemma is required for the proofs:

Lemma 1. After each sequential sample selection step $i = 1,\ldots,N$, we have $\Pr[C(i)] = F(i)$ and $P[\bar{C}(i)] = 1-F(i)$.

Proof. Reference is made to Table 1 for values of the conditional sequential probabilities. Note that at each step

$$\Pr[C(i)] = \Pr[C(i)|C(i-1)]\Pr[C(i-1)] + \Pr[C(i)|C(i-1)]\Pr[C(i-1)].$$

Consider first $i=1$ and recall that by convention $F(0) = 0$, $\Pr[\bar{C}(0)] = 1$, and $\Pr[C(0)] = 0$. Either case 1, 2, or 3 (Table 1) must apply depending on $F(1)$. If case 1 $(F(1) = 0)$ applies, then

$$\Pr[C(1)] = \Pr[C(1)|C(0)]\,\Pr[C(0)] = 0 \cdot 1 = 0.$$

Since $F(1) = 0$ for case 1, the Lemma is satisfied. If case 2 $(F(i) > F(i-1) \geq 0)$ applies, then

$$\Pr[C(1)] = [F(1) - F(0)]/[1 - F(0)]\cdot 1 = F(1) ,$$

and the Lemma is again satisfied. Case 3 $(F(i-1) \geq F(i) > 0)$ cannot apply for $i = 1$, because $F(0)$ is identically zero by convention. At step 1, we have thus shown that $\Pr[C(1)] = F(1)$.

The remainder of the proof follows by induction. Choose any i greater than 1 and assume that, $\Pr[C(i-1)] = F(i-1)$ and $\Pr[\bar{C}(i-1)] = 1 - F(i-1)$. As before, either case 1, 2, or 3 of Table 1 must apply. If case 1 $(F(i) = 0)$ applies, it follows immediately that $\Pr[C(i)] = 0$ and the Lemma is satisfied. If case 2 $(F(i) > F(i-1) \geq 0)$ applies, then

$$\Pr[C(i)] = [F(i) - F(i-1)]/[1 - F(i-1)]\cdot[1-F(i-1)]+ 1\cdot F(i-1) = F(i).$$

In case 3 $(F(i-1) \geq F(i) > 0)$,

$$\Pr[C(i)] = 0\cdot[1-F(i-1)] + [F(i)/F(i-1)]\cdot F(i-1) = F(i) .$$

Thus, Lemma 1 is satisfied for all 3 cases, and the proof is complete.

To prove that the *sample size* is always equal to n , Lemma 1 can be applied with $i = N$. By definition $I(N) = n$,

and so F(N) = 0 with Pr[$\bar{C}$(N)] = 1 by Lemma 1. Since the
event $\bar{C}$(N) corresponds to

$$\sum_{j=1}^{N} n(j) = n ,$$

the algorithm must produce a sample of size n with probability 1.

To show that the selection algorithm generates a PMR
design, it is first necessary to establish that Table 1 and Table
2 describe the same algorithm.

Lemma 2. Application of either the conditional sequential
probabilities of M(i) in Table 2 or the conditional sequential
probabilities for C(i) in Table 1 always yields the same sample.

Proof. In order to establish the required result, we will
show that the conditional probabilities given in Tables 1 and 2
for the events C(i) and M(i) yield the same samples. The
following relationships between F(i) , F(i-1), m(i), and f(i)
are required.

Case number (Table 2)	m(i)	f(i)
1a	I(i)-I(i-1)	0
1b	I(i)-I(i-1)-1	1-F(i-1)
2	I(i) - I(i-1)	F(i) - F(i-1)
3a	I(i)-I(i-1)-1	1 - F(i-1) + F(i)
3b	I(i) - I(i-1)	0

Each of the m(i) and f(i) values are obtained by determining
the integer and fractional portions of the difference
[I(i) + F(i)] - [I(i-1) + F(i-1)].

In addition Table 1 outcomes can be related to the n(i)
as follows.

Table 1 outcome	Resulting n(i)
C(i) and $\bar{C}(i-1)$	I(i) - I(i-1) + 1
$\bar{C}(i)$ and $\bar{C}(i-1)$	I(i) - I(i-1)
C(i) and C(i-1)	I(i) - I(i-1)
$\bar{C}(i)$ and C(i-1)	I(i) - I(i-1) - 1

Recalling that the event M(i) corresponds to n(i) = m(i)+1 , the correspondence of events in Tables 1 and 2 is as follows.

Outcome at unit i-1	Table 1 outcome at unit i	Table 2 outcome at unit i by case number				
		1a	1b	2	3a	3b
$\bar{C}(i-1)$	C(i)	0	0	M(i)	0	0
$\bar{C}(i-1)$	$\bar{C}(i)$	$\bar{M}(i)$	M(i)	$\bar{M}(i)$	M(i)	$\bar{M}(i)$
C(i-1)	C(i)	--	0	$\bar{M}(i)$	M(i)	$\bar{M}(i)$
C(i-1)	$\bar{C}(i)$	--	$\bar{M}(i)$	0	$\bar{M}(i)$	0

A zero in the above table indicates that the Table 1 outcome at unit i occurs with probability zero; a dash indicates that the unit i-1 outcome cannot occur for values of F(i-1) specified by the case number. The proof of Lemma 2 is completed by examining Tables 1 and 2 in connection with the correspondences defined above.

Corollary to Lemma 2. For the sequential selection algorithm with conditional probabilities for the events C(i) given in Table 1, the conditional probabilities for the events M(i) are as stated in Table 2. If the event M(i) does not occur, the only other possible event $\bar{M}(i)$ is that n(i) = m(i) with $Pr[\bar{M}(i)] = 1 - Pr[M(i)]$.

Lemmas 1 and 2 can be used to establish the PMR condition of the sample design. From Table 2, we get

$$Pr[M(i)] = Pr[M(i) \,|\, \bar{C}(i-1)] \, Pr[\bar{C}(i-1)] + Pr[M(i) \,|\, C(i-1)] Pr[C(i-1)].$$

Based on Lemma 1, this can be rewritten as

$$\Pr[M(i)] = \Pr[M(i)|\bar{C}(i-1)][1-F(i-1)] + \Pr[M(i)|C(i-1)]\ F(i-1).$$

The probability of the event M(i) can then be evaluated for each case in Table 2 as follows.

Case number	$\Pr[M(i)]$
1a	$0\cdot[1-F(i-1)]+0\cdot F(i-1) = 0$
1b	$1\cdot[1-F(i-1)]+0\cdot F(i-1) = 1-F(i-1)$
2	$[F(i)-F(i-1)]/[1-F(i-1)]\cdot[1-F(i-1)]+ 0\cdot F(i-1)$
	$= F(i)-F(i-1)$
3a	$1\cdot[1-F(i-1)]+F(i)/F(i-1)\cdot F(i-1)=1-F(i-1)+F(i)$
3b	$0\cdot[1-F(i-1)]+0\cdot F(i-1) = 0$

Note that, from the initial step in the proof of Lemma 2, the values above correspond to f(i) as determined by setting $m(i) + f(i) = I(i) + F(i) = [I(i-1) + F(i-1)]$. Since $\Pr[\bar{M}(i)] = 1 - f(i)$ from the corollary to Lemma 2, the PMR property of the design if established.

In order for the design to be PPS with respect to a size measure S(i) given that it is PMR , it is only necessary to require that $m(i) + f(i) = nS(i)/S(+)$ and to note that

$$En(i) = m(i)\cdot\Pr[\bar{M}(i)] + [m(i)+1]\Pr[M(i)] = m(i)[1-f(i)]+ [m(i)+1]f(i)$$
$$= m(i) + f(i).$$

3.3 A Modification for Unbiased Variance Estimation

Unbiased variance estimation requires each pair of units to have a positive chance of appearing in the same sample. In terms of the notation in this paper, this requirement can be stated as En(i)n(j) > 0 for all $i \neq j$. This discussion assumes that f(i) is positive for at least some units i. The following modification to the basic procedure is proposed.

1. Develop an ordered sampling frame of N sampling units.
2. Select a unit with probability proportional to its size to receive the label 1.

3. Continue labeling serially to the end of the sampling frame.

4. Assign the next serial label to the first unit at the beginning of the list and continue until all sampling units are labeled.

5. Apply the sequential PMR sample selection algorithm starting with the sampling unit labeled 1.

With this modification, an unbiased variance estimator can be obtained for sample designs with

$$\sum_{i=1}^{N} f(i) \geq 2 .$$

3.4 Estimation

The population total T is defined as

$$T = \sum_{i=1}^{N} Y(i)$$

where Y(i) is an observed value for sampling unit i. The sequential PMR selection algorithm always produces n selections, although they may sometimes be associated with fewer than n unique values of i. The sample can be represented in terms of an array of n labels $i_1,...,i_n$. The general form of estimator of T suggested for this design is

$$t = \sum_{i=1}^{n} Y(i_h)/En(i_h) .$$

This estimator can also be written in terms of the random variables n(i) , with summation over all labels i as

$$t = \sum_{i=1}^{N} n(i)[Y(i)/En(i)] .$$

Upon taking the expectation of the random variables n(i) , it is seen that E(t) = T . Since the n(i) are the only random variables in the above expression,

$$V[t] = \sum_{i=1}^{N} [Y(i)/En(i)]^2 Var[n(i)] + \sum_{i \neq j}^{N} \sum^{N} [Y(i)/En(i)]$$

$$[Y(j)/En(j)]Cov[n(i),n(j)].$$

Note that if this sample design is also PNR (i.e., if $En(i) < 1$ for each $i = 1,2,\ldots,N$), then the estimator t and its variance correspond to those developed by Horvitz & Thompson (1952).

An alternate expression for the variance corresponding to the one developed for PNR sample designs by Yates & Grundy (1953) can be written as

$$V(t) = \sum_{i<j}^{N} \sum^{N} [En(i)En(j) - En(i)n(j)] \; D^2(i,j)$$

where $D(i,j) = [Y(i)/En(i) - Y(j)/En(j)]$. Using the notation $\pi(i) = Pr[M(i)]$ and $\pi(i,j) = Pr[M(i),M(j)]$, we may express $V(t)$ in the form

$$V(t) = \sum_{i<j}^{N} \sum^{N} [\pi(i)\pi(j) - \pi(i,j)] \; D^2(i,j) \; .$$

An unbiased estimator of the variance can be written as

$$\hat{V}(t) = \sum_{g<h}^{n} \sum^{n} [\pi(i_g)\pi(i_h) - \pi(i_g,i_h)]/[En(i_g)n(i_h)]D^2(i_g,i_h) \; .$$

This estimator can also be written in terms of the random variables $n(i)$ as

$$\hat{V}(t) = \sum_{i<j}^{N} \sum^{N} [n(i)n(j)]/[En(i)n(j)][\pi(i)\pi(j) - \pi(i,j)]D^2(i,j) \; .$$

Upon taking the expection of the products $n(i)n(j)$, it is seen that $E\hat{V}(t) = V(t)$. As noted earlier, unbiased variance estimation requires that $En(i)n(j) > 0$ for all $i \neq j$. Note that for PMR designs $\pi(i) = f(i)$, and $En(i) = m(i) + \pi(i)$. The values of $\pi(i,j)$ and $En(i)n(j)$ satisfy the relationship

$$En(i)n(j) = m(i)m(j) + m(i)\pi(j) + \pi(i)m(j) + \pi(i,j) \; .$$

Since $m(i)$, $m(j)$, $\pi(i)$, and $\pi(j)$ are known, $En(i)n(j)$ can be determined by determining $\pi(i,j)$. Note that for $\pi(i)$ or $\pi(j)$ equal to zero, $\pi(i,j)$ is also zero. An algorithm for computing $\pi(i,j)$ when $\pi(i)$ and $\pi(j)$ are both positive is outlined in the next section.

3.5 Algorithm for Computing $\pi(i,j)$

The general approach for determining (i_u,i_v) with $i_u \neq i_v$ involves the following steps:

1. Note that each sampling unit   i   can be the starting point for the algorithm with probability $S(i)/S(+)$.

2. For a particular start i_s , relabel all sampling units as follows:

$$k = i + 1 - i_s \qquad \text{for} \quad i \geq i_s ,$$

$$k = N + i + 1 - i_s \quad \text{for} \quad i < i_s ,$$

$$i^* = \text{Min}[k_u,k_v], \quad \text{and} \quad j^* = \text{Max}[k_u,k_v]$$

where k_u and k_v (as relabeled by k) correspond to i_u and i_v , respectively, as labeled originally by i.

3. For each start i_s , compute $\pi(i^*,j^*|i_s) = \pi(i_u,i_v|i_s)$.

4. $\pi(i_u,i_v) = \sum_{i=1}^{N} \pi(i_u,i_v|i)S(i)/S(+))$.

5. Note that if $f(i_u) = 0$ or $f(i_v) = 0$, then $\pi(i_u,i_v)=0$.

Step 3 above must be expanded to allow for each possible fixed start i_s . Note that all steps below are conditional on a fixed start i_s , even though the notation is abbreviated by omission of reference to i_s . In addition, it is assumed that $f(i^*)$ and $f(j^*)$ are both positive.

1. Assume labeling by $k = 1,2,\ldots,i^*-1,i^*,i^*+1,\ldots,j^*-1,j^*,$ $j^*+1,\ldots,N$. Note that $\pi(i^*,j^*) = \Pr[M(i^*),M(j^*)]$.

2. By the properties of the algorithm $\Pr[C(i^*-1)] = F(i^*-1)$, $\Pr[\bar{C}(i^*-1)] = 1 - F(i^*-1)$.

3. Compute $Pr[M(i^*),C(i^*)]$ and $Pr[M(i^*),\bar{C}(i^*)]$. From Tables 1 and 2, the following results are established.

Case	Deterministic Conditions	$Pr[M(i^*), C(i^*)]$
(1)	$F(i^*) = 0$	0
(2)	$F(i^*) > F(i^*-1) \geq 0$	$F(i^*) - F(i^*-1)$
(3a)	$F(i^*-1) > F(i^*) > 0$	$F(i^*)$
(3b)	$F(i^*-1) = F(i^*) > 0$	0

In each case, $Pr[M(i^*),\bar{C}(i^*)] = \pi(i^*)-Pr[M(i^*),C(i^*)]$.

4. For each k such that $i^* < k \leq j^*-1$, iteratively compute $Pr[M(i^*),C(k)]$ and $Pr[M(i^*),\bar{C}(k)]$. From Table 1, $Pr[M(i^*),C(k)] = Pr[C(k)|\bar{C}(k-1)] Pr[M(i^*),\bar{C}(k-1)]$ $+ Pr[C(k)|C(k-1)] Pr[M(i^*),C(k-1)]$ and $Pr[M(i^*),\bar{C}(k)] = \pi(i^*) - Pr[M(i^*),C(k)]$.

5. Compute $Pr[M(i^*),M(j^*)]$. From Table 2, $Pr[M(i^*),M(j^*)] = Pr[M(j^*)|\bar{C}(j^*-1)]Pr[M(i^*),\bar{C}(j^*-1)]$ $+ Pr[M(j^*)|C(j^*-1)]Pr[M(i^*),C(j^*-1)]$.

3.6 Comments on the Selection Algorithm and the Resulting Sample Designs

The algorithm as modified for unbiased variance estimation (Section 3.3) does not pose a serious limitation on effective stratification if a closed or circular ordering of the sampling units can be achieved. Computerized methods for achieving such orderings with multiple stratification variables are discussed by Williams & Chromy (1980).

The choice of a PPS selection of a start unit in step 2 (Section 3.3) is arbitrary. Other probability assignment schemes may exist which yield designs with second order estimability and may require less computation.

Once the ordering and the start point are established, the selection procedure is implemented quite simply by determining the cumulative sample counts in a sequential manner. The PMR property of the resulting design eliminates the need to pre-identify and remove certainty or self-representing sample units. Estimates of the population total and of the variance of this estimate also require no special formulations for certainty units.

It appears reasonable to expect that reductions in variance usually associated with systematic, stratified, or zone sampling designs from meaningfully ordered lists should also accrue to the proposed sequential PMR design. The proposed design shares several properties with PPS systematic sampling using fractional intervals (Madow, 1949). Although such systematic sample designs are also PMR, no unbiased estimates of variance are possible.

Hartley & Rao's (1962) method utilizes a random ordering of the sampling frame before applying the PPS systematic sampling scheme. This permits unbiased variance estimation, but forsakes any potential gains from meaningful ordering of the list. The sampling method proposed in this paper preserves the frame ordering (and associated gains) by randomizing the starting point only. It also differs from Hartley and Rao's procedure because the actual selection method is more like a one-unit-per-stratum stratified sampling procedure than like a fixed-interval systematic sampling procedure.

4. SOME VARIANCE ESTIMATION ALTERNATIVES

4.1 General Comments

Although the ability to estimate variance is stated as a desirable property of the proposed design, other properties of the variance estimator must also be considered. This paper presents only some initial impressions concerning the unbiased estimator and some computational alternatives. Since PPS designs are often used at the first stage of multi-stage samples, the control of overall estimation bias (for all sampling stages) with a computationally feasible and implementable formula must be a major consideration in evaluating alternative variance estimation procedures.

4.2 Unbiased Single-Stage Estimator

The formula for the unbiased variance estimator with a
single-stage sample design is presented in Section 3.4. Based
on limited empirical work, it appears that the unbiased estimate
of variance will always be nonnegative. Example 1 below provides
a simple example for which certain feasible samples yield estimates
of variance which are identically equal to zero when based on the
unbiased estimator.

In order to study the behavior of $V(t)$ and $\hat{V}(t)$ under
PNR designs, we introduce the notation

$$\hat{V}_1(t) = \sum_{g<h}^{n} \sum^{n} [1-\gamma(i,j)]/\gamma(i,j)D^2(i,j)$$

where $\gamma(i,j) = \pi(i,j)/[\pi(i)\pi(j)]$. Note that for PNR designs
$En(i)n(j) = \pi(i,j)$.

Example 1. Consider an equal probability nonreplacement
sample of $n = 3$ sampling units from a sampling frame of $N = 9$
sampling units. Note that $En(i) = 1/3$ for all i. The behavior
of $\gamma(i,j)$ and $[1-\gamma(i,j)]/\gamma(i,j)$ for this example can be shown
in a matrix with rows indexed by i and columns by j. The
$\gamma(i,j)$ then appear in the upper right triangle of the below
matrix ;   the terms $[1-\gamma(i,j)]/\gamma(i,j)$ appear in the lower left
triangle.

-	1/3	2/3	1	1	1	1	2/3	1/3
2	-	1/3	2/3	1	1	1	1	2/3
1/2	2	-	1/3	2/3	1	1	1	1
0	1/2	2	-	1/3	2/3	1	1	1
0	0	1/2	2	-	1/3	2/3	1	1
0	0	0	1/2	2	-	1/3	2/3	1
0	0	0	0	1/2	2	-	1/3	2/3
1/2	0	0	0	0	1/2	2	-	1/3
2	1/2	0	0	0	0	1/2	2	-

Note that only units within 1 or 2 units of each other in the
closed ordering can provide any positive contribution to the variance or to the estimate of variance. Feasible samples under the
design, however, include the samples [1,4,7], [2,5,8], and [3,6,9].

The unbiased variance estimator yields variance estimates which are identically zero for all three of these samples.

In general, equal probability samples generated by the proposed algorithm with $k = N/n$ (an integer) produce variance estimates which are identically zero for $100(1/k)^{n-1}$ percent of the samples selected. When k is not an integer, a more regular behavior results, but some variance estimates will also necessarily be extremely small (rather than zero) regardless of the data. Example 2 illustrates this behavior.

Example 2. Consider an equal probability nonreplacement sample of $n = 3$ sampling units selected from a sampling frame of $N = 10$ units. The same matrix representation for $\gamma(i,j)$ and $[1-\gamma(i,j)]/\gamma(i,j)$ is shown in the following matrix (all entries have been rounded to two decimal places).

–	.33	.63	.90	.98	.98	.98	.90	.63	.33
2.02	–	.33	.63	.90	.98	.98	.98	.90	.63
.60	2.02	–	.33	.63	.90	.98	.98	.98	.90
.11	.60	2.02	–	.33	.63	.90	.98	.98	.98
.02	.11	.60	2.02	–	.33	.63	.90	.98	.98
.02	.02	.11	.60	2.02	–	.33	.63	.90	.98
.02	.02	.02	.11	.60	2.02	–	.33	.63	.90
.11	.02	.02	.02	.11	.60	2.02	–	.33	.63
.60	.11	.02	.02	.02	.11	.60	2.02	–	.33
2.02	.60	.11	.02	.02	.02	.11	.60	2.02	–

In this example, none of the variance estimator weights go to zero although some get very small.

The general property of $\gamma(i,j)$ approaching 1 as i and j diverge appears with other more general examples including those with unequal size measures. The fact that the unbiased estimator sometimes yields variance estimates which are identically zero provides the major motivation for considering other estimators of variance.

4.3 Unbiased Multi-Stage Estimators

The estimator for T in multi-stage designs can be written as

$$t = \sum_{h=1}^{n} \hat{Y}(i_h)/En(i_h) \ ,$$

where $\hat{Y}(i_h)$ is an estimate of $Y(i_h)$ based on second and subsequent stages of sampling. The variance of t can be written as

$$V(t) = \sum_{i<j}^{N} \sum^{N} [\pi(i)\pi(j)-\pi(i,j)]D^2(i,j) + \sum_{i=1}^{N} En(i)V[\hat{Y}(i)/En(i)]$$

$$= V_1 + \sum_{i=1}^{N} En(i)V_2(i).$$

An unbiased estimator of the variance of t is given by

$$\hat{v}_2(t) = \sum_{g}^{n}\sum_{h}^{n} [\pi(i_g)\pi(i_h)-\pi(i_g,i_h)]/[En(i_g)n(i_h)]\hat{D}^2(i_g,i_h)$$

$$+ \sum_{h=1}^{n} En(i_h)\hat{V}[\hat{Y}(i_h)/En(i_h)]$$

$$= v_1 + \sum_{h=1}^{n} En(i_h)v_2(i) \ ,$$

with v_1 and $v_2(i)$ having the obvious definitions and $\hat{D}(i,j) = \hat{Y}(i)/En(i)-\hat{Y}(j)/En(j)$.

Some of the comments regarding the single-stage estimator also apply to the multi-stage estimator. Since

$$Ev_1 = V_1 + \sum_{i=1}^{N} En(i)V_2(i) - \sum_{i=1}^{N} [En(i)]^2 V_2(i) \ ,$$

very small or zero values computed for v_1 for certain samples will also produce small estimated values for the second component of the variance of t regardless of the sample data.

4.4 An Assumed Replacement Sampling Estimator

A simpler estimator of the variance of t obtains if one assumes replacement sampling at the first stage. This estimator can be written as

$$\hat{V}_3(t) = 1/(n-1) \sum_{g<h}^{n} \sum^{n} \hat{D}^2(i_g, i_h) \ .$$

This estimator properly represents the contribution to variance from second and subsequent stages of sampling and conservatively overestimates for the first term of the variance. It is more appropriate for multi-stage designs than for single-stage designs.

4.5 Successive Difference Estimator

The successive difference estimator has been suggested for use with systematic samples (Kish, 1965). It also appears to be a close approximation to the variance of t based on the proposed PMR sequential design. The estimator can be written as

$$\hat{V}_4(t) = \frac{1}{2} \sum_{i=1}^{n-1} \hat{D}^2(i,i+1) + \frac{1}{2} \hat{D}^2(1,n) \ .$$

4.6 Collapsed Stratum Estimator

The collapsed stratum estimator is used as a conservative approximation with one unit per stratum stratified designs (e.g., Hansen, Hurwitz & Madow, 1953). For n even, it may be written as

$$\hat{V}_5(t) = \sum_{k=1}^{n/2} \hat{D}^2(2k, 2k-1) \ .$$

We can also write the collapsed stratum estimator as

$$\hat{V}_6(t) = \sum_{k=1}^{\frac{n-2}{2}} \hat{D}^2(2k, 2k+1) + \hat{D}^2(1,n)$$

based on the other possible collapsing of adjacent strata.

If we then choose $\hat{V}_5(t)$ or $\hat{V}_6(t)$ with equal probability, the result is an unbiased estimate of $\hat{V}_4(t)$. Under these conditions the expected value of the collapsed stratum estimator should be the same as that of the successive difference estimator. The successive difference estimator would be expected to be more stable.

One advantage of the collapsed stratum estimator is that it can be implemented with many available software packages for survey data analysis including those which utilize Taylor series, jackknife, or balanced repeated replication (BRR) techniques for handling nonlinear estimators (e.g., Kish & Frankel, 1970; Shah, 1980).

REFERENCES

Fan, C.T., Muller, M. E. & Rezucha, I. (1962). Development of sampling plans by using sequential (item by item) selection techniques and digital computers. *Journal of the American Statistical Association* 57, 387-402.

Hansen, M. H., Hurwitz, W. N. & Madow, W. G. (1953). *Sampling Survey Methods and Theory*. John Wiley & Sons, New York.

Hartley, H.O. & Rao, J. N. K. (1962). Sampling with unequal probabilities and without replacement, *Annals of Mathematical Statistics* 33, 350-374.

Horvitz, D. G. & Thompson, D. J. (1952). A generalization of sampling without replacement from a finite universe. *Journal of the American Statistical Association* 47, 663-685.

Kish, L. (1965). *Survey Sampling,* John Wiley & Sons, New York.

Kish, L. & Frankel, M. R. (1970). Balanced repeated replications for standard errors, *Journal of the American Statistical Association* 65, 1071-1094.

Madow, W. D. (1949). On the theory of systematic sampling II. *Annals of Mathematical Statistics* 24, 101-106.

Shah, B. V. (1980). SESUDAAN: Program for computing standard errors of standardized rates from sample survey data. Research Triangle Institute, Research Triangle Park, North Carolina.

Sunter, A. B. (1977). List sequential sampling with equal or unequal probabilities without replacement. *Applied Statistics* 26, 261-268.

Williams, R. L. & Chromy, J. R. (1980). SAS sample selection MACROS. *Proceedings of the Fifth Annual SAS Users Group International Conference.* 392-396.

Yates, F. & Grundy, P. M. (1953). Selection without replacement from within strata with probability proportional to size. *Journal of the Royal Statistical Society* B15, 235-261.

MAIL SURVEYS OF MIGRATORY GAME BIRDS IN NORTH AMERICA

A. R. Sen

Environment Canada

Two national surveys on migratory game birds
are annually conducted in U.S.A. and Canada. These
are (a) a mail questionnaire survey of waterfowl
hunters to measure size and distribution of kill
and (b) a species composition survey to estimate
species, age and sex composition of kill.

The present paper will briefly review the design
of the mail surveys and important sources of errors.
Both in U.S.A. and Canada the estimates of kill showed
marked skewness from normality. A method will be
described which resulted in approximate normality
and in considerable increase in precision of the
estimates of error.

1.   INTRODUCTION

Migratory game birds surveys are conducted annually in
Canada and in the U.S.A. to provide reliable estimates of kill and
activity, by geographical area and to assist management in the
setting of regulations. The surveys are the best available method
we have for estimating hunter kill which is a major component of
total annual waterfowl mortality. Migratory game bird harvest
estimates are useful (a) in making indirect preseason population
estimates, (b) in obtaining an index of band reporting rates,
(c) in providing information on the distribution of the recreation-
al opportunity due to hunting, (d) in determining characteristics
of waterfowl hunters for resource use and providing insight into

hunter behaviour, and (e) in providing research data needed for effective management of the resource.

Two large-scale surveys are being annually conducted in both the U.S.A. and Canada. These are (a) a mail questionnaire survey of waterfowl hunters to measure size and distribution of kill of ducks and geese by regions (provinces or states) and (b) a survey to collect duck wings and goose tails from birds bagged to estimate species, sex and age composition of the kill.

Methodological developments in the Canadian surveys have been discussed by Sen (1971, 1976), Smith (1975, 1978), Filion (1978, 1980) and Cooch et al. (1978). Developments in U.S. surveys have been described by Atwood (1956), Rosasco (1966) and Crissy (1967). We propose to describe these briefly and discuss a recent development in methodology which should provide more efficient estimates of the standard error of the kill estimates than being obtained at present.

2.   DESIGN OF THE MAIL SURVEYS

2.1   Canada

In 1966 it became obligatory for hunters, other than Indians and Eskimos, intending to shoot waterfowl in Canada to purchase a Canada Migratory game bird hunting permit before hunting in addition to such provincial licenses as may be necessary. The permits are numbered and the permit holder keeps one portion of the permit which has space for maintaining record on kill and activity; the other portion is sent to the Canadian Wildlife Service (CWS) headquarters in Ottawa where information on name, address, sex, whether the hunter purchased a permit during the previous season etc. is recorded on permit files. This information provides a sampling frame for the surveys.

After sorting the permit files by post office number, a

systematic random sample of hunters was selected from each of the
92 strata (23 geographical zones of Canada × 4 experience groups)
to be described later. Questionnaires are sent early in December
towards the end of the hunting season (September - December) to a
8-9 percent random sample of hunters (8.8 percent in 1976) with a
follow-up questionnaire to non-respondents in mid-January. Hunters
are asked about the number of ducks and geese taken and retrieved,
when and where taken and the days spent in shooting; approximately
1 in every 2 hunters contacted responds to the questionnaire.

The data collected from the questionnaire survey contained
two major sources of errors due to response and non-response. A
good deal of research has been done in Canada which are summarised
in Cooch et al. (1978) which tend to show that response bias may
lead to high exaggeration in kill estimates. No correction factors
could be developed to correct for response bias since we lack
national data to provide reliable estimates of bias. Several
changes have, however, been made in the questionnaire, in the
light of the studies which have reduced non-response and encouraged accurate reporting. During 1977-78 and shortly after the
mailing of the first questionnaire a postcard reminder was sent to
odd-numbered permits in the sample of permit holders in the harvest
survey. The even numbered permits in the sample served as the
control group. This raised the response rate by approximately
27 percent as shown by Couling and Smith (1980).

Before 1972, the sampling design consisted in selecting the
sample from the previous years (renewals) permits only. This
caused serious upward bias as demonstrated by Sen (1970) and was
estimated at 12.2 percent for Canada during 1975 by Cooch et al.
(1975). From 1972 the design was modified to include 'renewals'
as well as 'non-renewals'. Also, prior to 1972, only a hunter in
the sample was informed before the commencement of the season of
his being included in the survey and he received 'rules and
regulations' and a card for recording kill and activity. This

advance notice to members in the sample but not to others could
introduce bias and from 1972 onwards, all hunters, whether selected
for the survey or not were provided with record cards and hunters
in the sample were contacted only at the end of the season.

In 1972, the renewals within a zone were further stratified
into (i) hunters who bought a permit during the previous year and
(ii) who bought a permit during the preceding 2 years. The non-
renewals were stratified into (iii) resident hunters who did not
purchase a permit in the previous year and (iv) non-resident
hunters, usually Americans. In addition, an experimental sample
was selected during 1972 and 1973 surveys based on successive
sampling as given in Sen (1971) and Sen et al. (1975) for research
purposes and for use in simpler survey designs. In 1975 information from the 1970 survey was used to employ optimum allocation
(based on previous year's kill and sales) by strata within zones
which was aimed at providing estimates of provincial kill within
10 percent and national kill within 2.5 percent (with 95 percent
confidence).

The estimation of any subset of the hunting population e.g.,
kill or days hunted in a given area (A) is straight forward for
respondents from categories (iii) and (iv). Thus, if N_i be the
number of permits sold in the ith zone during the current year in
category (iii) say, and r_i the corresponding number of respondents who bagged k_{iA} birds in an area A , the estimate of total
birds shot in the area by hunters who bought permits in the ith
zone is $(N_i/r_i)k_{iA}$ where N_i/r_i is the weight assigned to each
respondent from the ith zone in the category. In particular A
may represent the ith zone.

In the renewal categories (i) and (ii) of a zone, some of
the hunters may not have purchased a permit or have changed zones
of purchase between years. If N_i be the number of permittees in
zone i of the category and r_i the number of respondents in the

sample during the preceding year, r_{ij} the number of respondents out of r_i buying permits in the jth zone during the current year, k_{ijA} the corresponding number of birds bagged in Area A, an estimate of total birds shot in the area by these hunters is given by

$$k_{ijA}(N_i/r_i)(M_j/\hat{M}_j) \quad \text{(Cooch et al. 1978)} \quad (1)$$

where $\hat{M}_j = \sum_i r_{ij}(N_i/r_i)$ and M_j is the actual number of permit holders in the jth zone during the current year by those who bought permits during the previous year. It would be seen that the weight assigned to each respondent in this case is $(N_i/r_i)(M_j/\hat{M}_j)$. The total kill in the area by all members of the category is obtained by summing (1) over i and j. Details for estimating standard errors of the estimates are given in Smith (1975).

2.2 United States

The United States Fish and Wildlife Service (USFWS) uses as its sampling frame a master list of post offices which sell duck stamps as cited by Rosasco (1966) and Crissy (1975). Unlike the Canadian system, the stamps are not numbered. Each state is stratified into geographical zones which, in turn, are further stratified into 3 groups of post offices (small, medium and large) based on the number of permits sold during the previous season. Basically, the plan is to select a stratified random sample of post offices each spring from each size group within a zone with allocation proportional to the duck stamp sales within a zone. The 1978 sample constituted 3 percent of the duck stamp purchasers in U.S.A. which provided flyway[1] estimates within 3% and estimates

---

[1] The U.S.A. is divided along the state lines into 4 bands running north and south called flyways.

for the entire U.S.A. within 1.7% (with 95% confidence). Each
person who purchased a duck stamp at one of those post offices is
requested to provide his name and address in one half of a special
card to the post clerk to be mailed to USFWS; the other half deals
with his participation in the survey and provides space for
recording kill and activity to facilitate in filling in the
questionnaire he would receive at the end of the season. The
practice of contacting purchasers in the sample at the beginning
rather than at the end of the season could lead to possible bias
in the estimates. By giving the name and address to the postal
clerk, the stamp purchaser places himself on the services' mailing
list. The number of addresses received may be quite low (35-40
percent) since, unlike the Canadian system, names and addresses
do not form part of the permit and stamp purchasers are not
required to provide them; also, postal clerks may forget at times
to hand over the cards to the duck stamp purchaser.

At the close of the hunting season, questionnaires are sent
to stamp buyers who provided the addresses; a follow-up is sent
to those who fail to return the questionnaire within about 3 weeks.
In all about 65 to 70 percent respond to the questionnaire though
the overall response rate would be much lower i.e., about 23 to
28 percent. The estimates of kill are assumed to be the same for
respondents and non-respondents which could be a dangerous
assumption. There is need for more work to examine the nature of
non-response bias and, in particular, at the address cards level.

Atwood (1956) demonstrated that there were serious upward
biases in responses to mailed questionnaires and developed adjustment factors to correct for the reported kill annually. The USFWS
has been using the 'correction factor' over years thus assuming
that response biases in hunting over years have not changed as
duck populations, regulations etc. have changed over years. This
assumption is untenable and there is need for fresh studies on
response bias to provide revised estimates of correction for bias

in the estimates in the light of changes in populations, regulations etc.

The data collected from the questionnaires are used to
estimate kill, hunting activity and success at the zonal, flyway
and national levels.

Let M be the number of hunters in the stratum known from
stamps sold during the season, y_i' the total reported kill in the
i^{th} post office of the stratum having N post offices, r_i the
total number of respondents and n the number of sample post
offices in the stratum. A biased but consistent estimator of the
population mean per hunter $(\bar{y})$ and of total kill (Y) are
respectively

$$\hat{\bar{y}} = \sum_{i=1}^{n} y_i' / \sum_{i=1}^{n} r_i \tag{2}$$

and

$$\hat{Y} = M\hat{\bar{y}} \tag{3}$$

An estimate of the approximate variance of $\hat{\bar{y}}$ is given by

$$v(\hat{\bar{y}}) = \frac{n(1 - \frac{n}{N})}{(\sum_{i=1}^{n} r_i)^2 (n-1)} (\sum_{i=1}^{n} y_i'^2 + \hat{\bar{y}}^2 \sum_{i=1}^{n} r_i^2 - 2\hat{\bar{y}} \sum_{i=1}^{n} y_i' r_i) \tag{4}$$

For estimating total kill in an area A (say) by members of the
population the results from the wing survey are used which provides
information on the number of birds taken, the time and place the
birds were taken by the hunter. We may recall that in the
Canadian survey this information is provided in the mail questionnaire itself. Let H_A and H_B be the estimated harvest of duck
stamp purchasers from zones A and B based on the mail survey
as given in (3) above and W_A, W_B the total birds in the sample
bagged by duck stamp purchasers in these zones from the wing
survey; also, let W_{AX}, W_{AY} be the number of wings received in

areas X and Y respectively by stamp purchasers in the sample in A and W_{BX}, W_{BY} the corresponding wings by stamp purchasers in B. Then, the total wings bagged in X by stamp purchasers in A and B are estimated by $(W_{AX}/W_A)H_A$ and $(W_{BX}/W_B)H_B$ respectively. Similarly, the estimates of total wings bagged in Y by stamp purchasers in A and B are respectively given by $(W_{AY}/W_A)H_A$ and $(W_{BY}/W_B)H_B$. Since the estimates W_{AX}/W_A and H_A are obtained from two separate surveys, there are independent of each other. Hence an unbiased estimate of variance of the estimators such as $(W_{AX}/W_A)H_A$ is given by

$$V(\frac{W_{AX}}{W_A} H_A) = p^2 v(H_A) + H_A^2 v(p) - v(H_A)\, v(p) \tag{5}$$

where $p = W_{AX}/W_A$ and $v(H_A)$ and $v(p)$ are unbiased estimates of $V(H_A)$ and $V(p)$.

Table 1. Use of wing collection and mail questionnaire data by zone of purchase.

Zones of 'duck stamp' purchase	Estimated Harvest	Wings Received	Harvest/Wing
A	H_A	W_A	H_A/W_A
B	H_B	W_B	H_B/W_B

Table 2. Use of wing collection data by origin of senders stamp and zone of harvest

Harvest Zone	Origin of Senders Stamp Purchase	
	A	B
	Wings	
X	W_{AX}	W_{BX}
Y	W_{AY}	W_{BY}

Table 3. Estimates of mean kill (K/H), standard error, per successful hunter, skewness (g_1) and kurtosis (g_2) for Provinces of Canada for 1967-70

Province	1967			1968			1969			1970		
	K/H	g_1**	g_2**	K/H	g_1**	g_2**	K/H	g_1**	g_2**	K/H	g_1**	g_2**
Nfld.	8.54 ±0.96	12.58	198.28	6.44 ±0.35	4.80	39.07	8.63 ±0.99	9.21	11.45	8.51 ±0.78	8.51	106.00
P.E.I.	10.98 ±1.23	3.37	17.57	8.99 ±0.64	2.82	10.48	8.33 ±0.68	2.07	4.46	11.26 ±0.83	2.55	8.45
N.S.	11.59 ±0.64	3.33	20.95	11.46 ±0.62	3.34	15.53	10.28 ±0.49	2.29	6.98	10.61 ±0.53	3.28	19.51
N.B.	8.96 ±0.48	3.53	19.94	9.86 ±0.55	5.43	45.30	9.73 ±0.45	1.98	4.42	9.12 ±0.44	2.83	11.24
Que.	4.58 ±0.62	3.43	16.12	13.56 ±0.64	3.07	16.24	15.41 ±0.67	2.51	8.32	14.11 ±0.68	3.67	20.56
Ont.	11.25 ±0.31	3.76	24.39	9.79 ±0.33	4.22	28.18	11.10 ±0.36	2.99	12.74	10.70 ±0.35	3.04	13.68
Man.	13.36 ±0.37	2.88	13.86	10.92 ±0.36	4.05	31.92	14.08 ±0.42	2.45	9.68	15.69 ±0.51	2.91	13.81
Sask	14.47 ±0.35	3.53	22.37	10.04 ±0.29	2.80	13.21	15.06 ±0.51	3.19	59.75	19.53 ±0.65	3.24	17.84
Alta.	16.89 ±0.44	3.53	23.60	13.04 ±0.41	3.79	25.94	19.34 ±0.90	3.68	19.58	19.17 ±0.82	3.98	25.24
B.C.	16.36 ±0.53	2.42	8.46	16.75 ±0.76	3.22	15.63	14.68 ±0.80	3.91	23.47	15.73 ±0.66	3.38	22.74

** $p < 0.01$

Table 4. Efficiency of the estimates of mean kill (K/H) per successful hunter and variance estimates with respect to estimates of mean (m) and variance (v) based on transformed data by provinces of Canada for 1967-70.

Province	Efficiency (percent)							
	Mean Kill (K/M)				Variance			
	1967	1968	1969	1970	1967	1968	1969	1970
Nfld.	87	90	85	87	16	21	13	14
P.E.I.	84	87	86	88	11	15	13	16
N.S.	84	85	87	85	11	12	16	13
N.B.	89	89	88	90	17	18	16	19
Que.	81	83	82	84	8	10	9	11
Ont.	83	84	84	84	10	11	10	11
Man.	88	88	88	88	16	17	16	17
Sask.	91	91	90	89	22	23	20	17
Alta.	87	88	87	84	15	16	15	11
B.C.	84	85	84	81	11	12	11	8

Table 5. Estimates of mean kill (K/H), standard error, skewness (g_1) and kurtosis (g_2) for 12 states of U.S.A. for 1971 and 1972.

Flyway	State	1971				1972			
		K/H	Std.Error	g_1	g_2	K/H	Std.Error	g_1	g_2
Atlantic	Florida	7.21	0.75	0.82*	0.07	7.43	0.67	1.05*	0.94
	Maryland	5.67	0.50	1.25**	1.90*	5.15	0.44	1.48	-0.32
	New York	4.01	0.30	2.07**	3.99**	4.37	0.28	1.91**	5.90
Mississippi	Arkansas	11.36	1.96	3.30**	12.74**	11.62	2.41	3.01**	8.96**
	Illinois	5.82	0.41	1.28**	1.72*	6.95	0.50	1.32**	3.34**
	Louisiana	13.78	1.50	1.75**	3.83**	15.20	1.53	1.68**	3.30**
	Minnesota	8.63	0.49	3.49**	17.52**	9.14	0.48	0.61*	0.16
	Missouri	7.06	0.62	0.82*	0.30	5.88	0.47	0.80*	0.87
	Wisconsin	5.86	0.46	1.18**	1.28*	5.96	0.49	0.79*	0.23
Central	Texas	7.62	0.64	1.47*	2.73**	12.57	1.50	3.45**	12.34**
Pacific	California	18.65	1.15	0.85*	1.07	15.97	1.13	0.76*	0.40
	Washington	10.41	0.85	1.38*	2.07*	11.66	1.20	2.06**	4.48**

* $P < 0.05$; ** $P < 0.01$

Table 6. Efficiency of the estimates of mean kill (K/H) and of variance estimates is the respect to estimates of mean (m) and variance (v) based on transformed data for 12 states of U.S.A. for 1971 and 1972.

Flyway	State	Efficiency (percent)			
		Mean Kill (K/H)		Variance	
		1971	1972	1971	1972
Atlantic	Florida	98	99	51	68
	Maryland	99	99	62	67
	New York	99	99	63	67
Mississippi	Arkansas	97	98	39	53
	Illinois	99	99	60	55
	Louisiana	98	99	54	61
	Minnesota	100	100	82	76
	Missouri	99	100	59	69
	Wisconsin	99	99	64	66
Central	Texas	98	99	54	62
Pacific	California	99	99	59	60
	Washington	100	99	73	67

3. RECENT DEVELOPMENTS

Preliminary analysis of data on kill and activity per hunter
from certain areas in U.S.A. and Canada show that the basic data
have highly skewed distributions and exhibit wide departures from
normality. The question, therefore, arose how wide are these
departures from normality and whether it is possible to find an
adequate distribution which would describe the population and
develop efficient procedures for estimation of its parameters,
e.g. mean and variance. We propose to examine this with regard
to estimates of kill from the migratory game bird surveys in
Canada and U.S.A.

3.1 Canada

The estimates of mean ducks shot per successful hunter,
standard error, skewness $(g_1 = m_3/m_2^{3/2})$ and kurtosis
$(g_2 = (m_4/m_2^4) - 3)$ for each province of Canada and for the four
years 1967-70 are presented in Table 3. It would be seen that kill
per hunter was highly and positively skewed in all the provinces
and in all the years which had the effect of increasing the
variance of the mean kill and decreasing its precision. Also, the
kurtosis was considerably high in all the provinces indicating
that estimates of variance were of low precision.

However, when the data was transformed by taking $x = \log_e(y)$
where y denotes the kill of the successful hunter, the distributions were approximately normal and deviations from normality
were significant $(P < 0.05)$ for only 2 of the 12 provinces.
We will estimate the gain in efficiency when the means of the x's
are transformed back into original variates. This is important
since the hunter's kill can be more readily interpreted than its
logarithm and is, therefore, more useful to management. Since
the samples are reasonably large the efficiency of the sample mean

kill per successful hunter $\bar{y}$ with respect to mean m is
approximately given by Finney (1941) as

$$\frac{\sigma^2 + \frac{\sigma^4}{2} + \frac{1}{2n}(\sigma^6 + \frac{\sigma^8}{4})}{e^{\sigma^2} - 1} \qquad (6)$$

where m is an efficient estimate of the mean of the y population and x is normally distributed with variance σ^2.

The efficiency of the direct sample means as given in Table
4 was uniformly high being over 80 percent for all the provinces
and all the years. Thus the mean was satisfactorily estimated at
the provincial level by the direct sample mean. Since efficiency
will be reduced with increasing values of σ^2, the above result
may not always be true and transformation may be necessary for
values of σ^2 exceeding 2.

The efficiency of the direct estimates of population variance
with respect to an efficient estimate v of the variance of the
y population based on the transformed data has been obtained by
Finney (1941). This is approximately given by

$$\frac{4\sigma^2(t-1)^2 + 2\sigma^4(2t-1)^2}{(t-1)^2(t^4 + 2t^3 + 3t^2 - 4)} \qquad (7)$$

where $t = e^{\sigma^2}$.

The efficiency given in Table 2 ranged between 8 and 23
percent showing that the use of the direct estimate of variance
of the y distribution is very inefficient for use in our kill
surveys.

3.2 U.S.A.

The data used in the study refer to kill per hunter for
1971-72 and 1972-73 seasons and for 12 of the largest states,
including at least one from each of the major migratory bird flyways which span the continent. These are Florida, Maryland and

New York (Atlantic Flyway); Arkansas, Illinois, Louisiana, Minnesota, Missouri and Wisconsin (Mississippi Flyway); Texax (Central Flyway); California and Washington (Pacific Flyway).

The mean ducks shot per hunter within each state based on post offices as units were used to determine g_1 and g_2. Post offices with less than 5 hunters reporting to have bought permits were omitted from the study. The estimates of mean kill, standard error, skewness and kurtosis are given in table 5 which shows that kill per hunter was positively and highly skewed in almost all the states. Also, the kurtosis was very high in about 50 percent of the states.

When the data was transformed by taking $x = \log_e(y+1)$ where y represents the mean kill for a post office in the sample, the distributions were again approximately normal; only 2 out of the 12 states showed significant skewness and kurtosis which was not consistent over the years. The efficiency of $\bar{y}$ with respect to the mean is given in Table 6 which shows that the direct sample mean was almost 100 percent efficient. The efficiencies of the direct estimates of the population variance with respect to the estimates based on the transformed data presented in Table 4 would show that a standard error obtained from the data without transformation would have 51 to 82 percent of the efficiency given by v. The efficiency was generally low for most of the states suggesting that the variance of the untransformed population will not be efficiently estimated by the variance of the original sample.

ACKNOWLEDGEMENT

The author wishes to thank Gail Butler of the Biometrics Division for some helpful comments.

REFERENCES

Atwood, E.L. (1956). Validity of mail survey data on bagged
    waterfowl. *J. of Wildlife Management* 20, 1-16.

Cooch, F.G., Wendt, S., Smith, G.E.J. and Butler, G. (1978).
    The Canada migratory game bird hunting permit and associated
    surveys. Canadian Wildlife Service Report Series Number 43,
    Environment Canada, Ottawa pp. 8-39.

Couling, L. and Smith, G.E.J. (1980). Impact of a post card
    follow-up on harvest survey returns. Canadian Wildlife
    Service, Progress Note No. 3. Environment Canada, Ottawa.

Crissy, W.F. (1967). Aims and methods of waterfowl research in
    North America. *Finnish Game Research* 30, 1-9.

Crissy, W.F. (1975). Determination of appropriate waterfowl
    hunting regulations. U.S. Bur. Sport Fish, Wildlife Admin.
    Rep. 87 pp.

Filion, F.L. (1978). Increasing the effectiveness of mail
    surveys. *Wildlife Society Bulletin* 6, 135-141.

Filion, F.L. (1980). Human surveys in wildlife management.
    In: *Wildlife Management Techniques*. (Schemnitz, S.D.
    editor) 4th edition. The Wildlife Society, Washington, D.C.

Finney, D.J. (1941). On the distribution of a variate whose
    logarithm is normally distributed. *J. Royal Stat. Soc.*7(B),
    155-161.

Rosasco, M.E. (1966). Waterfowl kill survey. U.S. Fish and
    Wildlife Service, Special Scientific Report No.99, 17-20.

Sen, A.R. (1970). On the bias in estimation due to imperfect
    frame in the Canadian Waterfowl Surveys. *J. of Wildlife
    Management* 34, 703-706.

Sen, A.R. (1971). Increased precision in Canadian waterfowl
    harvest surveys through successive sampling. *J. of Wildlife
    Management* 35, 664-668.

Sen, A.R. (1971). Some recent developments in waterfowl sample
    survey techniques. *J. of Royal Stat. Soc.* 20, 139-147.

Sen, A.R. (1976). Developments in migratory game bird surveys.
    J. of Am. Stat. Ass. 71, 43-48.

Smith, G.E.J. (1975). Sampling and estimation procedures in the 1973-74 Canadian waterfowl harvest survey. Canadian Wildlife Service, Biometrics Section Manuscript No. 12, Environment Canada, Ottawa.

ON THE PROBLEM OF VARIANCE ESTIMATION
FOR A DESEASONALIZED SERIES

Kirk M. Wolter and Nash J. Monsour

U.S. Bureau of the Census

The problem of variance estimation for a deseasonalized time series is discussed. It is assumed that the
series is generated from a sequence of sample surveys
conducted at regular intervals of time. Two concepts
of variability are presented: the first assumes the
classical finite population model and treats the true
but unobserved series as fixed, while the second assumes
the true series is generated by a time series process.
Results are given in the form of plots showing the
variance as a function of time. It is shown that the
variance of a *year-ahead* seasonally adjusted observation
is larger than the variance of the original, unadjusted
observation.

1. INTRODUCTION

Let $\{\hat{X}_t\}_{t=1}^{T}$ be an observed time series, where $\hat{X}_t$ is an
estimate of a finite population parameter at time t that is
based on the data from a sequence of sample surveys conducted at
the various time points. For example, $\hat{X}_t$ may be an estimate of
total retail sales or of the unemployment rate in the U.S. in
month t. A common problem in the analysis of such data is to
adjust the observed series $\{\hat{X}_t\}$ for seasonal variations, trading
day variations, holiday variations, and other movements that
repeat themselves in a more or less regular fashion from year to
year. The Census X-11 algorithm (Shiskin et al., 1967) is one
technique for carrying out this adjustment and has been widely

used throughout the world. Many additional techniques are available, some based on regression models, some on Autoregressive-
integrated-Moving-Average (ARIMA) models, some on ratio-to-moving-
average methods, and some on a combination of these techniques.
See Lovell (1963), Box, Hillmer, and Tiao (1978), Dagum (1978),
and the references cited by these authors.

This paper is concerned with the problem of variance
estimation for a deseasonalized (or seasonally adjusted) time
series. Statistical bureaus usually publish reliability statements
and estimates of sampling variability associated with the original
observations $\hat{X}_t$, but not for the deseasonalized observations.
It is usually assumed either explicitly or implicitly that the
variance associated with the adjusted series is equal to that of
the original series. We shall discuss whether or not this assumption is valid, and propose methods for estimating the variances
associated with the deseasonalized series.

The need for variance estimates for a deseasonalized series
is not new, but rather has been recognized for at least 20 years.
In the final report of the President's Committee to Appraise
Employment and Unemployment Statistics (the Gordon Committee)
(1962), it was suggested that standard errors for an adjusted
labor force series are more important than those for the original
series because of the increased reliance of policy makers on the
deseasonalized series, and that research be undertaken on how to
estimate and publish such standard errors. Unfortunately, this
suggestion did not lead to any major advances or to practical
methods for estimating and publishing the standard errors.
Recognizing this fact, the National Commission on Employment and
Unemployment Statistics (1979) recently reaffirmed the Gordon
Committee's findings. In part, their recommendations read as
follows:

> "The commission reemphasizes the importance of standard
> errors for seasonally adjusted statistics and urges the
> Census Bureau to undertake research to develop them."

Finkner and Nisselson (1978) characterized the need for variance estimates for a deseasonalized series as one of the main statistical problems with continuing cross-sectional surveys.

As of the writing of this paper, most official publications of U.S. and Canadian statistical bureaus continue to provide the user with little information about the level of error in the deseasonalized statistics. Both the U.S. publication Employment and Earnings (January, 1980) and the Canadian publication The Labour Force (January, 1980) give moderately complete discussions of the reliability of the original data. Employment and Earnings also warns the reader that "Seasonally adjusted estimates have a broader margin of possible error than the original data on which they are based, since they are subject not only to sampling and other errors, but in addition, are affected by the uncertainties of the seasonal adjustment process itself." Neither publication presents standard errors for the deseasonalized statistics. Both publications are based, in part, on data from household surveys. Publications giving estimates prepared from establishment surveys have similar deficiencies. For example, see the U.S. publication Current Business Reports: Monthly Wholesale Trade (January, 1980) and the Canadian publication Retail Trade (December, 1979).

It is useful to distinguish between at least three different kinds of deseasonalized estimators, because the variance will depend on the kind of estimator. For ease of exposition, we relate this discussion to the Census X-11 algorithm, though similar remarks could be made about any seasonal adjustment methodology. In the U.S., it has been customary to project seasonal factors one year in advance using data through December of the previous year. As the observations are made for the new year, they are deseasonalized by using the projected seasonal factors. The resulting seasonally adjusted estimates are the first such estimates to be published for the new year. We shall call them the *year-ahead* deseasonalized estimates. A second kind of deseasonalized estimate

is called the *end-term* estimate. This is prepared by using a
seasonal factor that is computed from the observed series up to
and including the observation being adjusted. The third kind of
deseasonalized estimate, called the *central-term* estimate, is
prepared by using a seasonal factor that is based on 145 observations of the series, with 72 of the observations being on each
side of the observation being adjusted.[1] Clearly this adjustment
can only be made when the series is of length $T \geq 145$ and the
observation being adjusted is at least 72 terms from the ends of
the series. There are, in principle, 71 additional kinds of
deseasonalized estimates that could be computed for a given time
point t. They are the asymmetric adjustments created when the
seasonal factors are based on the 73 observations for periods
t-72, t-71,..., t-1, t and one additional observation (t+1), two
additional observations (t+1 and t+2),..., and 71 additional
observations (t+1, t+2,..., t+71). For the X-11 program, no
additional changes occur in the deseasonalized estimate for a
given time period when more than 72 additional observations are
available after the period. In the U.S., it has been customary
to run the X-11 program periodically computing the end-term adjustment, the central-term adjustment, and some of the intermediate
adjustments. On occasion, a revised publication is issued, where
the year-ahead deseasonalized estimates are replaced by later
adjustments. Implicit in this discussion, of course, is a certain
minimum length of the series. For series shorter than this minimum,
X-11 operates in a different way, leading to further categories
of deseasonalized estimates.

Although standard errors for deseasonalized estimates have
apparently not been published nor the topic of widespread research,
some limited research has been accomplished since the Gordon

---

[1] The default version of X-11 uses 145 observations for the
central-term adjustment. More or fewer observations would be
used for various X-11 options or with other seasonal adjustment
programs.

Committee's report. During the late 1950's and early 1960's the
U.S. Bureau of the Census used 20 half-sample replicates for
estimating the variances of the unadjusted estimators from the
Current Population Survey (CPS). See, e.g., Hanson (1978). By
individually seasonally adjusting the 20 replicate series created
by the half-sample replicates and computing the variability
between the 20 resulting deseasonalized values, it is possible to
estimate the sampling variance of the published (i.e. combined
over the 20 replicates) deseasonalized series. This work was
never completed and was not published. Kaitz (1974), replying on
Young's (1968) linear approximation to the X-11 algorithm, estimated variances for some deseasonalized CPS labor force series
by making simplifying assumptions about the correlation structure
of the original series. He concluded that the standard errors of
the year-ahead seasonally adjusted observations are approximately
10-15 percent larger than those of the original observations, but
that the standard errors of the central-term adjusted data are
approximately 10-15 percent smaller than those of the original
observations. Monsour (1975) extended Kaitz' work, in part by
using more accurate estimates of the variances of the original
series. This work showed that variances are roughly in the order

Central-term adjusted $\leq$ End-term adjusted $\doteq$ Original $\leq$ Year-ahead
adjusted.

In the remainder of this article, we discuss various
approaches that we have been exploring for the estimation of the
variances of a seasonally adjusted series. Section 2 presents
two general models or concepts of the variance. One is the classical finite population model where the true series, i.e. the
series that would result from a complete enumeration each month,
is treated as fixed, and the variability of the deseasonalized
values arises solely from survey errors. The second assumes a
time series model for the true series, and the variability of the
deseasonalized value arises from both survey errors and the

irregular component in the series. Some results for the first
model are presented in Section 3. Some business series (from
establishment surveys) and some labor force series (from a household survey) are discussed there. The second model is presented in
Section 4, and we close the paper with a general summary in
Section 5.

2. ALTERNATIVE CONCEPTS OF VARIABILITY

Let $\{X_t\}_{t=1}^{T}$ denote the unobserved series of "true" values.
For example, X_t may be total retail sales or the total number of
employed persons in the U.S. at time t. Suppose that $\hat{X}_t$ is an
estimator of X_t. Then we have the model

$$\hat{X}_t = X_t + u_t , \qquad (2.1)$$

where u_t denotes the survey error. In this section we discuss
two concepts of variability associated with this model.

2.1 First concept

Assume that the realization of the time series $\{X_t\}$ is
given, or equivalently, that the X_t are fixed values with no
formal relationship one to another. Then (2.1) is the classical
finite population (or survey sampling) model. The variance of
the estimator $\hat{X}_t$ is

$$\text{Var} \{\hat{X}_t\} = \text{Var}_2 \{u_t\} , \qquad (2.2)$$

where the subscript 2 signifies that the expectation is with
respect to the sampling design. Only sampling variability and
variability due to other survey errors enter into this concept of
the variance.

2.2 Second concept

Now assume that $\{X_t\}$ is the realization of a time series

process. In other words, X_t is not regarded as fixed, but is
itself a random variable that is connected to other values in the
series by a formal time series model. Scott, Smith, and Jones
(1977) have suggested this approach as a means of developing
improved estimators of X_t . The variance of the estimator $\hat{X}_t$
now satisfies

$$\text{Var}\{\hat{X}_t\} = \text{Var}_1\{X_t\} + E_1\text{Var}_2\{u_t\} \qquad (2.3)$$

where the subscript 1 denotes the expectation with respect to the
time series model. The variability in $\hat{X}_t$ arises from two
sources:

    (i)   variability in the true observation X_t about its
              expectation

    (ii) variability due to sampling and other survey errors.

    If the trend-cycle and seasonal components of X_t can be
regarded as deterministic, then the first source of variability
arises from the irregular component of the true series. In this
circumstance, the irregular component of the observed series is a
combination of the survey error series $\{u_t\}$ and the irregular
component of the true series.

2.3 Variance of seasonally adjusted data

    Assume that the seasonal adjustment is of the form

$$\hat{A}_t = \sum_{j=-L_t}^{M_t} \omega_{t,t+j}\hat{X}_{t+j} \;, \qquad (2.4)$$

where $\hat{X}_t$ is either the original observation, the log of the
observation, or some other transformed value. Note that the
limits of summation, L_t and M_t , and the weights $\omega_{t,t+j}$ depend
on the observation being adjusted. They also depend on the length
of the series and the adjustment algorithm, but for convenience
these variables will not be used in the notation. Most adjustment

algorithms are of the form (2.4). For example, Young (1968) has
shown that the additive version of X-11 is of this form and that
the multiplicative version is approximately of this form for the
log of the original series.

The variance of the deseasonalized value $\hat{A}_t$ is

$$\text{Var}\{\hat{A}_t\} = \sum_{j=-L_t}^{M_t} \sum_{i=-L_t}^{M_t} \omega_{t,t+j}\omega_{t,t+i} \text{Cov}\{\hat{X}_{t+j}, \hat{X}_{t+i}\} \qquad (2.5)$$

where $\text{Var}\{\hat{X}_{t+j}\}$ is defined according to (2.2) or (2.3). The
covariance between adjacent values is

$$\text{Cov}\{\hat{A}_t, \hat{A}_{t-1}\} = \sum_{j=-L_t}^{M_t} \sum_{i=-L_{t-1}}^{M_{t-1}} \omega_{t,t+j}\omega_{t-1,t-1+i} \text{Cov}\{\hat{X}_{t+j}, \hat{X}_{t-1+i}\} \quad (2.6)$$

where covariances such as $\text{Cov}\{\hat{X}_{t+j}, \hat{X}_{t-1+i}\}$ also depend on
whether the first or second concept of error is specified. Let
$\underset{\sim}{V}$ denote the $(T \times T)$ covariance matrix of the $\hat{X}_t$ series, and
let $\underset{\sim}{\omega}_t$ be the $(1 \times T)$ row vector with elements $\omega_{t,t+j}$ for
$j = -L_t, \ldots, M_t$ and zeros elsewhere. Then (2.5) and (2.6) may be
expressed in matrix notation as

$$\text{Var}\{\hat{A}_t\} = \underset{\sim}{\omega}_t \underset{\sim}{V} \underset{\sim}{\omega}_t' \qquad (2.7)$$

and

$$\text{Cov}\{\hat{A}_t, \hat{A}_{t-1}\} = \underset{\sim}{\omega}_t \underset{\sim}{V} \underset{\sim}{\omega}_{t-1}' . \qquad (2.8)$$

Similarly, the variance of month-to-month change is

$$\text{Var}\{\hat{A}_t - \hat{A}_{t-1}\} = \underset{\sim}{\omega}_t \underset{\sim}{V} \underset{\sim}{\omega}_t' + \underset{\sim}{\omega}_{t-1} \underset{\sim}{V} \underset{\sim}{\omega}_{t-1}' - 2\underset{\sim}{\omega}_t \underset{\sim}{V} \underset{\sim}{\omega}_{t-1}' . \qquad (2.9)$$

Variances (2.7) and (2.9) may be estimated by substituting
an estimate $\hat{\underset{\sim}{V}}$ for $\underset{\sim}{V}$. Clearly, the estimator must depend on
whether the first or second concept of variability is specified.
Variance estimation methodology based on the first concept is
discussed in Section 3, while Section 4 presents some methodology
for the second concept.

For purposes of the year-ahead adjustment, it is necessary to augment the covariance matrix by 12 additional rows and columns to form the $(T+12 \times T+12)$ matrix $\underset{\sim}{V}^*$. Also let $\underset{\sim}{\omega}_t^*$ denote the $(1 \times T+12)$ vector of weights associated with the deseasonalized value $\hat{A}_t$ for $t = T+1, T+2, \ldots, T+12$, where these denote the year-ahead months. Variances for the year-ahead months are then given by (2.7) and (2.9) with $\underset{\sim}{V}^*$ and $\underset{\sim}{\omega}_t^*$ replacing $\underset{\sim}{V}$ and $\underset{\sim}{\omega}_t$, respectively. Variances may be estimated by substituting an estimate $\hat{\underset{\sim}{V}}^*$ for $\underset{\sim}{V}^*$.

2.4 Discussion

The first concept of variability is appealing because it makes no model assumptions about the $\{X_t\}$ series. It is the standard concept used by statistical bureaus in making reliability statements about the observed series $\{\hat{X}_t\}$. An advantage to this approach is that it puts the variance of the adjusted series on the same basis (i.e. model-free) as the variance of the original series. Because of this consistency of concepts, comparisons between the two variances are straightforward and easily interpretable.

There is, however, a philosophical problem with the first concept that the second concept of variability avoids. The problem is that the very act of seasonally adjusting the observed series $\{\hat{X}_t\}$ implies a time series structure for the true series $\{X_t\}$. Trend-cycle, seasonal, and irregular movements are usually regarded as properties of the $\{X_t\}$ series, not (or only partial-ly) of the $\{u_t\}$ series, and the process of adjustment presumably removes the seasonal movements. If the first concept of variance is accepted for the seasonally adjusted series, then this is to deny or contradict the stochastic nature of the $\{X_t\}$ series implied by the very process of seasonal adjustment. The second concept of variability clearly avoids this problem. It has the disadvantage, however, of not being strictly comparable with the

classical finite population variances usually used for the original series.

3. RESULTS FOR THE CLASSICAL MODEL

In this section we present some illustrative results for
the first concept of variability. Two deseasonalization algorithms
are considered. The first is the Census X-11 algorithm, which is
based on ratio-to-moving-average methods. The second is based on
a simple class of regression models.

3.1 Linear representation of the deseasonalized series

To facilitate the discussion, let

$$\hat{\underset{\sim}{X}} = (\hat{X}_1, \hat{X}_2, \ldots, \hat{X}_T)'$$

$$\hat{\underset{\sim}{X}}^* = (\hat{\underset{\sim}{X}}', \hat{X}_{T+1}, \ldots, \hat{X}_{T+12})'$$

$$\hat{\underset{\sim}{A}} = (\hat{A}_1, \hat{A}_2, \ldots, \hat{A}_T)'$$

$$\hat{\underset{\sim}{A}}^* = (\hat{\underset{\sim}{A}}', \hat{A}_{T+1}, \ldots, \hat{A}_{T+12})' \, ,$$

where T is the current month (or the last month that is available
for input to the deseasonalization algorithm) and T+1, ..., T+12
denote the year-ahead months. Then (2.4) may be written as

$$\hat{\underset{\sim}{A}} = \underset{\sim}{\Omega} \, \hat{\underset{\sim}{X}} \, , \tag{3.1}$$

where $\underset{\sim}{\Omega}$ is the (T × T) matrix with typical row $\underset{\sim}{\omega}_t$. For
purposes of the *year-ahead* seasonal adjustments, we have

$$\hat{\underset{\sim}{A}}^* = \underset{\sim}{\Omega}^* \, \hat{\underset{\sim}{X}}^* \, , \tag{3.2}$$

where $\underset{\sim}{\Omega}^*$ is the (T+12 × T+12) matrix with typical row $\underset{\sim}{\omega}_t^*$.

Following Young (1968), the weight matrix $\underset{\sim}{\Omega}$ for the X-11
algorithm may be expressed by

$$\underset{\sim}{\Omega} = \underset{\sim}{I} - \underset{\sim}{S}$$

(3.3)

$$\underset{\sim}{S} = \underset{\sim}{H}(\underset{\sim}{I}-\underset{\sim}{G}(\underset{\sim}{I}-\underset{\sim}{F}(\underset{\sim}{I}-\underset{\sim}{D}))) \ ,$$

where each of the matrices is $(T \times T)$ and $\underset{\sim}{S}$ is the seasonal factor weight matrix. The matrices $\underset{\sim}{D}, \underset{\sim}{F}, \underset{\sim}{G}, \underset{\sim}{H}$ represent the standard moving average filters used in the X-11 method developed by Shiskin et al. (1967). For the center terms they are the centered 12-term (2×12) trend moving average, the 5-term (3×3) seasonal moving average, the 13-term Henderson trend moving average, and the 7-term (3×5) seasonal moving average, respectively. Somewhat different filters are used at the beginning and end of the time series. The additive version of X-11 operates on the original untransformed series and uses the representation (3.3). The multiplicative version is approximated by applying (3.3) to the log of the original series, the approximation arising because arithmetic moving averages are used instead of geometric averages and sets of 12 consecutive seasonal factors are forced to sum to 12. An additional approximation arises in both versions of X-11 because (3.3) does not represent the outlier and trading day adjustments found in X-11.

To obtain $\underset{\sim}{\Omega}^*$, we first obtain the $(T+12 \times T+12)$ matrix of seasonal factor weights, say $\underset{\sim}{S}^*$. In the X-11 method, a seasonal factor for the year-ahead is obtained by summing the corresponding seasonal factor for the current year plus one-half the difference between the seasonal factor for the current year and the seasonal factor for the prior year. Letting s_{tj} and s_{tj}^* denote typical elements of $\underset{\sim}{S}$ and $\underset{\sim}{S}^*$, respectively, we have

$$s_{tj}^* = s_{t-12,j} + \frac{1}{2}(s_{t-12,j} - s_{t-24,j}), \ t=T+1,\ldots,T+12 \ j=1,\ldots,T \ ,$$

$$s_{tj}^* = s_{tj}, \ t = 1,\ldots,T, \ j = 1,\ldots,T \ ,$$

and

$$s_{tj}^* = 0, \ t = 1,\ldots,T+12, \ j = T+1,\ldots,T+12 \ .$$

The weight matrix is then given by $\underset{\sim}{\Omega}^* = \underset{\sim}{I}^* - \underset{\sim}{S}^*$, where $\underset{\sim}{I}^*$ is the (T+12 × T+12) identity matrix.

We may also find weight matrices $\underset{\sim}{\Omega}$ corresponding to regression based deseasonalization algorithms. As an example, we consider the class of models presented by Rosenblatt (1965). The general model considered by Rosenblatt is

$$\underset{\sim}{\hat{X}} = \underset{\sim}{\Phi} \underset{\sim}{\beta} + \underset{\sim}{v} \tag{3.4}$$

where $\underset{\sim}{\Phi}$ is a T × 28 matrix with typical element ϕ_{ti} , where

$$\phi_{t,2k-1} = \cos(2\pi kt/12) , \qquad k = 1,\ldots,6$$

$$\phi_{t,2k} = \sin(2\pi kt/12) , \qquad k = 1,\ldots,5$$

$$\phi_{t,12} = t \sin(2\pi t/12)$$

$$\phi_{t,13} = t \cos(2\pi t/12)$$

$$\phi_{t,14} = t^2 \sin(2\pi t/12)$$

$$\phi_{t,15} = t^2 \cos(2\pi t/12)$$

$$\phi_{t,16+k} = t^k , \qquad k = 0,1,\ldots,6$$

$$\phi_{t,22+2k-1} = \cos(2k\pi t/M) \qquad k = 1,2,3$$

$$\phi_{t,22+2k} = \sin(2k\pi t/M) \qquad k = 1,2,3 .$$

The first 15 columns of $\underset{\sim}{\Phi}$ are trigonometric polynomials which allow for various types of fixed and moving seasonality, while the last 13 columns of $\underset{\sim}{\Phi}$ allow for nonseasonal movement. Letting $\underset{\sim}{S}$ be the (T × T) matrix of seasonal factor weights, we have

$$\underset{\sim}{S} = (\underset{\sim}{\Phi}_s, 0)(\underset{\sim}{\Phi}'\underset{\sim}{\Phi})^{-1} \underset{\sim}{\Phi}'$$

where $\underset{\sim}{\Phi}_s$ is the submatrix containing the first 15 (seasonal) columns of $\underset{\sim}{\Phi}$. The deseasonalized weight matrix $\underset{\sim}{\Omega}$ is then given by

$$\underset{\sim}{\Omega} = \underset{\sim}{I} - \underset{\sim}{S} .$$

For purposes of the year-ahead adjustments, define

$$S^* = (\underset{\sim}{\Phi}{}^*, \; 0)(\underset{\sim}{\Phi}{}'\underset{\sim}{\Phi})^{-1}(\underset{\sim}{\Phi}{}', \; 0)$$

where $\underset{\sim}{\Phi}{}^*$ is the (T+12 × 28) matrix associated with $\underset{\sim}{X}{}^*$, of which the first 15 columns are $\underset{\sim S}{\Phi}{}^*$, and $\underset{\sim}{0}$ is an appropriately dimensioned matrix of zeros. The corresponding deseasonalization matrix is then

$$\underset{\sim}{\Omega}{}^* = \underset{\sim}{I}{}^* - \underset{\sim}{S}{}^* .$$

3.2 Variances for the deseasonalized series

Using the deseasonalized weight matrices developed in Section 3.1, we may write the covariance matrix of $\hat{\underset{\sim}{A}}{}^*$ as

$$\underset{\sim}{\Sigma}{}^* = \underset{\sim}{\Omega}{}^* \; \underset{\sim}{V}{}^* \; \underset{\sim}{\Omega}{}^{*'} . \qquad (3.5)$$

If the seasonal adjustment is based on the log of the original observation for month t, then $\exp\{\hat{A}_t\}$ is the deseasonalized estimator of monthly level based on the original scale. In this case, $\underset{\sim}{V}{}^*$ is (to a Taylor series approximation) the relative covariance matrix of the original series and $\underset{\sim}{\Sigma}{}^*$ is the relative covariance matrix of the deseasonalized statistics $\exp\{\hat{A}_t\}$.

The variance of a month-to-month change $\hat{A}_t - \hat{A}_{t-1}$ was stated in (2.9). If the adjustment was carried out on the log series, then $\exp\{\hat{A}_t\}/\exp\{\hat{A}_{t-1}\}$ is the deseasonalized estimator of change based on the original scale and (2.9) is the form of its relative variance.

In the next section, we present some illustrative variances for the X-11 and for the regression algorithm. Throughout this work, the multiplicative version of X-11 is used and the log of the series is used with the regression algorithm. It is assumed that the weight matrix in (3.3) applied to the log of the series is an adequate approximation to the multiplicative X-11. Thus, in all cases, $\underset{\sim}{V}{}^*$ represents the covariance matrix of $\hat{\underset{\sim}{X}}{}^*$ and

the relative covariance matrix of the original series. Similarly,
Σ^* represents the covariance matrix of $\overset{*}{\underset{\sim}{A}}$ and the relative
covariance matrix of the exp $\{\hat{A}_t\}$.

3.3 Examples of variances of deseasonalized series

In this section we present graphically variances for some
typical business deseasonalized time series. For all series used
in this section the month T , defined in Section 3.1 as the last
month used in computing the seasonal adjustment, will correspond
to December 1978. This will make the year 1979 correspond to the
year-ahead months T+1, ..., T+12. The value of T will be 72,
120, or 145. The value T = 145 is particularly interesting
because T must be at least this large for the X-11 *central-term*
adjustment (Young, 1968). In this case the central weights only
appear in the 73rd row of $\underset{\sim}{\Omega}$.

Throughout the section, we employ the first concept of
variability, and in terms of this concept assume that u_t is
correlation stationary. Thus we assume autocorrelations are of
the form

$$\rho_2(u_t, u_{t+k}) = \rho(k) .\qquad(3.6)$$

We further assume that the variance is of the form

$$\text{Var }\{\hat{X}_t\} = \text{Var}_2\{u_t\}$$
$$= a + b/Y_t ,\qquad(3.7)$$

where a and b are specified parameters and $Y_t = \exp\{X_t\}$ is
the true value of the original series. For the business series
we assume b = 0 so that the variance is equal to the constant a.
This is equivalent to assuming that the relative variance of the
original series is constant. See Wolter (1979) for some discussion of this assumption. For U.S. labor force series, it is also
customary to employ models of the form (3.7). See Hanson (1978)

and Jones (1977).

The structure in (3.6) and (3.7) is used throughout the section to specify the matrix $\underset{\sim}{V}^*$. In all cases, the Y_t and parameters a and b are estimated from recent data. The $\rho(k)$ are also estimated from recent data, but usually for $k \leq 24$. For $k > 24$, we define the $\rho(k)$ in a manner that seems to be the natural extrapolation of the pattern in the range $0 \leq k \leq 24$.

The first series we examine is the U.S. women's apparel sales series. Both the original series and the deseasonalized series are presented in Figure 1.1. The deseasonalized series is estimated using the linear approximation to the X-11 seasonal adjustment method with $T = 145$. The correlogram for the original apparel sales series is Figure 1.4, and the variance parameter is $a = .000955$. The peaks at lags 12, 24, and 36 are due to the seasonality of the series. In Figure 1.2 the relative variance of the deseasonalized series increases as the months move away from the center portion of the series. This is particularly noticeable for the year-ahead period (1979). Figure 1.3 shows a clear increase in the relative variance of the estimated month-to-month change for the year-ahead period, with a large jump at the first month of the year-ahead period. The symmetry of the relative variances of the deseasonalized estimates in Figures 1.2 and 1.3 about the center months is due to the symmetries existing in the weight matrix $\underset{\sim}{\Omega}$ and to the fact that the relative variance of original series is assumed constant.

The U.S. wholesale grocery inventory series is examined in Figures 2.1 - 2.4. The seasonally adjusted series results from the linear approximation to the X-11 with $T = 120$ and $a = .000331$. In many respects the relative variance of the deseasonalized series behaves the same as for the women's apparel sales

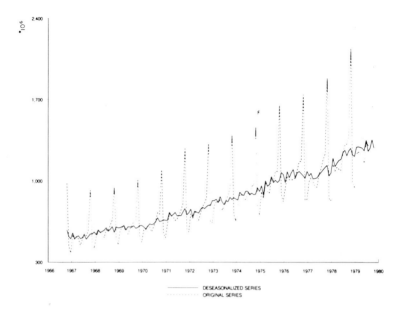

Fig. 1.1 Women's Apparel Sales Versus Time

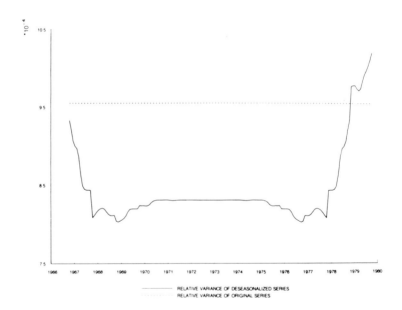

Fig. 1.2   Relative Variance of Level Versus Time
          for Women's Apparel Sales

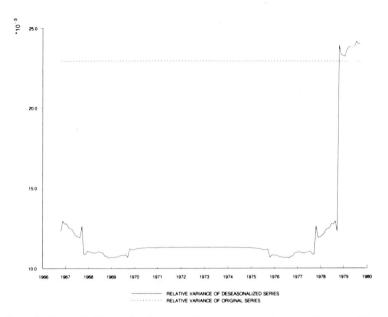

Fig. 1.3   Relative Variance of Mo-to-Mo Change Versus Time
           for Women's Apparel Sales

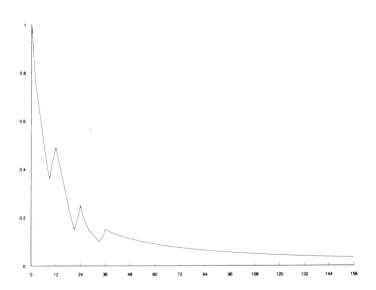

Fig. 1.4 Correlogram for Women's Apparel Sales

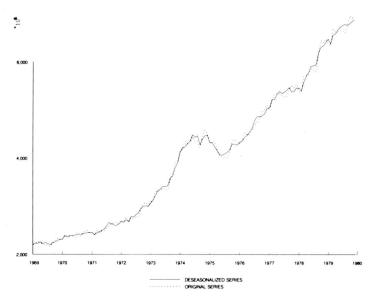

Fig. 2.1 Wholesale Grocery Inventories Versus Time

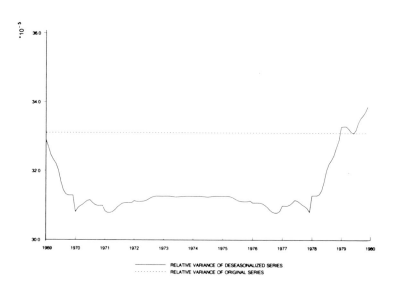

Fig. 2.2 Relative Variance of Level Versus Time
         for Wholesale Grocery Inventories

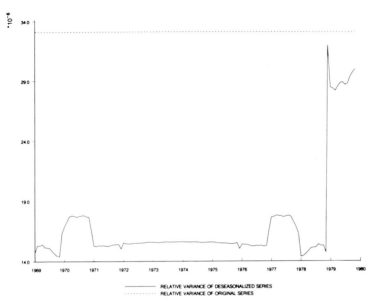

Fig. 2.3   Relative Variance of Mo-to-Mo Change Versus Time
           for Wholesale Grocery Inventories

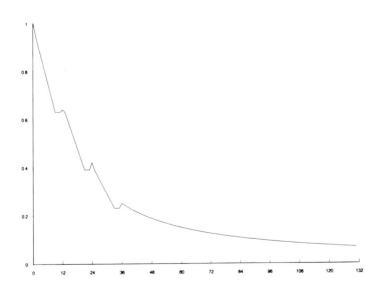

Fig. 2.4   Correlogram for Wholesale Grocery Inventories

deseasonalization series. Note again the large jump in the
relative variance of the estimated month-to-month change for the
first month of the year-ahead period. The fact that the deseasonalized series has its relative variance of month-to-month change
less than that of the original series even for the year-ahead
period (see Figure 2.3) may be due to the relatively high auto-
correlation of the original series and the fact that it damps
down rather slowly (see Figure 2.4).

In Figures 3.1-3.2, we examine once more the wholesale
grocery inventory series. Again T = 120, and the same set of
autocorrelations as given in Figure 2.4 is used. This time the
deseasonalized series results from the submodel of (3.4) that uses
the first six terms (ϕ_{tj} for $j = 1,\ldots,6$) for stable seasonality,
the six terms for a fifth degree polynomial trend (ϕ_{tj} for
$j = 16,\ldots, 21$), and the six low frequency sine and cosine terms
for trend (ϕ_{tj} for $j = 23,\ldots, 28$). We shall call this Model 1.
Several submodels of (3.4) were considered for these data, with
Model 1 providing a very reasonable fit (the R^2 was in excess
of .98). Comparing the corresponding Figures 2.2-2.3 with 3.1-
3.2, we note their similarity and some differences. The relative
variances using X-11 are smaller particularly for the central month
months of the series. Also it is interesting that in Figure 2.3
the relative variance of the month-to-month change for the
deseasonalized series is less than the corresponding relvariance
of the original series for 1979 (the year-ahead portion of the
series), but that in Figure 3.2 this is not true.

3.4 Practical variance estimation methods

In this section two practical methods of estimating the
variance of a deseasonalized series are discussed. These are
methods which could be applied to published series such as those
used as examples in Section 3.3, provided certain summary data
existed for past months of the series.

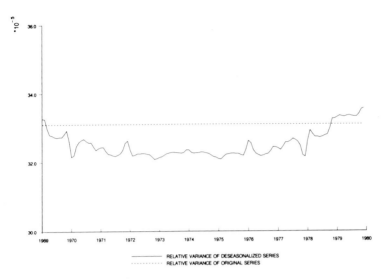

Fig. 3.1   Relative Variance of Level Versus Time
for Wholesale Grocery Inventories

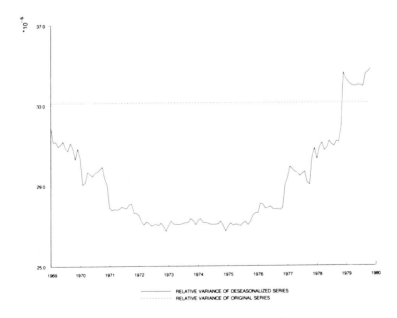

Fig. 3.2   Relative Variance of Mo-to-Mo Change Versus Time
for Wholesale Grocery Inventories

The first method simply involves computing an estimate, say $\hat{V}^*$, of the covariance matrix of $\{\underset{\sim}{x}_t\}_{t=1}^{T+12}$. A variety of standard variance estimating methodologies are available for this purpose, including Taylor series methods, jackknife methods, random group (or replication) methods, balanced fractional sample methods, or some combination of the above. The method of choice will certainly be heavily dependent upon the special circumstances of the computer environment and the series itself, including the estimators used and the sampling design. Having computed the estimate $\hat{V}^*$, the covariance matrix of the deseasonalized series $\{\hat{A}_t\}_{t=1}^{T+12}$ is then estimated by $\hat{\underset{\sim}{\Sigma}}^* = \underset{\sim}{\Omega}^* \underset{\sim}{V}^* \underset{\sim}{\Omega}^{*'}$, where the matrix of weights $\underset{\sim}{\Omega}^*$ is chosen to correspond to the particular deseasonalization algorithm. If all computations to this point have been based on the log series (including the multiplicative X-11), then $\{\exp(\hat{A}_t)\}$ is the deseasonalized series on the original scale and its Taylor series estimated covariance matrix is $\underset{\sim}{E}^* \hat{\underset{\sim}{\Sigma}}^* \underset{\sim}{E}^{*'}$, where

$$\underset{\sim}{E}^* = \text{diag}(\exp\{\hat{A}_1\}, \ldots, \exp\{\hat{A}_{T+12}\}) .$$

Note that periodic sample revisions can be accommodated by this estimation procedure. In this case $\hat{V}^*$ is a block-diagonal covariance matrix: the diagonal blocks correspond to the periods between revisions and are estimated as described above. The off-diagonal blocks are zero.

If random groups, balanced fractional samples, jackknife, or some other form of replication are used in the estimation, then it may be possible to avoid the explicit computation of $\hat{\underset{\sim}{V}}^*$. Rather, this computation is made implicitly by individually deseasonalizing each of the replicate series. The variance is then estimated by the variability between the deseasonalized replicate values.

As an example of this use of replication, we consider a labor force series, "employed nonagricultural industries", from

the U.S. Current Population Survey. The original series as well
as each of 20 half-sample replicate series were seasonally
adjusted by the X-11 method with December 1963 as the current
month and the months of 1964 as the year-ahead period. Using
these data the relative variance of level and of month-to-month
change were computed. Results for the relative variance of level
.are shown in Figure 4.1. It is immediately evident that the
estimated relative variance is subject to high variability. Some
characteristics are discernable though. First, for the year-ahead
period (1964, in this case), the deseasonalized series has higher
variability than the original series. Second, the relative
variance of the original series is for the most part higher than
the relative variance of deseasonalized series, particularly for
the central years.

The second method for variance estimation, which we shall
call the factor method, is based on the approach discussed in
Section 3.3. Here $\hat{V}^*$, and subsequently $\hat{\Sigma}^*$, is based on an
assumed model, such as (3.7), for the variability in the original
series. The ratio factors

$$f_t = \hat{\sigma}_{tt}/\hat{v}_{tt}$$

are computed from these results, where $\hat{\sigma}_{tt}$ and $\hat{v}_{tt}$ are the
t-th diagonal elements of $\hat{\Sigma}^*$ and $\hat{V}^*$, respectively. The
variance of a deseasonalized observation $\hat{A}_s$ is then estimated
by the product $f_t\hat{Var}\{\hat{x}_s\}$, where $\hat{Var}\{\hat{x}_s\}$ is a sample based
estimator of the variance of $\hat{X}_s$. It is not necessary, though it
is certainly sufficient that s = t. All that is required is
that f_t be based on the same deseasonalization filter as is
used in $\hat{A}_s$. Periodically the factors f_t should be updated,
using the most recent data.

In practical applications, the first method is appealing
because it provides a direct variance estimate that is based on
the sample data themselves. The disadvantage is that the

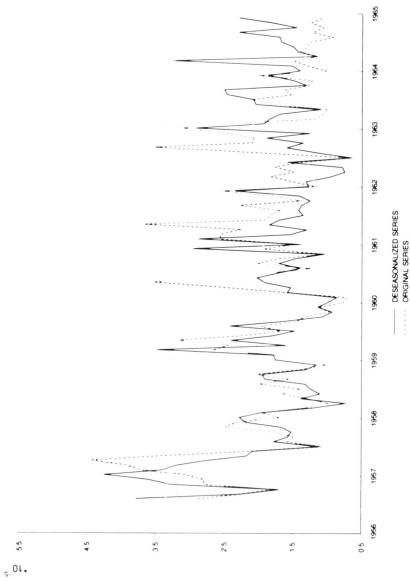

Fig. 4.1 Relative Variance of Level versus Time Based on 20 Half-Sample Replicates
for Employed Nonagricultural Industries

computation of $\hat{V}^*$ requires certain summary data for all past months in the series. For large surveys of the kind discussed in this article, it may not be feasible to retain such data at reasonable cost. Another disadvantage is that the estimated variances will themselves be subject to large variance, as was seen in Figure 4.1. The factor method clearly avoids the difficulties of the first method. The main disadvantage is that it is based on an assumed model for the variances and covariances of the original series. It also requires that the model remain relatively stable through time.

4. COMBINING TIME SERIES AND SURVEY VARIABILITY

In this section we develop the second concept of variability, where $\{x_t\}_{t=1}^T$ is no longer treated as fixed but is a time series process. Throughout the development, X_t will represent a finite population total at time t, and $\hat{X}_t$ will be a design unbiased estimator of X_t. For convenience, we shall assume monthly time intervals.

4.1 Model and assumptions

We proceed by stating the model and assumptions that we shall be using. Although we are fairly explicit about our assumptions, we feel that the methodology to be described is valid under somewhat more general conditions. We are active in further research to determine the most general conditions under which the methods are valid.

We shall consider a finite population of N units where N remains constant through time. The population parameter at time t for the characteristic of interest is $X_{t\cdot} = (X_{t1}, \ldots, X_{tN})$. We assume that a probability sample, s_t, is observed at time t, and that the sampling and observational schemes used to generate the sequence of samples $\{s_t\}$ are such that the $\{u_t\}$ series may be regarded as covariance stationary. In practice, this means

that the sampling and observational schemes are the same for each
month and that the sample rotation pattern (or pattern of overlap)
remains constant through time. The series of estimators $\{\hat{X}_t\}$ is
generated from the observed data for the sequence $\{s_t\}$. The
finite population total at time t is

$$X_t = \sum_i^N X_{ti} \, ,$$

and this forms the "true" but unobservable series $\{X_t\}$.

   With regard to the time series structure of the true series,
we shall assume that

$$X_t = \underset{\sim}{\Phi}_{t\cdot} \cdot \underset{\sim}{\beta} + z_t \, ,$$

where $\underset{\sim}{\Phi}_{t\cdot}$ is a $(1 \times r)$ vector whose elements are fixed functions
of time, $\underset{\sim}{\beta}$ is an $(r \times 1)$ vector of unknown parameters, and $\{z_t\}$
is a stationary time series with zero mean and absolutely summable
autocovariance function. We shall assume that each unit in the
population is consistent with this model. Thus

$$X_{ti} = \underset{\sim}{\Phi}_{t\cdot} \cdot \underset{\sim}{\beta} + z_{ti} \, , \tag{4.1}$$

where

$$z_t = \sum_1^N z_{ti} \, .$$

In addition, we shall assume that

$$z_{ti} = \sum_{j=0}^{\infty} \alpha_j e_{t-j,i} \tag{4.2}$$

and

$$z_t = \sum_{j=0}^{\infty} \alpha_j e_{t-j} \, ,$$

where $\{\alpha_j\}$ is absolutely summable and

$$e_{t-j} = \sum_i^N e_{t-j,i} \, .$$

The e_{ti} are $(0, \sigma^2)$ random variables which are independent over time, but not necessarily independent between units at the same time. Thus we allow nonzero covariance between e_{ti} and e_{tj} for $i \neq j$.

Using (2.1) we may express the observed series as

$$\hat{X}_t = \underset{\sim t \cdot \sim}{\Phi}_{t \cdot} \beta + w_t , \tag{4.3}$$

where $w_t = z_t + u_t$. The total variance of $\hat{X}_t$ is then

$$Var\{\hat{X}_t\} = Var\{w_t\}$$
$$= Var_1\{z_t\} + E_1 Var_2\{u_t\} , \tag{4.4}$$

which is consistent with the expression (2.3). The first term on the right side of (4.4) represents the variability arising from the time series structure of the true series, while the second term is the contribution due to survey errors.

Given our assumptions on the time series error z_t and on the survey error u_t , w_t is covariance stationary. We denote the autocovariance function by

$$\gamma_w(k) = Cov \{w_t, w_{t-k}\} \tag{4.5}$$
$$= Cov_1\{z_t, z_{t-k}\} + E_1 Cov_2\{u_t, u_{t-k}\} .$$

Clearly, the covariance structure of the $\{\hat{X}_t\}$ series is the same as that of the $\{w_t\}$ series. As in Section 2, let $\underset{\sim}{V}$ denote the covariance matrix of $\{\hat{X}_t\}_{t=1}^T$. Then the elements of $\underset{\sim}{V}$ are

$$v_{ij} = \gamma_w(|i-j|) . \tag{4.6}$$

As in previous sections, we consider a seasonal adjustment of the form of (2.4). The variance of a deseasonalized estimator is then given by (2.7) or (2.9) where $\underset{\sim}{V}$ is defined in (4.6).

If seasonal factors are projected a year in advance for the purpose of making a *year-ahead* seasonal adjustment, then it is necessary to augment the $(T \times T)$ covariance matrix with 12 additional rows and columns. Denote the augmented matrix by $\underset{\sim}{V}^{*}$ with elements

$$v_{ij}^{*} = \gamma_{w}(|i - j|) . \qquad (4.7)$$

Let $\underset{\sim}{\omega}_{t}^{*}$ denote the $(1 \times T+12)$ vector of weights associated with the *year-ahead* adjustment of month t, $t = T+1, \ldots, T+12$. Then, the variance of the deseasonalized estimators $\hat{A}_{t}$ and $\hat{A}_{t} - \hat{A}_{t-1}$ are given by (2.7) and (2.9) with $\underset{\sim}{V}^{*}$ replacing $\underset{\sim}{V}$ and the $\underset{\sim}{\omega}_{t}^{*}$ replacing the $\underset{\sim}{\omega}_{t}$.

4.2   Methodology for variance estimation

For estimating the variance of the deseasonalized estimator of level, $\hat{A}_{t}$, or of month-to-month change, $\hat{A}_{t} - \hat{A}_{t-1}$, we suggest substituting an estimator of $\underset{\sim}{V}$, say $\hat{\underset{\sim}{V}}$, into (2.7) – (2.9). The natural estimator for model (4.3) is $\hat{\underset{\sim}{V}}$ with elements

$$\hat{v}_{ij} = \hat{\gamma}_{w}(|i-j|) \qquad (4.8)$$

where

$$\hat{\gamma}_{w}(k) = \frac{1}{T} \sum_{t=1}^{T-k} w_{t} w_{t+k}$$

for $k = 0, 1, \ldots, T-1$. Unfortunately, there are two problems with this estimator: first, the errors w_{t} will be unknown in practice, and second, even if the w_{t} were known, the sample autocovariances would be quite unstable for large k .

In view of these problems, we suggest the following strategy for variance estimation:

1. Fit (4.3) using ordinary least squares. Let the estimated residuals be denoted by

$$\hat{w}_{t} = \hat{x}_{t} - \underset{\sim}{\Phi}_{t \cdot} \hat{\underset{\sim}{\beta}} ,$$

where

$$\hat{\underset{\sim}{X}} = (\hat{X}_1, \ldots, \hat{X}_T)' ,$$

$$\underset{\sim}{\Phi} = (\Phi'_{\underset{\sim}{1.}}, \ldots, \Phi'_{\underset{\sim}{T.}})' ,$$

and

$$\hat{\underset{\sim}{\beta}} = (\underset{\sim}{\Phi}'\underset{\sim}{\Phi})^{-1} \underset{\sim}{\Phi}'\underset{\sim}{X} .$$

2. Compute the sample autocovariances

$$\hat{\gamma}_{\hat{w}}(k) = \frac{1}{T} \sum_{t=1}^{T-k} \hat{w}_t \hat{w}_{t-k}$$

for $k = 0,1,\ldots,K$, where $K = T/3$ or some similar cutoff. Define

$$\hat{\gamma}_{\hat{w}}(k) = 0$$

for $k = K+1, \ldots, T-1$.

3. Define the estimated covariance matrix $\hat{\underset{\sim}{V}}$ by using the $\hat{\gamma}_{\hat{w}}(k)$ instead of the $\hat{\gamma}_w(k)$.

4. Estimators of the variances for the deseasonalized series are then given by

$$\hat{Var} \{\hat{A}_t\} = \underset{\sim}{\omega}_t \hat{\underset{\sim}{V}} \underset{\sim}{\omega}'_t \tag{4.9}$$

$$\hat{Var}\{\hat{A}_t - \hat{A}_{t-1}\} = \underset{\sim}{\omega}_t\hat{\underset{\sim}{V}}\underset{\sim}{\omega}'_t + \underset{\sim}{\omega}_{t-1}\hat{\underset{\sim}{V}}\underset{\sim}{\omega}'_{t-1} - 2\underset{\sim}{\omega}_t\hat{\underset{\sim}{V}}\underset{\sim}{\omega}'_{t-1} \tag{4.10}$$

where the $\underset{\sim}{\omega}_t$ are the vectors of weights discussed in Section 2.3. If the seasonal adjustment was performed on the basis of the log series, then $\exp\{\hat{A}_t\}$ and $\exp\{\hat{A}_t\}/\exp\{\hat{A}_{t-1}\}$ are the deseasonalized values of level and month-to-month change on the original scale and (4.9) - (4.10) are their estimated relative variances.

5. If seasonal factors are projected one year in advance, then define the augmented covariance matrix $\hat{\underset{\sim}{V}}^*$ as in (4.7) using $\hat{\gamma}_{\hat{v}}(k) = 0$ for $k = K+1,\ldots,T+11$. Estimated variances

for $\hat{A}_t$ and $\hat{A}_t - \hat{A}_{t-1}$ are then given by

$$\hat{\text{Var}}\,\{\hat{A}_t\} = \underset{\sim t}{\omega^*}\,\hat{\underset{\sim}{V}}^*\,\underset{\sim t}{\omega^*}{}' \qquad\qquad (4.11)$$

$$\hat{\text{Var}}\,\{\hat{A}_t - \hat{A}_{t-1}\} = \underset{\sim t}{\omega^*}\,\hat{\underset{\sim}{V}}^*\,\underset{\sim t}{\omega^*}{}' + \underset{\sim t-1}{\omega^*}\,\hat{\underset{\sim}{V}}^*\,\underset{\sim t-1}{\omega^*}{}' - 2\underset{\sim t}{\omega^*}\,\hat{\underset{\sim}{V}}^*\,\underset{\sim t-1}{\omega^*}{}'. \qquad (4.12)$$

The choice of cutoff K in Step 2 is guided by two considerations. First, K should be chosen large enough so that the true auto-covariances $\gamma_v(k)$ have damped down to zero for $k \geq K+1$, or at least approximately so. Fortunately, our method of variance estimation is not extremely sensitive to this consideration because the weights $\omega_{t,t+j}$ typically damp down to zero for large $|j|$. Second, K should be chosen small enough so that the estimated autocovariances $\hat{\gamma}_v(k)$ are reasonably stable for $k \leq K$.

The theoretical justification for the variance estimation procedure suggested in 1 - 5 is based on the result that $\underset{T \to \infty}{\text{plim}}\,\hat{\beta} = \beta$. The proof of the result is available from the authors.

4.3 An example: U.S. wholesale grocery inventories

We illustrate these methods using the wholesale grocery inventories series. This series deviates somewhat from the assumptions stated in Section 4.1. Two of the principal differences are that the wholesale population has births and deaths (i.e. the population N does not remain exactly fixed through time), and there are periodic sample revisions (i.e. the $\{u_t\}$ series is not exactly stationary). We feel, however, that the methods of Section 4.2 may be applied in a wider range of circumstances than was originally stated, including the circumstances of this series.

In Section 3, models of the form (4.3) were fit to the log of these data, and the final model (referred to as Model 1) used r = 18 with six terms representing the seasonal; six terms for a fifth degree polynomial with intercept representing the trend-cycle and six low frequency sine and cosine terms also representing the trend-cycle. Sample autocorrelations $\hat{\rho}_{\hat{w}}(k) = \hat{\gamma}_{\hat{w}}(k)/\hat{\gamma}_{\hat{w}}(0)$ were computed from the fitted residuals $\hat{w}_t$. The estimated variance was $\hat{\gamma}_{\hat{w}}(0) = .000277$. In this example, the autocorrelations were computed for $k = 0, \ldots, K$, with $K = 39$. For $k \geq K+1 = 40$, they were set to zero. Although not presented in this paper, the estimated correlogram suggests that $\{w_t\}$ is not a white noise process.

Based on (4.11) and (4.12), relative variances (or variances for the logged series) for level and month-to-month trend are presented in Figures 5.1 and 5.2 respectively. In both cases the weights ω were those associated with Model 1, $T = 120$, and December 1978 was treated as the current month. As usual, 1979 was treated as the year-ahead.

The relative variance plots for both level and month-to-month change have the same basic shape as the plots in Section 3. Relative variances for the deseasonalized series are generally smaller than for the original series for $1 \leq t \leq T$. The size of the difference is particularly large in the neighborhoods of $12 \leq t \leq 24$ and $T-24 \leq t \leq T-12$. When seasonal factors are projected for the *year-ahead* adjustment, the relative variances of the deseasonalized series are larger than those of the original series. This behavior differs somewhat from that observed in Section 3, where the relative variance of the *central-term* adjustment was clearly smaller than that of the original observation.

Figures 5.1 and 5.2 may be compared with Figures 3.1 and 3.2. All of these figures present data for the grocery inventories series as adjusted by Model 1; 3.1 and 3.2 employ the first concept of variability while 5.1 and 5.2 employ the second concept. Notice that the variances are larger in Figures 3.1 and 3.2 than in 5.1 and 5.2. Our a priori expectation was the reverse of this behavior, that the variances in 5.1 and 5.2 should be larger.

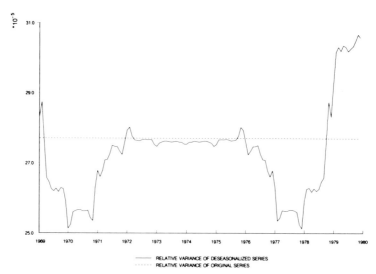

Fig. 5.1   Relative Variance of Level Versus Time
           Based on Second Concept of Variability
           for Wholesale Grocery Inventories

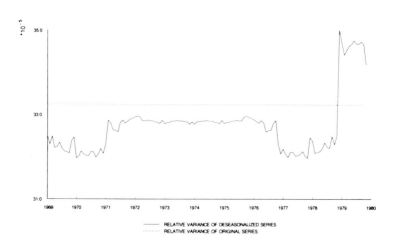

Fig. 5.2   Relative Variance of Mo-to-Mo Change Versus Time
           Based on Second Concept of Variability
           for Wholesale Grocery Inventories

This expectation was based on the fact that the first concept of
variability only considers survey errors, while the second concept
includes survey errors and also time series errors.

The rotating panel design of the survey that generates
these data may offer an explanation for the observed behavior of
the variances. As is described in Isaki et al. (1976) and Wolter
et al. (1976), the survey uses three rotating panels, each reporting four times per year at three month intervals. Because wholesale sales and inventories are highly correlated over time, panels
tend to maintain a reasonably fixed relationship with the true
series $\{X_t\}$. That is, if the panel reporting in month t is
high or low in relation to the true total for the month (i.e.
$\hat{X}_t > X_t$ or $\hat{X}_t < X_t$), then the panel will tend to remain high or
low the next and successive times that it is enumerated.[2] Thus,
the panel rotation scheme tends to induce a seasonal behavior on
the survey errors. The panel effect behaves almost like a
deterministic seasonal effect. The deseasonalization filters, of
course, can not distinguish this induced seasonal component from
the seasonal component in the true $\{X_t\}$ series. As a result,
both components tend to be filtered out in the deseasonalization
process. The residual $\hat{w}_t$ that remains after fitting a model of
the form (4.3) may, therefore, be less variable than the survey
error u_t, because a large portion of u_t is picked up as a
seasonal component. Under conditions such as these, it is not
surprising that the variances are lower in Figures 5.1 and 5.2
than in 3.1 and 3.2.

A possible model for the panel effect is

$$u_{tp} = \bar{u}_t + e_{tp},$$

where u_{tp} is the survey error associated with the panel p
reporting in month t,

---

[2] To some extent this effect is dissipated because X_t is a
composite estimate in this survey.

$$\bar{u}_t = \frac{1}{3} \sum_{p=1}^{3} u_{tp} \ ,$$

and

$$e_{tp} = u_{tp} - \bar{u}_t \ .$$

In terms of (4.3), the e_{tp} component of survey error tends to be picked up by the $\phi_{\underset{\sim}{t} \cdot \underset{\sim}{\beta}}$ term, leaving a residual

$$w_t = z_t + \bar{u}_t$$

that is less variable than the residual $w_t = z_t + u_t$ that was anticipated.

5. SUMMARY

In this article we have discussed two concepts of variability associated with a deseasonalized time series. The first concept only includes variability arising from survey errors, and some example results are presented in Section 3. Variances for the two parameters of greatest practical interest, monthly level and month-to-month change, were investigated. The main finding of this work was that the variance, both for level and month-to-month change, of the deseasonalized series is higher than that of the original series during the year-ahead months, but smaller during the central months of the series. The second concept was presented in Section 4 and considered both survey and time series variability. The main finding of this section was similar to that of Section 3.

We consider the investigations reported in this article to be of a preliminary nature, and several extensions are already being planned. For example, the methods easily extend to the optional filters used in X-11 and to other deseasonalization algorithms, including Statistics Canada's X-11 ARIMA. We hope to investigate several of these alternatives. Another area that requires much further development is the estimation for the second

concept of variance. In Section 4 our results were somewhat
unsatisfactory because the deseasonalization filter picked up part
of the sampling error. We hope to develop improved methods of
estimation which completely account for all of the various sources
of variability. Another important topic is the sensitivity of
the calculations to the autocorrelations $\rho(k)$ (see (3.6)) for
large lags k. In most practical applications, the statistical
information available about such correlations will be very sketchy
and some assumptions will have to be made. We hope to show that
one does not require very precise estimates of the $\rho(k)$ for
large k , because the weights $\underset{\sim}{\omega}$ themselves damp down quickly
toward zero.

Until further research is completed, we feel that the
factor method for the first concept of variance (see Section 3.4)
is the only feasible method of estimation for large-scale surveys.
The ratio factors can be computed as described in Section 3, and
their application to the variances of the original series is quite
straightforward. It would not be a computational burden to perform
such calculations for the many kinds of seasonally adjusted
estimates (e.g. year-ahead, end-term, etc.) and the numerous
characteristics associated with most large-scale surveys.

6. ACKNOWLEDGEMENTS

As is evident from the graphic results, a great deal of
computer work was required in the preparation of this paper. We
are especially grateful to Quentin Ludgin and Zigmund Krivitsky
for the computations and graphs, only a few of which are presented
here. We thank Joseph Stith and Mitchell Trager for their assistance with the computations in Section 4, and past members of the
Bureau's technical staff for creating and retaining the 20 half-
samples used in Section 3.4. We also acknowledge Lillian Principe
for careful typing of various drafts.

REFERENCES

Box, G.E.P., Hillmer, S. and Tiao, G.C. (1978). Analysis and
    Modeling of seasonal time series. In *Seasonal Analysis of
    Economic Time Series*, A. Zellner (ed.). U.S. Government
    Printing Office, Washington, D.C.

Dagum, E.B. (1978). Estimation of changing seasonal variations
    in economic time series. In *Survey Sampling and Measure-
    ment*. N.K. Namboodiri (ed.). Academic Press, New York.

Finkner, A.L. and Nisselson, H. (1978). Some statistical problems
    associated with continuing cross-sectional surveys. In
    Survey Sampling and Measurement. N.K. Namboodiri (ed.).
    Academic Press, New York.

Hanson, R.H. (1978). The current population survey. Technical
    paper 40, U.S. Bureau of the Census, Washington, D.C.

Isaki, C.T., Wolter, K.M., Strudevant, T., Monsour, N., and
    Trager, M. (1976). Sample redesign aspects of the Census
    Bureau's monthly business surveys. *Proceedings of the
    American Statistical Association, Business and Economic
    Statistics Section*. American Statistical Association,
    Washington, D.C., 90-98.

Jones, C.D., (1977). CPS variances - new standard errors for
    monthly estimates of levels, percentages and participation
    rates for the CPS labor force data for the 461 area design.
    Unpublished Memorandum to E.J. Gerson, U.S. Bureau of the
    Census, Washington, D.C.

Kaitz, H.B. (1974). Analysis of current economic data on employ-
    ment and unemployment. Unpublished manuscript prepared
    for the Committee on National Statistics, National Research
    Council, Washington, D.C.

Lovell, M.C. (1963). Seasonal adjustment of economic time series
    and multiple regression analysis. *Journal of the American
    Statistical Association* 58, 993-1010.

Monsour, N.J. (1975). On the analysis of the standard error of
    seasonally adjusted series. Unpublished memorandum to
    H. Nisselson, U.S. Bureau of the Census, Washington, D.C.

National Commission on Employment and Unemployment Statistics (1979)
    Counting the Labor Force. U.S. Government Printing Office,
    Washington, D.C.

President's Committee to Appraise Employment and Unemployment
     Statistics (1962). *Measuring Employment and Unemployment*.
     U.S. Government Printing Office, Washington, D.C.

Rosenblatt, H.M. (1965). Spectral analysis and parametric methods
     for seasonal adjustment of economic time series. Working
     paper 23, U.S. Bureau of the Census, Washington D.C.

Scott, A.J., Smith, T.M.F. and Jones, R.G. (1977). The application
     of time series methods to the analysis of repeated surveys.
     International Statistical Review 45, 13-28.

Shiskin, J., Young, A.H. and Musgrave, J.C. The X-11 variant of
     the census method 11 seasonal adjustment program. Technical
     paper 15, U.S. Bureau of the Census, Washington, D.C.

Statistics Canada (1979). *Retail Trade*. Statistics Canada,
     Ottawa.

Statistics Canada (1980). *The Labour Force*. Statistics Canada,
     Ottawa.

U.S. Bureau of the Census (1980). *Current Business Reports:
     Monthly Wholesale Trade*. U.S. Government Printing Office,
     Washington, D.C.

U.S. Bureau of Labour Statistics (1980). Employment and Earnings.
     U.S. Government Printing Office, Washington, D.C.

Wolter, K.M., Isaki, C.T., Sturdevant, T., Monsour, N., and Mayes,
     F. (1976). Sample selection and estimation aspects of the
     Census Bureau's monthly business surveys. *Proceedings
     of the American Statistical Association, Business and
     Economic Statistics Section*. American Statistical
     Association, Washington, D.C., 99-109.

Wolter, K.M. (1979). Composite estimation in finite populations.
     Journal of the American Statistical Association 74, 604-613.

Young, A.H. (1968). Linear approximation to the Census and BLS
     seasonal adjustment methods. *Journal of the American
     Statistical Association* 63, 445-471.

VARIANCE ESTIMATION: GENERAL DISCUSSION

A.R. SEN, *Environment Canada* (to P.S.R.S. Rao): You said that when the sample size n increases the three estimators, the classical one and the two others, are more or less the same so far as efficiency goes. Can you tell me roughly for what size of n is this true?

P.S.R.S. RAO, *University of Rochester*: n is thirty. The sampling fraction need not even be 30%. Even at about 20% this holds.

A.R. SEN, *Environment Canada*: The second point is a small one. At the end you have put the values of the expected mean square error of the estimators for $t = 0$ and $t = 2$. What is the position for $t = 1$?

P.S.R.S. RAO, *University of Rochester*: When t is equal to 0 and 1, the weighted least square estimator has larger expected mean square error than the classical estimator. When t is equal to 2, there have to be some conditions on the sample size.

J.N.K. RAO, *Carleton University* (to P.S.R.S. Rao): Sam, I object to your calling $\hat{V}_C$ the classical estimator and $\hat{V}_R$ not the classical estimator. Both are equally classical. Another point I would like to make is that the conditional biases and conditional mean square errors you have given are also given in my paper with Kuzik, in the Appendix.

P.S.R.S. RAO, *University of Rochester*: The thing I would
like to tell you Jon is that $\hat{V}_R$ is not the classical estimator.
As I pointed out, the one we know in the literature is this $\hat{V}_C$
multiplied by the reciprocal of this, because $\overline{X}$ (capital $\overline{X}$)
comes in the dominator there. You see this one I just got as an
approximation of the Jackknife one. But, don't worry about the
names, call it classical or whatever, they are all there in the
literature. Whether I multiply here with $\overline{X}/\overline{x}$ or $\overline{x}/\overline{X}$, does
make some difference. Also, I'm aware of the fact that you had
the conditional mean square errors in your Kuzik paper, but here
I have done a little more analysis and derived them myself.

D. KREWSKI, *Health and Welfare Canada* (to P.S.R.S. Rao):
I have a little trouble with the interpretation of these conditional biases and mean square errors as measures of accuracy and
precision relative to these same quantities defined in terms of
the distribution induced by random sampling. Certainly these
conditional results are going to depend on the model. For example,
if you throw in an intercept term, it's going to make quite a
difference and this model cannot be reliably determined on the
basis of small sample sizes. If you must use a conditional
approach however, why not average the conditional results over all
possible samples, in order to give what was referred to yesterday
as anticipated biases and mean square errors.

As a second point, I wonder if you could comment on the
rationale for this new modification to the jackknife estimator?
The Kish and Frankel modification simply involves replacing the
jackknife estimator of the ratio in the formula of a variance, by
the ratio estimator itself. Is there some simple interpretation
to the modification that you are now making?

P.S.R.S. RAO, *University of Rochester*: Let me take the
last question first. The reason for my modification is that I do
not trust the computer results that come out when we try to put in

some equations to evaluate the jackknife estimator. As you know, there are too many equations, and the more equations you put in, the less accuracy you get. I think I told you that in one place I used ninety decimal precision to get two good digits. Computer scientists get upset with that, but in any case, my modification was out of the frustration that I don't know whether I should believe the computer results. There's no way I can verify them on the desk calculator. I got to that final modification by seeing whether I could simplify the estimator and not lose the main feature. I'm happy that, in fact, its mean square error is very close to the original jackknife estimator.

Now, addressing your first question, it is true that all of these things are model based, but my results here at least do not insist on taking $t = 1$. I examined the estimators for the entire range of t, although I gave you results for $t = 0$, 1 and 2. When I'm comparing the sample mean with the ratio estimator, the result does not depend on the value of t. Secondly, even with the balancing, if you have an intercept term the ratio estimator is still unbiased. So, although the results do depend on the linear model, the final conclusions I gave you were averaged over the x_i.

P.S.R.S. RAO, *University of Rochester* (to J. Chromy): Your PMR should have some advantages over the Horvitz-Thompson estimator in terms of efficiency and nonnegativeness of the variance estimator. You have pointed out some of these things, but would you elaborate a little more on this.

J. CHROMY, *Research Triangle Institute*: I hope it has an advantage. The first thing you have to do is worry about self-representing units and how to subsample them. One of the conveniences of the whole thing is that you can just put in all your size measures and this automatically gives you the self-weighting design where the number of subsamples you draw per primary unit is determined by the way you determine the $n(i)$. I think it has

some kind of convenience features in it. It's not really an
alternative to the Horvitz-Thompson estimator, it is more or less
a method of selection that I think keeps the pairwise probabilities
well behaved except that I can get zeros. I don't think zero
variance estimates are much better than negatives. I don't like
either one of them.

J.N.K. Rao, *Carleton University* (to J. Chromy): Did you
see a paper by Alan Sunter in "Applied Statistics" where he does
this sequential sampling from a file for unequal probability
selection?

J. CHROMY, *Research Triangle Institute*: No, I haven't.

J.N.K. RAO, *Carleton University*: And secondly, Poisson
sampling is also a very convenient way of sampling from a file.
The trouble with Poisson sampling is that the sample size is a
random variable. Recently Ken Brewer has proposed a modification
which he calls "colocated sampling". It has the features of
Poisson sampling, and also reduces the variability in the sample
size, so I don't think it has the problem of zero variance
estimator.

J. CHROMY, *Research Triangle Institute*: What I would have
liked to have said is that I am not proposing the use of the
Horvitz-Thompson estimator for this. I think you need to find an
alternative that gets away from those zeros, but that is a problem
that is not really resolved yet. We don't know what the best
variance estimator is.

M. HIDIROGLOU, *Statistics Canada* (to J. Chromy): Have you
ever run into situations where you actually got negative estimates
of variance with this algorithm?

J. CHROMY, *Research Triangle Institute*:   No I haven't.   I
don't think I will.   I haven't been able to prove analytically
that I wouldn't but the way the algorithm handles those pairwise
probabilities, it doesn't appear that you get them.

M. HIDIROGLOU, *Statistics Canada*:   For those who want to
use it, is there a computer algorithm available?

J. CHROMY, *Research Triangle Institute*:   Yes there is.   We
developed what's called a macro within the SAS statistical analysis
system package.   We are not selling it, but we are trying to get
it documented for internal use.   We have used it on several sample
selection schemes.   We consider it a good alternative certainly to
systematic sampling, and to zone sampling; it has some features
that we like a little better.

P.S.R.S. RAO, *University of Rochester* (to K. Wolter): Kirk,
your second concept indicates to me either of two things.   One,
your series is like a double sampling procedure, so that the u_t's
are a sample from an infinite superpopulation.   Secondly, if I
look at it, $\hat{X}_t$ is $X_t + u_t$ which is like a classical variance
component model.   If you carefully look at it, what you are doing
is variance of the conditional expectation plus expectation of the
conditional variance.   That means my time series is a sample from
a superpopulation like a two stage double sampling procedure.
Otherwise, why would I get variances of conditional expectations
plus expectations of conditional variances?   Have you used either
of these concepts in estimating the variance?

K. WOLTER, *U.S. Bureau of the Census*:   No we haven't but
you are correct in that the X_t series is a realization of a time
series.

P.S.R.S. RAO, *University of Rochester*: That's right. In
all of the sampling literature, if we have a process like that we
use the between sums of squares and the within sums of squares and
combine them properly to estimate the variance. What I want to
know is whether you have used a similar approach to estimate the
variance of your time series? In your work, $\hat{X}_t$ is $X_t + u_t$ and
the variance of $\hat{X}_t$ is the variance of X_t plus the variance of
u_t. That means you are treating X_t as a random variable. That
is our classical variance components model. If I look at it that
way, we have several ways of estimating these variances. I want
to know whether you have used either of those concepts to exploit
the estimation procedure in a more efficient way?

K. WOLTER, *U.S. Bureau of the Census*: Well, the answer is
again no. As far as variance components go, the difference
between this problem, I guess, and the classical variance components models that I know about, is that this is time series data.
There are correlations over time that you don't have in classical
variance components models. As far as your first comment goes,
you're quite right, the variance for the second concept is an
expected value of the conditional variance plus the variance of a
conditional expectation. The terminology of between and within
doesn't really make sense here: between what? What falls out of
what we've done, is that the last variance estimation procedure
that I've described to you will estimate the total variance of a
deseasonalized value. If you wanted somehow to separate the
sampling variance component from the time series variance component, you could do that, just as you do in multiple stage sampling.
You do it by getting an estimate of the total variance as I
describe here, estimating sampling variability as we do following
standard practice, and then by subtraction you would get an estimate of the variability associated with the time series error. In
that sense, this is analogous to what you're talking about.

VI IMPUTATION TECHNIQUES

THE EFFECTS OF PROCEDURES WHICH IMPUTE FOR MISSING ITEMS: A SIMULATION STUDY USING AN AGRICULTURAL SURVEY

Barry L. Ford, Douglas G. Kleweno, and Robert D. Tortora

U.S. Department of Agriculture

This simulation study compares the effects of six procedures which impute values for missing items. Using data from an agricultural survey, this experiment covers a range of conditions which account for the method of designating which values are missing and the rate at which they are missing. An analysis of the mean square errors, the effects on the correlations, and the costs show that two versions of a ratio procedure give the best results for sample sizes which are very large.

1. INTRODUCTION

The problem of incomplete data, i.e. missing values, is one of the most common problems of survey work. Incomplete data is of two types -- missing units and missing items. Missing units are the result of nonresponse for a sample unit and, thus, consist of refusals and inaccessibles. Missing items refer to those units which have missing values but also have some reported values. For example, the respondent answers some questions but not others, or he answers some questions incorrectly. The problem of missing units is the subject of previous studies at the U.S. Department of Agriculture (Ford, 1976). *The purpose of this study is to compare six procedures which impute for missing items.*

The basic research tool of this study is simulation. Using a complete data set (no missing values) from a current survey, the

authors simulate which values are missing. Six missing item procedures are then applied, and the imputed values are compared to
each other and to the original values. Although simulation
experiments are in a sense artificial, they do allow analysis over
a wide range of conditions and a comparison against "true" values.

The simulations in this study are over various levels of
two effects: 1) the randomization mechanism used to designate
which values are missing and 2) the rate at which values are
missing. For each level of these two effects, there are several
incomplete data sets simulated from the original data set. The
original data set is divided into three replicates, and this
replicate structure is carried over into each simulated data set.
Missing item procedures are applied to each replicate independently in order to obtain unbiased estimates of standard errors.

The original data set used in this study is from one
stratum of a hog survey conducted by the U.S. Department of
Agriculture. There are 201 complete sample units which are
divided into three replicates of 67 units each. Each unit has
15 quantitative variables -- 14 survey variables and 1 control
variable which has been used to stratify the population. For
the purposes of this study the authors confine the simulation
of missing values to two major survey variables, $y^{(1)}$ and $y^{(2)}$,
of the 14 survey variables. Values for either or both of these
variables can be designated as missing. All imputations must
obey the edit check $y^{(1)} + y^{(2)} \leq w$, where w is another of
the 14 survey variables.

The distributions of $y^{(1)}$ and $y^{(2)}$ are both highly
skewed. Figure 1.1 gives two bar graphs to show the general
shape of the distributions. The mean of $y^{(1)}$ is 22.11 and the
variance if 509.32; the mean of $y^{(2)}$ is 21.45 and the variance
is 502.22. Thus, $y^{(1)}$ and $y^{(2)}$ are similar in distribution.
The correlation between w and $y^{(1)}$ is 0.82 and between w and
$y^{(2)}$ is 0.81. Both $y^{(1)}$ and $y^{(2)}$ have integer values greater

than or equal to zero.

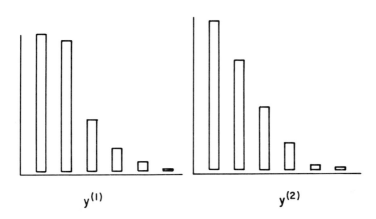

$y^{(1)}$ $y^{(2)}$

Figure 1.1: Bar graphs to show the general shape of the distributions of $y^{(1)}$ and $y^{(2)}$.

2. THE PROCEDURES

This study compares the effects of six procedures which impute for missing items. This section gives a description of each procedure, a description which includes the estimation techniques and assumptions used by the procedures. The descriptions are written in general terms of how the procedures would impute for a data set which has both complete and incomplete units.

2.1 The Ratio Procedure (Variations 1 and 2)

The ratio procedure examined in this study imputes a value for each missing value by using the equation:

$$y_{ratio} = \hat{R} \, x^*$$

where:

$\hat{R}$ is the estimated ratio between the variables x and y

x^* is the value of an x variable for a sample unit which has a missing y value

y_{ratio} is the value imputed for the missing y value.

An estimate of R is based on the sample units which are complete. If x' and y' are totals for the complete units in the sample, then $\hat{R} = y'/x'$. Thus, this estimator of R assumes that the ratio for the complete units is a good estimate of the ratio for the incomplete units.

Although called an auxiliary variable, the x variable may be a survey variable or the control variable. When a y value is missing, the ratio procedure uses as the x variable that variable which is most highly correlated with the y variable. If the value of the most highly correlated variable is missing from the unit, the procedure uses the next most highly correlated variable. If that value is also missing, then the procedure continues in the same fashion until a reported value is found. Correlations are estimated by using only the complete sample units of a data set which has missing values.

This study uses two variations of the ratio procedure. These two variations arise because of the linear restriction imposed on the two variables -- $y^{(1)} + y^{(2)} \leq w$. The first variation simply imputes independently for $y^{(1)}$ and $y^{(2)}$ and then checks to see whether $y^{(1)} + y^{(2)} \leq w$. When $y^{(1)} + y^{(2)} > w$, then the procedure adjusts any imputed values so that $y^{(1)} + y^{(2)} = w$. The second variation uses the constructed variable $z = y^{(1)} + y^{(2)}$ as though it is a survey variable. If z is missing (either $y^{(1)}$ or $y^{(2)}$ is missing), the procedure: 1) finds an x variable by using correlations with z where the correlations are estimated from the complete units, 2) imputes a value for z , 3) makes z = w if z > w , and 4) imputes for

missing values of $y^{(1)}$ and/or $y^{(2)}$ so that $y^{(1)} + y^{(2)} = z$.
If both $y^{(1)}$ and $y^{(2)}$ are missing, z is split into $y^{(1)}$ and
$y^{(2)}$ proportionally by using relationships from the complete
units in the data set.

2.2 The Array Procedure

The array procedure is not a procedure in general use but a
procedure designed within the U.S. Department of Agriculture
(Beller and Bynum, 1971) and proposed as a method of imputing for
missing values on the Department's hog survey. Although not
designed by the authors, the array procedure is included among the
test procedures because its effects have never been assessed.

The array procedure uses a two-way table to impute for
missing values. Two survey variables, a_1 and a_2 , are chosen
to define the table. If these two variables have c_1 and c_2
classes respectively, then the array procedure would form a table
for $y^{(1)}$, as an illustration, of the form:

Variable a_2

Cell values must be initialized with an estimate of the ratio
$y^{(1)}/a_3$ where a_3 is another survey variable. As the procedure
processes the units in a sample, each unit is classified into a
cell of the table by the values of a_1 and a_2 for that unit.
If $y^{(1)}$ is reported, the ratio $r = y^{(1)}/a_3$ from the unit is
added into a cell by using the weighted formula:

$$\frac{2(\text{previous value for the cell}) + r}{3}$$

The purpose of this weighted formula is to prevent the imputation of extremely large values, i.e. outliers. If $y^{(1)}$ is missing, the value of a_3 from the unit is multiplied by the ratio from the appropriate cell and imputed for the value of $y^{(1)}$. Obviously, the ordering of the data has some importance for the estimates from the array procedure. Although data from surveys by the U.S. Department of Agriculture are often in a roughly geographic order, the data of this study were in a random order except that complete units were processed before incomplete units.

The array procedure is similar to the ratio procedure because the array procedure also uses a type of ratio to impute values. However, the array procedure is a more complex method of obtaining the ratio and a more rigid process. For example, the array procedure can use the information from *three* auxiliary variables -- a_1, a_2 and a_3. However, these three variables must be chosen before applying the procedure to the data set and are not allowed to have missing values. Another difference is that once the array procedure processes all complete units in a data set, then the procedure can also use incomplete units to change the ratio values in the cells as long as a_1, a_2, a_3 and $y^{(1)}$ are not missing. The ratio procedure, as used in this study can only use estimates of ratios and correlations from the complete units in a data set.

In this study the a_1 and a_3 variables are the same variable, w. The variable a_2 is another survey variable which is highly correlated with $y^{(1)}$ and $y^{(2)}$.

2.3   The ESTMAT Procedure

The ESTMAT procedure is an iterative solution to the problem of finding the maximum likelihood estimates for a multivariate data set in which some values are missing (Hartley and

Hocking, 1971). The ESTMAT procedure imputes by using multivariate
regressions as defined by the reported values. As long as the same
regression relationships apply to both reported and missing values,
the ESTMAT procedure should be able to impute accurately even if
the reported and missing values have different means.

The ESTMAT procedure represents an extension of the double
sampling regression estimator to a multivariate setting. However,
the ESTMAT procedure can take into account many different patterns
of missing data in the data set. For example, once the data is
collected for two variables, there are four possible patterns of
missing data -- both variables are reported, only the first
variable is reported, only the second variable is reported, or
both variables are missing. With k variables there are 2^k
possible patterns if one also counts as a pattern the set of
complete units.

The estimation formulas which the ESTMAT procedure uses are
complex and are not given in this paper. However, they can be
found in the references. Convergence of the iteration process
used by ESTMAT is not assured in general, but in practical applications the convergence has usually taken less than ten iterations.

The two major assumptions of the ESTMAT procedures are:
1) values follow a multivariate normal distribution, and 2) the
values are missing at random. The first assumption is necessary,
of course, for the derivation of the maximum likelihood estimators
used in the ESTMAT procedure. One example to show robustness to
the normality assumption has been given (Hocking, Huddleston and
Hund, 1974), but no one has made a thorough study. The second
assumption is unlikely to hold when the data are missing because
of refusals, inaccessibles, editing, etc. The second assumption
emphasizes the fact that the ESTMAT procedure seems more appropriate for survey situations in which the missing values are
planned -- double sampling schemes, triple sampling schemes, etc.
(Ford and Hocking, 1974). However, if the procedure is robust to

the randomness assumption, then applying multivariate regressions seems as reasonable as applying the ratio of a ratio procedure. The data set in this study does not obey either of the two assumptions for the ESTMAT procedure.

The ESTMAT procedure was *not* initially designed to impute individual values but to estimate directly the mean vector of the population. However, the procedure also estimates the variance-covariance matrix, and this estimate allows the computation of multivariate regression equations which can be used to impute individual values. These imputed values lack what Pregiborn (1976) calls "commutativity" with the estimated mean vector. In other words, if one averages the reported and imputed values in a data set, this average does not equal the mean estimated directly by the ESTMAT procedure. Thus, the reader must be aware that the results of the ESTMAT procedure in this study are affected by an imputation process which may not be a part of other ESTMAT applications.

2.4   The Zero Spike Procedure

The zero spike procedure takes its name from the fact that zeros often dominate the response space of many surveys -- thus resulting in a "spike" of zeros when one draws a histogram of the distribution. The data set of this study has this characteristic. The first bar in each of the graphs of Figure 1.1 represents the zeros in the data set. For $y^{(1)}$ 33 percent of the 201 original values are zeros, and for $y^{(2)}$ 38 percent of the original values are zeros.

The zero spike procedure forms an indicator matrix for each unit in the sample. For each variable, there is an element in the vector. The value of this element is "0" if the value of the variable is zero, "1" if the value is positive, and "2" if the value is missing. If unit A has a "2" in its indicator vector, then the "2" is changed to a "0" or "1" using probabilities based on S -- that subset of the sample units which: 1) is complete, and 2) matches the indicator vector of unit A for those variables reported on A. For example, if there are two variables, then the complete units can form four groups -- (0,0), (0,1), (1,0) and (1,1). If a unit has the form (0,2) then the "2" is changed to a "0" with probability $n_{(0,0)}/[n_{(0,0)} + n_{(0,1)}]$ or changed to a "1" with probability $n_{(0,1)}/[n_{(0,0)} + n_{(0,1)}]$ where $n_{(i,j)}$ is the number of complete units in the (i,j) group; i,j = 0,1. If a "2" is changed to a "0", the missing value becomes zero. If a "2" is changed to a "1", the missing value becomes a positive number of the form $\hat{R}x$, where x is the most highly correlated variable which also has a "1" in the indicator vector of A . R is the ratio which relates x to y and is estimated from the units in S .

Pregiborn actually recommends the use of any, even subject-ive, information to estimate the probabilities for assignments of "0" and "1" and not just the use of units in the sample. Thus, his recommendations allow a Bayesian approach to the imputation through the estimation of the probabilities. Also, Pregiborn (1978) notes that there are many possible methods -- hot decks, regressions, averages, etc. -- to decide what positive value to impute for a missing value. This study uses a ratio method because the first three procedures described also use a ratio or regression method in some way. Thus, in the comparisons of estimates from the procedures, any differences for the zero spike procedure are not mainly a result of the method used to determine positive values but mainly a result of the "zero-positive" structure employed.

2.5   The Princomp Procedure

This procedure uses the first principal component when im-puting for missing values. The first principal component is applied as a distance measure to select the complete unit which is

most like a unit with a missing value. The reported value for
this complete unit is then substituted for the corresponding
missing value. The first principal component is a linear combination of all reported variables and has the maximum variance of all
possible linear combinations of these variables. It is the line
of closest fit in the sense that it minimizes the sum of squares
of distances from data points to the line (note that a regression
line minimizes the sum of squares in particular directions).

For this study the princomp procedure: 1) constructs four
subsets of the data -- S1 contains the complete units, S2 contains
those units with the variable $y^{(1)}$ missing, S3 contains those
units with the variable $y^{(2)}$ missing, and S4 contains those units
with both variables $y^{(1)}$ and $y^{(2)}$ missing; 2) computes the
first principal component for S2 by using all 15 variables except
$y^{(1)}$ and then computes the value of the first principal component
for each unit in S1 and S2; 3) for each unit in S2, finds the
S1 unit which has a principal component value closest (minimum
absolute deviation) to the unit in S2 and substitutes the corresponding values of $y^{(1)}$ from the S1 unit into the missing values
of $y^{(1)}$ in the S2 unit; 4) repeats steps 2 and 3 to substitute
reported values from S1 for missing values in S3 and S4 by using
the principal component that corresponds to each subset.

The princomp procedure is essentially a hot deck procedure
(a hot deck procedure is defined as a procedure which substitutes
reported values for missing values) which substitutes by the
minimization of a distance function rather than substituting
randomly. There are many distance measures which could have been
tested, but the authors felt that only one procedure of this type
could be added to the experiment due to time and cost constraints
and that the princomp procedure is a distribution-free method
which has the potential for accurate imputation.

3. ANALYSIS

The goal of this analysis is to identify the "best" procedure of the six described procedures which impute for missing items.
There are five criteria for selection of the "best" procedure:
1) the accuracy of estimated means, 2) the standard errors,
3) the accuracy of imputations on a unit level, 4) the effect
on correlations between variables, and 5) costs.

3.1 Experimental Design

Three methods designate units which have missing items:
1) a random designation, 2) a 15 percent designation of
incomplete units below the median and 85 percent above, and
3) an 85 percent designation of incomplete units below the median
and 15 percent above. (The median of $z = y^{(1)} + y^{(2)}$ is used in
these designations.) For each of these three methods, there are
two rates to designate how many units have missing items -- 10
percent and 30 percent. The combined effect of the type of
designation and the rate of designation results in six different
situations in which means are estimated for the entire population.

Five data sets are simulated for each level of bias. Thus,
a total of 30 data sets are generated from the original data set.
Each data set consists of three replicates, and each procedure is
run independently on each replicate to provide unbiased estimates
of the standard errors. Within each data set the group of units
with missing values contains 40 percent of the units with $y^{(1)}$
missing, 40 percent with $y^{(2)}$ missing, and 20 percent with both
$y^{(1)}$ and $y^{(2)}$ missing.

The structure of the simulations corresponds to an analysis
of variance model. This model is:

$$Y_{ijk\ell m} = \mu + \alpha_i + \beta_j + \gamma_k + T_\ell + (\alpha\beta)_{ij} + (\alpha T)_{i\ell} + (\beta T)_{j\ell} + \varepsilon_{ijk\ell m}$$

where:

α_i = the effect of the ith designation method

β_j = the effect of the jth rate of designating how many units have missing items

γ_k = the effect of the kth replicate

T_ℓ = the effect of the ℓth missing item procedure

$(\alpha\beta)_{ij}$ = the interaction between designation method and rate

$(\alpha T)_{i\ell}$ = the interaction between designation method and missing item procedure

$(\beta T)_{j\ell}$ = the interaction between rate and missing item procedure

$\varepsilon_{ijk\ell m}$ = the error of the model associated with $y_{ijk\ell m}$

$Y_{ijk\ell m}$ = the value of the dependent variable associated with designation method i, rate j, replicate k, missing item procedure ℓ, and observation m

i = 1, 2, 3

j = 1, 2

k = 1, 2, 3

ℓ = 1, 2, ..., 6

m = a number which varies with the definition of the dependent variable.

An example of a dependent variable is the difference for a sample unit between the imputed value and the corresponding original value. If an analysis of variance shows a significant difference due to an effect, then Duncan's multiple range test is used to identify which levels of the effect caused the differences. All tests are at a five percent level of significance.

3.2 Results

An analysis of variance shows significant differences among the six missing item procedures when the dependent variable is the average difference between the imputed values and the "true" values. Table 3.2.1 gives the results of Duncan's multiple comparison test and the patterns that are characteristic of each

procedure. The ratio 1 and ratio 2 procedures are usually significantly different from the other procedures but not from each
other. The princomp and zero spike procedures also tend to be
different from the other procedures but are not significantly
different from each other. The array procedure does not show
consistent trends but tends to group with the princomp and zero
spike procedures. The ESTMAT procedure tends to be by itself.
Apparently the ESTMAT procedure is not robust to its normality
and random error assumptions because the estimated means from this
procedure are not very accurate under the random designation of
missing values. All procedures tend to underestimate the mean --
even when values are randomly missing. This underestimation may
not only be a result of biases inherent in the procedures but also
a result of the skewness in the underlying data.

The interaction between the designation methods and the
procedures is a significant effect. However, as Table 3.2.1 shows,
this significance is a result of the fluctuation of the ESTMAT
procedure in relation to the other procedures. The remaining
tables in this paper give overall results across designation
methods and rates. These overall results do not imply that the
interactions are insignificant, but, as in Table 3.2.1, they are
not important enough in this study to warrant the complexity of
presenting the results in each cell. Table 3.2.2, for example, is
much simpler and clearer than Table 3.2.1 and does not lose much
information.

Table 3.2.2 gives overall results for Duncan's multiple
comparison test in terms of average difference and relative bias.
In this table the relative bias is the average difference in
imputed and original values divided by the "true" mean of the
sample. Across both variables the ratio 1 and ratio 2 procedures
give the best results. It is disturbing that the ESTMAT procedure
can give the best results for $y^{(1)}$ and the worst for $y^{(2)}$.
This result may be an effect of the imputation part of the ESTMAT

Table 3.2.1: Results of Duncan's multiple range test* when the dependent variable is the average difference between the imputed value and the corresponding original value.

Variable	Designation Method					
	Random		15% Below Median/85% Above		85% Below Median/15% Above	
	Average Difference	Procedure	Average Difference	Procedure	Average Difference	Procedure
$y^{(1)}$	-0.133	Ratio 2	3.781	ESTMAT	5.781	ESTMAT
	-1.281	Array	-5.719	Ratio 2	-1.315	Ratio 2
	-2.359	Ratio 1	-5.922	Ratio 1	0.041	Ratio 1
	-5.285	ESTMAT	-10.715	Array	-4.104	Zero Spike
	-5.933	Zero Spike	-14.170	Princomp	-4.337	Princomp
	-5.756	Princomp	-15.870	Zero Spike	-5.759	Array
$y^{(2)}$	-0.852	Ratio 2	-7.567	Ratio 2	3.463	Ratio 2
	-1.500	Array	-8.711	Ratio 1	1.204	Ratio 1
	-2.104	Ratio 1	-14.378	Array	-2.641	Princomp
	-5.156	ESTMAT	-17.219	Princomp	-3.552	Array
	-5.815	Zero Spike	-18.330	Zero Spike	-4.285	Zero Spike
	-6.026	Princomp	-24.748	ESTMAT	-12.189	ESTMAT

* Any two means connected by the same bracket are not significantly different at α = 0.05.

Table 3.2.2:   Overall results of Duncan's multiple comparison test[*]

Variable	Procedure	Average Difference in Imputed Values and Original Values	Effect on Mean Estimates of Entire Population (Relative Bias)
$y^{(1)}$	ESTMAT]	1.426	+0.3%
	Ratio 2 ⌉	-1.512	-0.3%
	Ratio 1 ⌋	-2.747	-0.6%
	Array]	-5.918	-1.2%
	Princomp ⌉	-8.421	-1.7%
	Zero Spike ⌋	-8.636	-1.7%
$y^{(2)}$	Ratio 2 ⌉	-1.652	-0.4%
	Ratio 1 ⌋	-3.204	-0.8%
	Array]	-6.477	-1.3%
	Princomp ⌉	-8.629	-1.8%
	Zero Spike ⌋	-9.447	-2.0%
	ESTMAT]	-14.031	-2.9%

[*]
Any two means connected by the same bracket are not significantly different at $\alpha = 0.05$.

procedure since direct estimates from ESTMAT showed a relative bias of -1.3 percent and -0.2 percent for $y^{(1)}$ and $y^{(2)}$ when estimating the mean for the entire population -- a result which seems more reasonable. Thus, *imputations* using the ESTMAT procedure appear to be unreliable.

To judge the accuracy of the imputations at a unit level, Table 3.2.3 gives the total of the absolute differences between the imputed values and "true" values. The optimum procedure should minimize this total. Table 3.2.3 confirms the superiority of the ratio 1 and ratio 2 procedures and explains the contradictory results in Table 3.2.2 between $y^{(1)}$ and $y^{(2)}$ for the ESTMAT procedure. The ESTMAT procedure gives the lowest difference for

$y^{(1)}$ in Table 3.2.2 because of offsetting extremes in positive and
negative directions. Thus, when absolute differences are calculated, the ESTMAT procedure gives the largest totals for both
variable in Table 3.2.3.

Table 3.2.3: Total of the absolute differences between each
              imputed value and the corresponding "true" value

Variable	Procedure	Absolute Difference
$y^{(1)}$	Ratio 2	6,683
	Ratio 1	6,711
	Array	9,648
	Zero Spike	9,909
	Princomp	10,129
	ESTMAT	14,643
$y^{(2)}$	Ratio 2	7,132
	Ratio 1	7,439
	Array	9,862
	Princomp	10,873
	Zero Spike	10,922
	ESTMAT	14,137

Table 3.2.4 gives the coefficient of variation of the
estimated mean for the entire population. The coefficient of
variation is the standard error (an unbiased estimate calculated
using replicates) for a procedure divided by the "true" mean of
the sample. The coefficients of variation in Table 3.2.4 are
similar in size except that the ESTMAT procedure is larger for
$y^{(1)}$.

An overall measure of the quality of the procedures is
the root mean square error. This measure is defined as:

Table 3.2.4: Coefficients of variation for the estimated mean
of the entire population

Variable	Procedure	Coefficient of Variation
$y^{(1)}$	Zero Spike	0.062
	Princomp	0.063
	Array	0.065
	Ratio 1	0.065
	Ratio 2	0.070
	ESTMAT	0.100
$y^{(2)}$	Princomp	0.065
	Ratio 1	0.065
	Zero Spike	0.068
	Ratio 2	0.068
	ESTMAT	0.070
	Array	0.070

$$\sqrt{MSE'} = \sqrt{(\text{Relative Bias})^2 + (\text{Coefficient of Variation})^2}$$

The $\sqrt{MSE'}$ is sensitive to the sample size since the sample size
affects the magnitude of the coefficient of variation and sometimes the magnitude of the relative bias. Assuming, however, the
relative bias is not affected by the sample size, Table 3.2.5
displays $\sqrt{MSE'}$ for several sample sizes by using the relative
biases in Table 3.2.2 and the coefficients of variation in Table
3.2.4. Only for sample sizes larger than 1000 does the relative
bias component dominate the root mean square error rather than the
component due to the coefficient of variation. Thus, for very
large sample sizes, such as those often used in government surveys,
the two ratio procedures give the best results. For smaller sample
sizes, however, there is little difference in the procedures
except that the ESTMAT procedure is substantially larger for $y^{(1)}$.

Table 3.2.5: Root mean square error relative to the "true"
              sample mean

Variable	Procedure	Sample Size				
		50	100	1,000	10,000	∞
		(%)	(%)	(%)	(%)	(%)
$y^{(1)}$	Ratio 1	13.0	9.2	3.0	1.0	0.6
	Ratio 2	14.0	10.0	3.1	1.0	0.3
	Array	13.1	9.3	3.2	1.5	1.2
	Princomp	12.7	9.1	3.3	1.9	1.7
	Zero Spike	12.5	8.9	3.3	1.9	1.7
	ESTMAT	20.1	14.2	4.5	1.4	0.3
$y^{(2)}$	Ratio 1	13.0	9.2	3.0	1.2	0.8
	Ratio 2	13.6	9.6	3.1	1.0	0.4
	Array	14.1	10.0	3.4	1.6	1.3
	Princomp	13.2	9.4	3.4	2.0	1.8
	Zero Spike	13.8	9.8	3.6	2.2	2.0
	ESTMAT	14.3	10.3	4.3	3.1	2.9

When a data set contains imputed values, estimates of
standard errors are often calculated by ignoring the imputation
process and treating the imputed data set as though all the values
are reported. This method may lead to biases in the estimates of
standard errors. Table 3.2.6 gives the ratio of the variance
calculated by using the conventional formula and the variance
calculated by using replicates. The conventional formula treats
the imputed values as though they are original, reported values:

$$V_c(\bar{y}) = \frac{\sum_{i=1}^{n} (y_i - \bar{y})^2}{n(n-1)}$$

where y_i is the value for observation $i, \bar{y} = \sum_{i=1}^{n} y_i/n$, and n

is the number of observations both reported and imputed in a
simulated data set. An unbiased estimate of variance can be
calculated by using replicates:

$$V_r(\overline{y}) = \frac{\sum\limits_{i=1}^{n'} (y_i' - \overline{y}')^2}{n'(n'-1)}$$

where y_i' is the mean of replicate $i,\overline{y}' = \sum\limits_{i=1}^{n'} y_i'/n'$, and n' is
the number of replicates. Table 3.2.6 shows there can be large
biases in either direction for almost any of the procedures using
$V_c(\overline{y})$ as an estimate of the standard error.

Another important aspect of imputation is the effect on the
correlation structure of the data set. Although correlations are
not important for estimates of univariate statistics such as means
and standard errors, correlations are important when the data set
is used to explore and assess relationships among variables
through regression analysis, principal components, or other multivariate techniques. Table 3.2.7 gives an example of the effects
of the missing item procedures on the correlation structure. This
table shows the correlations between w and $y^{(1)}$ and between
w and $y^{(2)}$. Most of the procedures tend to lower the correlations, but the ratio 1 and ratio 2 procedures tend to inflate the
correlations.

The cost of each procedure for imputing data is shown in
Table 3.2.8. This cost is based on imputation for all 30 data
sets for the two variables $y^{(1)}$ and $y^{(2)}$. The system resource
units (SRU's) -- a measure of computer usage -- required by each
procedure are reasonably close except for the ESTMAT procedure.
ESTMAT requires more SRU's than the other five procedures combined.
This requirement is because of the complexity of the procedure.
Thus, cost alone imposes a severe restriction on the use of the
ESTMAT procedure. The other five imputation techniques are very
similar in cost with the ratio 1 and ratio 2 procedures costing

Table 3.2.6: Ratio of estimated variances of estimated means -- variance estimate assuming imputed values are reported values divided by unbiased variance estimate using replication

Variable	Designation Method	Imputation Procedure						
		Ratio 1	Ratio 2	Array	Zero Spike	Princomp	ESTMAT
$y^{(1)}$	Random	0.922	1.049	0.967	0.867	0.806	1.057
	15% Below Median/ 85% Above	1.084	1.172	0.970	0.902	0.746	1.316
	85% Below Median/ 15% Above	1.283	1.242	1.172	1.217	1.103	1.387
	Overall	1.096	1.154	1.036	0.995	0.885	1.253
$y^{(2)}$	Random	0.889	0.961	1.226	0.933	0.819	1.009
	15% Below Median/ 85% Above	0.869	0.933	0.980	1.063	0.838	1.136
	85% Below Median/ 15% Above	1.244	1.325	1.233	1.330	1.262	1.144
	Overall	1.000	1.073	1.146	1.109	0.973	1.096

the least.

Table 3.2.7: Correlations between w and $y^{(1)}$ and between
w and $y^{(2)}$ for six missing item procedures.

Procedure	Variable					
	$y^{(1)}$			$y^{(2)}$		
	Random	15% Below Median/ 85% Above	85% Below Median/ 15% Above	Random	15% Below Median/ 85% Above	85% Below Median/ 15% Above
(Actual)	.82	.79	.84	.81	.72	.77
Ratio 1	.97	.94	.97	.89	.66	.93
Ratio 2	.88	.86	.94	.89	.72	.94
Array	.57	.73	.83	.54	.62	.76
Zero Spike	.80	.53	.74	.67	.42	.62
Princomp	.72	.60	.77	.68	.28	.74
ESTMAT	.79	.59	.81	.72	.33	.20

Table 3.2.8: Processing costs of six missing item procedures.

Procedure	SRU's [1]	Cost [2]
Ratio 1	937	$ 145
Ratio 2	938	$ 150
Array	1278	$ 205
Zero Spike	1215	$ 194
Princomp	1103	$ 177
ESTMAT	9604	$1537

[1] SRU: System resource unit
[2] Cost projected at 16¢ per SRU

4. SUMMARY

This study serves as an example of using a simulation
experiment to make a preliminary assessment of the impact of
imputation procedures on a specific survey. The scarcity of theory
and guidelines about imputation procedures in the statistical
literature causes the need for simulation experiments which assess
the effects of these procedures. The *preliminary* nature of the
simulation experiment in this study deserves emphasis. For large
scale, repetitive surveys the objective of a simulation study is
to winnow the many possible procedures and their variations to a
few procedures. Further research under actual survey conditions,
efficient computer programming, and other more costly requirements
should be used to narrow these few procedures to one operational
procedure.

This study compares the effects of six procedures which
impute for missing items -- two versions of the ratio procedure,
the array procedure, the ESTMAT procedure, the zero spike procedure, and the princomp procedure. The comparison of these
procedures is from an experiment in which a complete data set has
values deleted to simulate an incomplete data set. Simulations
are over a range of conditions which account for the method of
designating missing values and the percentage of missing values.
Comparisons of the procedures are made with respect to: 1) the
accuracy of the estimated means, 2) the standard errors,
3) the accurary of imputations on a unit level, 4) the effect on
correlations between variables, and 5) costs.

The two versions of the ratio procedure perform the best
for very large sample sizes (at least as large as 1000). For
smaller sample sizes all of the procedures except the ESTMAT
procedure have approximately the same mean square error. The
main disadvantage of the ratio procedure is an inflation of the
correlations between variables in the data set.

The ESTMAT procedure emerges as the least attractive procedure because it does not impute very accurately and it has an extremely high cost relative to the other procedures. This result only applies to the ESTMAT procedure as an *imputation* process and not as a missing data procedure in general. For example, the ESTMAT procedure is probably a suitable method for sample designs in which missing data is planned -- in other words, a survey design in which one plans to collect only partial information on some designated units.

Standard errors of the estimates from any missing item procedure should account for the fact that data is imputed. Replication is a method to obtain unbiased estimates of standard errors. Estimates of standard errors which treat the imputed values as though they are original, reported values may be biased up or down depending on the procedure and the particular situation.

Finally, the reader is cautioned that the results of the study are based on one data set. The variables of this data set have skewed distributions dominated by zero values. These distributions are characteristic of much survey data but not all. Generalizations must wait until other survey organizations apply and compare procedures on their own data so that an empirical body of knowledge about imputation procedures can be created.

ACKNOWLEDGEMENT

The assistance of Ann Adams with parts of the computations in this paper is gratefully acknowledged.

REFERENCES

Beller, N.D. and Bynum, H.E. (1971). Multiple Frame Hog Survey;
    Nebraska's Hot Deck Edit Procedure, Version 2. Documenta-
    tion of computer program.

Ford, B.L. (1976). Missing data procedures: a comparative study,
    Proceedings of the Social Statistics Section. American
    Statistical Association, Washington, D.C., 324-329.

Ford, B.L., Hocking, R.R. and Coleman,A.(1978). Reducing respondent
    burden on an agricultural survey, *Proceedings of the Section
    on Survey Research Methods*. American Statistical Associa-
    tion, Washington, D.C., 341-345.

Hartley, H.O. and Hocking, R.R. (1971). The analysis of incomplete
    data, *Biometrics,* 27, 783-823.

Hocking, R.R., Huddleston, H.F. and Hunt, H.H. (1974). A procedure
    for editing survey data. *Journal of the Royal Statistical
    Society (Series C),* 23, 121-133.

Pregiborn, D. (1976). Incomplete survey data: estimation and
    imputation. *Survey Metholology,* 2, 70-103.

Pregiborn, D. (1978). Discussion of papers by Huddleston and
    Hocking and Patrick, *Proceedings of the Section on Survey
    Research Methods*. American Statistical Association,
    Washington, D.C., 492-493.

DATA ADJUSTMENT PROCEDURES IN THE 1981 CANADIAN CENSUS OF POPULATION AND HOUSING

Christopher J. Hill and Henry A. Puderer

Statistics Canada

This paper is a general discussion of the methods of adjustment for missing and inconsistent data that will be applied to the data from the 1981 Canadian Census of Population and Housing. The scope of the discussion is deliberately broad including all methods of adjustment rather than limiting itself to the narrowly defined topic of the main edit and imputation operation. A Census because it is on so large a scale and is all inclusive allows for the possibility of applying certain methods not normally available to other surveys. Given the general nature of the discussion, the generalized system known as CAN-EDIT that was developed for the 1976 Census and is being reused in the 1981 Census will only be discussed in general terms.

1. INTRODUCTION

The complete estimation process for a survey or census can be considered as comprising two aspects, estimation to take account of the survey design and estimation or adjustment for errors or discrepancies from this design. Errors or discrepancies from the design may arise either as a result of mistakes made by the survey staff or as a result of respondent behaviour. Both these sources of error are impossible to eliminate, costly to reduce below some reasonable minimum and difficult to accurately estimate. The result of these errors may in turn result in either data problems that are directly identifiable in the micro-data as missing or

inconsistent responses or as errors that may only be detected by
means of an external evaluation.

One may recognize two broad classes of methods of adjustment,
imputation methods and weighting methods. Both methods will be
applied to the 1981 Census of Population and Housing. It is also
possible to apply methods which are mixtures of the two approaches
or cannot easily be classified as falling into either class. It
is planned that methods of this kind will also be used in Census[1].

The discussion to follow is presented in three sections.
Section 2 discusses the factors that necessitate adjustment procedures in the Census, namely Survey Design related factors and
Deviations from Survey Design related factors. As part of this
discussion, and for the purpose of orienting the later discussion
of adjustment procedures which follows in sections 3 and 4, a
typology of deviations from the Survey Design is presented. In
Section 3 the methods of adjustment to be applied to the 1981
Census are discussed. This includes a short discussion of the
generalized system known as CAN-EDIT that was developed for the
1976 Census and is being reused in the 1981 Census. Anyone interested in a more detailed discussion can find this reported elsewhere
(Fellegi & Holt, 1976; Graves, 1976; Hill, 1978). Section 4 is
devoted to a more detailed discussion of the edit and imputation
step in the adjustment process. Due to the complexity of the
adjustment procedures, the focus of sections 3 and 4 is that of

---

[1]
  Imputation may be defined as a method of adjustment in which
  incompleteness or inconsistency in a data record is eliminated
  by assigning new values to one of the variables in that record
  and where these values are obtained by means other than reference back to the source of the data.

  Weighting may be defined as a method of adjustment in which
  incompleteness in data records are taken account of in the final
  estimates by weighting up the complete data. This method may
  also be applied to inconsistent data if some of the variables
  contributing to the inconsistency are treated as equivalent to
  missing data items.

an overview, with appropriate references to more detailed reports.

2. FACTORS THAT NECESSITATE ADJUSTMENT PROCEDURES IN THE CENSUS

2.1 Survey Design Related Factors

The 1981 Census Data will be collected by means of two
questionnaires, a short form sent to four fifths of all households,
and long form sent to one fifth of all households. The short form
consists of dwelling questions and a set of questions asked of all
persons permanently resident in the household. These questions
are a set of basic demographic questions on age, sex and marital
status, a question on each person's relationship to a household
reference person and a question on their mother tongue. The long
form includes all the short form questions together with additional
housing questions, and questions on education, residential mobility,
cultural characteristics and a variety of economic variables. Most
of these questions are only requested of persons age 15 and over.

Two features of the survey design for the 1981 Census give
rise to specific adjustment requirements, namely, the use of
sampling (1/5 household sample) and the 'de jure' principle of the
Canadian census.

The collection of data for some variables on a sample basis
has been applied since the 1961 Census. In the 1981 Census, as in
previous Censuses, this data will be published as weighted estimates rather to the raw counts. Further information on this is
given below.

The question of adjusting data to conform to the 'de jure'
principle does however require some discussion. A 'de jure' Census
is a Census in which a person is counted at their usual or legal
place of residence. The alternative is a 'de facto' Census in
which a person is counted at their location on Census night.
Given a 'de jure' Census, persons found 'de facto' must be matched
back to their 'de jure' place of residence. On finding a person

'de facto' who may or may not have been also found 'de jure' the
Census has three options:

(1) Ignore these people resulting in omitting that proportion
     that have been missed in the 'de jure' coverage.

(2) Attempt some strategy of 'trace back' to include all these
     persons at their 'de jure' residence.

(3) Adopt an estimation procedure to adjust for these persons.

The Canadian Census follows the third strategy which is therefore
within the scope of this discussion.

2.2 Deviations from Survey Design Related Factors

     For the purpose of this discussion, it is useful to first
start with a broad classification of the types of deviation error.
An important distinction can be made between those deviations from
the survey design that cannot be detected within the survey process
and those which can. In the case of the former, detection is
usually the result of a post survey evaluation. In general the
results of these evaluations have no effect vis-à-vis adjusting
the statistics of the current survey. Rather, their impact is
often in the form of appropriate warning notes to users, restrictions on the release of certain statistics and input to future
surveys - often in the form of new adjustment procedures or
modifications to existing adjustment procedures. In contrast,
those deviations which can be detected may be directly linked to
procedures which adjust the statistics of the current survey before
their release.

     In terms of errors that represent deviations from the survey
design, we have identified three broad types - coverage, geographic
and response. (This includes both errors occurring during collection and errors introduced during processing.) Coverage error
occurs when a unit of observation is missed, incorrectly included
or counted more than once. Geographic errors occur when the
element being measured is positioned in the incorrect geographic

location. Response and/or processing error includes three sub-
types. These are invalid code (code outside the range of acceptable values); non-response (either total non-response from an
element, in this case a household or person record, or partial
non-response) and classification error. Classification error may
be either inconsistent or consistent. In the former instance
values between two or more response and/or control fields are by
definition not possible, while in the latter case although an
incorrect value(s) is present it is still by definition within the
range of acceptable values, i.e., consistent.

We now turn out attention to a general discussion of the
adjustment procedures to be applied to the 1981 Census with respect
to those required as a consequence of the survey design and those
required as a consequence of the deviations from the survey design.

3. METHODS OF ADJUSTMENT TO BE APPLIED TO THE 1981 CENSUS

The methods of adjustment to be applied to the 1981 Census
may be classified into four groups:

(1) Adjustments made during a preliminary grooming operation[2],

(2) A weighting adjustment applied to sampled data,

(3) Adjustment procedures to correct for special populations,

(4) Edit and imputation procedures.

The weighting adjustment applied to sample data and a special
population adjustment procedures to assign 'de facto' enumerated
population to their 'de jure' residence are required to take

---

[2] Preliminary grooming refers to a series of adjustments which
occur prior to the creation of a census data base. The distinction is of importance with respect to the adjustment procedures.
Preliminary grooming involves some form of source re-observation
(original source re-observation excluded) while post Census data
base adjustment procedures are procedures which effectively
exploit the information provided by the collected data.

account of the survey design. Adjustment for deviations from the
survey design will utilize all the above methods.

3.1 Adjustment procedures to Take Account of Survey Design

3.1.1 *Weighting for sampling*

The total estimation procedures for a survey must take into
account adjustments of sample data as dictated by the survey design.
The method of adjustment for sampling used by the Census is to
determine weights by a raking ratio method to be assigned at the
individual record level (Arora and Brackstone, 1977). As discussed
in 3.2.3. adjustment for total nonresponse to sample questionnaires
is treated as part of this process.

3.1.2 *Adjustment for persons temporarily absent from home*

The Canadian Census is a 'de jure' Census. A small percentage of the population however are away from home on Census night
and are enumerated as temporary residents where they have been
found. These persons may or may not have been counted at their
usual place of residence. A corrective action should be taken for
those who have been missed at their usual place of residence. In
principle all these persons could be re-assigned to their 'de jure'
residence. The cost and time delay associated with this re-assign-
ment is however considered unjustified. A sample of persons is
therefore traced and an adjustment is made based on this sample.
This method includes an estimate of the number of additional
persons to be added for each geographic area and a method of
estimating for their characteristics[3]. Given a certain number of

---

[3] Post 1971 censuses evaluations identified a geographic error in
the adjustment procedures for the temporarily absent population.
This error was a consequence of too simple a procedure for
assigning the temporarily absent population on a de jure basis.
Procedures for 1976 were subsequently amended and are to be
employed again for 1981. This sequence of events reflects the
manner in which evaluations of deviations which cannot be detected within the survey process can affect adjustment procedures
in succeeding surveys.

persons are to be added this is achieved by selecting donor persons
to represent the additions where the donors are selected to conform
to a previously determined distribution of basic demographic
characteristics. This representation is achieved in practice by
a method of Weighting or Hot Deck. This adjustment procedures
increases the population count by about 0.50% compared with an
estimated undercoverage of 2.01% (Statistics Canada, 1976).

The population enumerated outside of Canada is also matched
back to a 'de jure' residence. The procedures differ somewhat
from those applied to the temporarily absents population in that
no sampling is performed (i.e., a 100% match back is applied) and
the records (household and person) are placed on the data base in
contrast to the weighting method utilized for the temporarily
absent population.

3.2 Adjustment Procedures to take Account of Deviations from Survey Design

3.2.1 Coverage error

Evaluation studies undertaken during previous Censuses
have indicated that there is an underestimate of persons resulting
from the misclassification of occupied dwellings as unoccupied.
This misclassification accounts for an undercoverage of persons of
0.26% (Theroux, 1976). The plans for 1981 include a procedure to
adjust for this undercount, similar to the one being used for
persons temporarily absent from home.[4] The adjustment procedure
involves a sample follow-up to estimate the proportion of occupied
dwellings misclassified as vacant, and their distribution in terms
of type of dwelling and the number of persons in household, by
geographic area. These estimates will then be reflected in an
adjustment at the record level by selecting households to represent

---

[4] This is another example of how post survey evaluations of
deviations from the survey design which cannot be detected
within the survey can lead to improved procedures for future
surveys.

the misclassified dwellings. These households will be weighted
accordingly. To balance this addition an equivalent number of
vacant dwellings will be excluded from the population of dwellings.
It is appropriate to note at this point that no procedures are in
place for 1981 Census of Housing and Population to adjust the
final estimates for population undercoverage. Current plans call
for only a continuance of the evaluation studies similar to the
1976 studies which serve to estimate the degree of undercoverage
(Statistics Canada, 1976). However, work is continuing with
regard to the implications of adjusting for population undercoverage. Considerable investigation is currently in progress with
respect to this issue by the Demography Division and the Census
and Household Surveys Methodology Division of the Social Statistics
Field of Statistics Canada.

3.2.2 *Geographic Error*

An adjustment of the preliminary grooming type is performed
when it is determined that the Census Representative (CR) did not
restrict his/her enumeration activity to the designated enumeration
area (EA). The adjustment involves manually reassigning question-
haires to their correct EA.

3.2.3 *Response and/or Processing Error*

Adjustment for response or processing error initially
involves re-observation techniques during the preliminary grooming
operations whereas the final adjustment involves estimation using
both edit and imputation and weighting - procedures which exploit
existing information on the data base.

Adjustments during the preliminary grooming operations for
response and/or processing deviations from the survey design include
corrections due to range checks on certain coded variables (e.g.,
industry, occupation, geographic variables as related to mobility st
and place of work); probability edits which help to trap a classific
tion (consistent) error (e.g., income) and various quality control

procedures the objective of which are to restrict processing
error to predefined limits. Throughout the preliminary grooming
operations the adjustment technique generally involves some form
of re-observation. For example, verification of the output (which
fails edit) to the input source, followed by adjustment, when verification procedure determines incomparability between output and
input.

The adjustments during the preliminary grooming operations
also include a series of ad hoc procedures which are a mixture of
clerical corrective actions and resolutions of inconsistencies
that are essentially imputation actions. The primary function of
these adjustments is to fix coverage by resolving whether or not a
person or a household exists and to freeze the allocation of persons
to households and both persons and households to geographic locations. This process therefore, determines, subject to certain
adjustments for special populations, the official population count.
The impact of these adjustments is generally slight and will not
be discussed further except to say that given ambiguous information
the procedures veer on the side of assuming that a person exists.

As noted previously both weighting and edit-imputation
procedures consistute the final series of planned adjustments for
Response and/or Processing Error in the 1981 Census of Housing
and Population. In case of a sampled household, where a *response*
to the sample questions has not been obtained for any household
member, the plan in 1981 is to incorporate a weight adjustment
for this type of household rather than impute values for the
missing data. In previous Censuses a household was determined to
be in the sample by design. In 1981 however households will be
treated as not in the sample if de facto they have not responded
to the sample questionnaires. In other words the weighting procedure will be used to adjust for total non-response to sample
items rather than imputing for these records.[5]

The final adjustment for deviations from the sample design

vis-à-vis response and/or processing error will be accomplished
via CAN-EDIT procedures (similar to those used in 1976) complement-
ed by a series of linked custom written system modules that may be
known as 'E & I System '81.' A more detailed description of the
approach followed by these edit and imputation systems in detecting
response and/or processing errors and subsequently adjusting the
statistics, is discussed in Section 4.

4. THE 1981 EDIT AND IMPUTATION SYSTEM

4.1 General Background

The operation of editing and imputing the data is concerned
with two quite distinct objectives that may be only partially
reconciled. The specific objective of edit and imputation is to
ensure that the survey data conforms to a conceptual framework.
A more general objective,one which is common to most survey
methods,is the minimization of the mean square error. Clearly
in the case of a multi-purpose survey we are concerned with a
principle of generally minimizing the mean square error.

These two objectives may be formulated into the following
principles.

(1) The data in each record shall be made to satisfy all the edits.

(2) In order to ensure that the data satisfies all the edits,
erroneous records may be modified so that either

2.1 They emulate 'clean' data records

OR

2.2 They fall into a data space predefined as correct

5
The Census of Canada has in previous censuses treated them as
separate operations. Previously, households determined to be
in the sample but in which sample questions were not answered,
had values imputed during edit and imputation.

by a logical deduction from the editing specifications.

(3)   In order to optimize the quality of the data after correcting

3.1 The amount of data changes shall be minimized

AND

3.2 As far as possible the imputation will preserve the
distributional properties of the data.

The preservation of the distributional properties of data is a
deliberately vague formulation. Ideally, of course, one wishes
to approach closer to the true distributions. In practice however
the best that can be achieved is to minimize the change from the
original data.

4.2 CAN-EDIT

A general consideration of the realization of these principles in the 1981 Census edit and imputation system will now be
given. The Fellegi and Holt methodology was not only implemented
in the system known as CAN-EDIT but has also been a major focus
of much subsequent thought on the problems of edit and imputation.
They proposed that a system should be developed using the following criteria.

(1)   The data in each record should be made to satisfy all edits
by changing the fewest possible items of data (fields);

(2)   As far as possible the frequency structure of the data file
should be maintained;

(3)   Imputation rules should be derived from the corresponding
edit rules without explicit specification.

The task of developing a generalized system was made possible by:

(1)   A methodology for ensuring that the corrective action would
pass the edits without the requirements that the edits be
re-invoked.

(2) A methodology for ensuring that the logical consistency of
     the edits can be verified during a pre-edit stage.

(3) The application of two general assumptions about the nature
     of errors;

    (i)   that the variables in error are the least number of
          fields that must be changed to satisfy the edits.

    (ii) that within a set of records matched on certain
          variables the erroneous records are a randomly selected
          sub-set. This in turn implies that the information in
          a   variable to be changed does not contribute
          additional information that could predict the true
          value for that variable and that the donor values
          can be found by means of a hot-deck search.

This methodology meant that imputation was controlled by general
principles together with the edit rules that are a variable part
to the system. The only additional control the specifier has over
the imputation action is the ability to give additional matching
conditions. Both the edit rules and the additional matching
conditions are user parameters.

     The CAN-EDIT system may be considered as consisting of four
distinct processes. The first of these processes stands along,
the other three are linked. These processes are:

(1) A pre-edit process that involves an analysis of the logic of
     the edit and imputation processes.

(2) An edit process that detects erroneous conditions in the data.

(3) A process that selects which variables need to be imputed.

(4) An imputation process that selects values to be donated to
     the selected variables.

Pre-edit: The automation of the pre-edit stage of analysing the
logic of the edits and the imputation actions is a recent develop-

ment. This stage consists of two operations:

(i)   the analysis of the edit and imputation rules,

(ii) the incorporation of these rules into a computer system.

In the past these operations have traditionally proved to be a
high risk phase of the operation, depending upon the specifier to
give complete and unambiguous specifications and the systems
development staff to render a correct realization of these rules.
As Fellegi and Holt (1976) suggest, this approach

"..... offers both an assurance that the rules are logically
correct and allows flexibility to rapidly change the rules during
the development stage".

The edit: The edit stage of the process may be considered as a
totally passive operation. Its functions is purely a diagnostic
process of identifying erroneous conditions. At this stage no
attempt is made to select between alternative corrective actions.
It should be noted that in the past edit and imputation procedures
have often intermingled edits and imputations with confusing
results.

The selection of the variables to impute: The first stage of the
imputation process is the selection of which variables should be
changed during imputation. The identification of this process as
a distinct and important step is sometimes overlooked. It may be
treated as part of an edit stage that determines not only the
erroneous condition but also which variables are judged to be in
error or an imputation stage that determines which variables
should be changed and to what value. We believe however that it
is very important to identify this process as distinct even though
it may in some instances be conveniently combined with the second
stage of imputation that, of selecting a value.

Imputation: The final stage of the process is the selection of
values for the variables to be imputed. There are clearly many
different methods by which this may be done. In the case of

CAN-EDIT this is done entirely by means of a selection from a hot-
deck i.e., a set of records of which the erroneous record is
normally a member. IT can be noted at this point that other more
common methods of imputation which will be applied to the 1981
Census but which fall outside the scope of CAN-EDIT are:

   (i)   Selection at random from a pre-defined distribution,

  (ii)   Selection by a set of pre-defined rules from other records
       linked to the record to be imputed e.g., imputing the
       value of a variable from that variable for other persons
       in the same household;

(iii)   Direct assignment based upon the other variables in the
       record.

The CAN-EDIT system was developed and applied in 1976 to
edit and impute most of the variables. These variables were:

(1)   Age, sex, marital status and relationship to head (100%
      person variables);

(2)   The household variables type of dwelling and tenure;

(3)   Education and labour force status (sample variables).

In 1981 CAN-EDIT will be re-used for these variables and will in
addition be applied to:

(4)   The dwelling variables collected on the sample questionnaire;

(5)   Fertility;

(6)   A classification of Occupation and Industry at the major group
      level.

Given the availability of a generalized system and the
resulting economy of development costs some explanation is required
concerning why CAN-EDIT was not applied more generally. CAN-EDIT
was not used in 1976 for either Mother Tongue or Mobility Status.
Again, in 1981 it will not be used for these variables or for
Nuptuality, Place of Work, Income, Cultural Characteristics or

Shelter Costs. The following reasons were given for not using CAN-EDIT for Mobility Status in 1976.[6]

(i) CAN-EDIT by design gave rise to the imputation action directly from the edit rules precluding any error specific data corrections. In the case of mobility status data, where previous evaluations indicated certain error patterns had very specific causes demanding very specific corrections, this feature of CAN-EDIT was found to be an unacceptable restriction.

(ii) The mobility status imputation strategy requiring a within household search for a suitable donor record could not be performed using CAN-EDIT.

(iii) Mobility status edit imputation involved lengthy code lists with i.e. over 128 codes in the case of previous and current place of residence while CAN-EDIT capability was limited to relatively short value sets.

The limitations of CAN-EDIT for certain variables can be broadly defined as:

(1) the specific imputation methodology is inappropriate.

(2) The variable requires a lengthy code list outside the scope of CAN-EDIT.

The first limitation could be one circumvented by a front end program that does various custom made fixes or imputations. Such an approach however is only cost justified if the major part of the imputations pass through CAN-EDIT and the fix-up program is not made unduly complicated in order to fit with CAN-EDIT. The second limitation in CAN-EDIT is one of practice rather than principle. In order to develop a generalised system all variables are treated as equivalent,

---

[6] This was taken from an internal document by H.A. Puderer, *An Overview of the 76 Edit and Imputation of Mobility Status*, December, 1978.

equally likely to be used as matching conditions when imputing
or as variables to be imputed. In these circumstances, were
the code lists to be unrestricted, exact matching would
become rare and the potential number of edit rules infinite.
Other solutions to the problem were therefore needed.

4.3 E & I System '81

In 1976 the approach taken to variables handled outside
CAN-EDIT was to treat each by a separate program. In 1981 the
plan is to develop a single system that will utilize generalized
modules wherever possible. As these are still in the development
stage it is only possible to give a general picture of what is
planned:

(1) We clearly recognize the virtue of having a pre-processor
     that will analyse the specific rules to be entered into the
     system. The system in 1981 will analyse both edits and
     imputation actions in the form of decision tables to ensure
     that they are logically correct. These will then be entered
     into the system as parameters.

(2) Depending upon the error conditions the actions taken by the
     system may be either a deterministic imputation or a resort
     to a hot-deck search. Where specific actions are not required
     a minimum change algorithm will be invoked to limit the amount
     of change in the data.

(3) All errors not resolved by a specific action will be imputed
     by a common hot-deck imputation module.

This system loses the generality of CAN-EDIT in that:

(1) A distinction is made between variables defining edit
     conditions and controlling matching conditions, and variables
     to be imputed.

(2) The inputs to the system are no longer simply edit rules, as
     user parameters, but are decision tables that involve the

active participation of system personnel in order that they may be entered into the system.

5. SUMMARY

The adjustments to be applied to the 1981 Census of Population and Housing can be summarized as follows.

A. Procedures to take account of Survey Design:

    (i)   A raking ratio weighting procedure to scale up sample items to population total.

    (ii) A special adjustment procedure for assigning 'de facto' enumerated population to their 'de jure' residence.

B. Procedures to take account of Deviations from Survey Design:

    (i)   A special adjustment procedure for undercoverage caused by the misclassification of occupied dwellings as unoccupie dwellings.

    (ii) A modification of the sample weighting procedures (Ai) above to include as sampled households those in which no household member responded to the sample questions.

    (iii) A series of quality control and edit procedures to fix the count of households and persons to a specific geographic location prior to creation of a Census data base.

    (iv) A series of quality control and edit procedures to ensure that data processing was within predetermined limits of deviation from originally observed values.

    (v) An Edit and Imputation procedures to make final adjustments for response and/or processing error.

REFERENCES

Fellegi, I.P. and Holt, D. (1976). A systematic approach to
    automatic edit and imputation. *Journal of the American
    Statistical Association*   71, 17-35.

Graves, R.B., (1976). Can-Edit. A Generalized Edit and Imputation
    System in a Data Base Environment. A report to the working
    party on electronic data processing, Conference of European
    Statisticians, 1976.

Hill, C.J. (1978). The application of a systematic method of
    automatic edit and imputation to the 1967 Canadian census
    for population and housing. *Survey Methodology* 4, 178-202.

Statistics Canada (1976). Quality of Data, 1976 Census, Series 1:
    Coverage Errors in the 1976 Census of Population and
    Housing. Statistics Canada, Ottawa, Catalogue No. 99-840.

Theroux, G. (1976). Parametric Evaluation Project, Vacancy Check,
    Final Results, Part 1. Statistics Canada, Ottawa.

Arora, H.R. and Brackstone, G.J., (1977). An investigation of the
    properties of raking, ratio estimators with simple random
    sampling. *Survey Methodology* 3, 62-83.

ISSUES OF NONRESPONSE AND IMPUTATION IN THE SURVEY OF INCOME AND PROGRAM PARTICIPATION[*]

Graham Kalton

University of Michigan

Daniel Kasprzyk

Department of Health and Human Services

Robert Santos

University of Michigan

This paper describes the extent and nature of the household-, person- and item-level nonresponse that the U.S. Survey of Income and Program Participation (SIPP) is likely to encounter, based on experience obtained from the 1978 Income Survey Development Program Research Panel. Some issues involved in developing nonresponse adjustment and imputation strategies for the SIPP are reviewed, and two promising imputation procedures are examined. Alternative, closely related, information was collected for certain questions which were expected to experience sizeable degrees of nonresponse; one procedure uses this information in imputing for the missing responses to those questions. The other procedure uses responses to a question on one wave of the panel in imputing for missing responses to the same question on another wave.

---

[*] This material is based in part on work performed under contract no. HEW-100-79-0127 with the U.S. Department of Health and Human Services. Views expressed are those of the authors and do not necessarily represent the official position or policies of the Department.

1. INTRODUCTION

The Survey of Income and Program Participation (SIPP) is a
U.S. national longitudinal survey program which will begin in 1982.
It is a program of the Departments of Health and Human Services
(HHS) and of Commerce designed to obtain detailed intra-year
information on household composition, money and in-kind income,
assets, liabilities, program eligibility criteria and program
participation, labor force participation, taxes, and selected
topics of policy interest. The SIPP will collect information
through personal interviews in panel households, and will link the
information to data from various administrative record systems.
The goals and objectives of the SIPP are described in more detail
by Lininger (1980).

In preparation for the SIPP, a development program - the
Income Survey Development Program (ISDP) - was established in 1975
to examine and resolve content, operational and technical issues
for the survey (see Ycas and Lininger, 1980, for a review of the
ISDP). Two SIPP prototype national household panel surveys, known
as the 1978 and 1979 ISDP Research Panels, have been conducted in
this program to aid in the SIPP design decisions. Since the
extensive amount and sensitivity of the data to be collected in
the SIPP and the panel aspect of the SIPP design could give rise
to a greater degree of nonresponse than would be encountered in a
simple one-time survey, the issues of nonresponse, nonresponse
adjustments and imputation are of special concern to the ISDP.
These issues are being examined by a team from the Survey Research
Center at the University of Michigan in conjunction with HHS and
the Bureau of the Census.

The purpose of this paper is to describe the extent and
nature of the nonresponse likely to occur in the SIPP, based on
experience obtained in the 1978 ISDP Research Panel, and to review
some of the issues involved in deciding on appropriate nonresponse

adjustment and imputation procedures for the SIPP. As yet, work
on procedures for dealing with nonresponse in the SIPP is in its
early stages. In consequence, this paper aims only to examine
various aspects of the problem, and to indicate some possible
solutions; further research will be carried out before decisions
are reached on the procedures to be adopted.

The rest of the paper is organized in four sections. Section
2 provides a brief description of the 1978 ISDP Research Panel,
and Section 3 outlines the nonresponse encountered in that Panel,
including both whole-unit and item nonresponse. Section 4 examines
the utility of two specific imputation schemes for item nonresponse
and Section 5 reviews some general issues that need to be considered in developing a nonresponse adjustment and imputation strategy
for the SIPP.

2.   THE 1978 ISDP RESEARCH PANEL

The 1978 ISDP Research Panel was designed to be a nationally
representative feasibility test for the SIPP. It was carried out
in 60 Census Primary Sampling Units (PSU's) with an initial sample
of 2,358 sample units, consisting of 1,947 area probability sample
housing units and 411 persons drawn from Supplemental Security
Income (SSI) files of the Social Security Administration. The
area sample was defined in terms of the households occupying
sampled residential units and the SSI sample was defined as the
households in which the sampled persons were found to be living.
When the list addresses of SSI sampled persons turned out to be
incorrect (as frequently happened), attempts were made to trace
the persons to their current addresses. The initial sample of
persons was defined as all those aged 16 or over living in the
sampled households.

Five interviews were conducted with each panel household at
quarterly intervals, the first wave of interviews taking place in

April, 1978 and the last in April, 1979. All members of the
initial sample of persons were included in the Panel. Field procedures specified that movers would be followed to their new
addresses; however, for cost reasons, those moving fifty miles
beyond the sample areas were not followed. Persons who moved in
with members of the initial sample (or vice-versa) became part of
the sample.

   The initial interview of the Panel collected a detailed
income profile, which was then updated on subsequent waves.
Additional information was also collected at each wave on a non-
recurring basis. The general contents of each quarterly interview are summarized below:

> *April, 1978:* Detailed income profile and a series of
> attitudinal questions for self-respondents on satis-
> faction with life and perceived adequacy of past, present
> and future income.

> *July, 1978:* Income update; items on the value of assets
> that are used to screen for program eligibility and items
> on expenses that are used to adjust the recipient's gross
> income when determining eligibility and benefit amounts;
> and an extension to questions on educational and medical
> expenses to gather information on school enrollment and
> receipt of in-kind income.

> *October, 1978:* Income update; items on disability to
> provide additional information on program eligibility;
> "personal history" questions on marital history, job
> history, educational attainment, and migration; and,
> for self-respondents, attitudinal questions similar to
> those asked in April, 1978.

> *January, 1979:* Income update; values of assets and
> liabilities - including a net worth worksheet sent to
> respondents in advance of the interviews; and information
> on life cycle earnings.

> *April, 1979:* Annual round-up of income and taxes for the
> calendar year 1978, using records such as W-2 forms where
> available; persons who had not been interviewed on one or
> more waves were asked briefly about income during the
> missing quarters; where sale of assets had been reported
> earlier, data on the resulting capital gains were collected;
> and tax data were collected to adjust gross into disposable
> income.

The results given here are mainly restricted to the April
and July waves of the 1978 ISDP Research Panel. Before presenting
the results, some general comments on them need to be made. First,
they should be viewed as preliminary; more detailed analyses are
underway. Minor inconsistencies may occur in the results of
different analyses both because the 1978 Panel data were unedited
and because the analyses were not all based on the same data
source. Secondly, all the results are unweighted, that is no
adjustments have been made to compensate for households' unequal
selection probabilities. A major cause of the inequality in
selection probabilities is the Panel's combination of the two
component samples, the area sample and the SSI sample. Many of
the analyses presented are restricted to either one or the other
of these samples, usually to the larger area sample. Thirdly, it
should be noted that because of the experimental and developmental
nature of the 1978 Panel, the levels of nonresponse it experienced
may be somewhat higher than those that will be encountered in the
ongoing SIPP program.

3. NONRESPONSE IN THE 1978 ISDP RESEARCH PANEL

3.1. Household and Person Nonresponse

Nonresponse occurred at the household level when no data
were collected for any household member. One source of this type
of nonresponse arose from adults in original sample households
changing residences. The sample design called for following all
adults in the original sample households, but for cost reasons
the fieldwork procedures stipulated that adults moving beyond
fifty miles of the Panel's primary sampling units would not be
followed. This feature, and other movers who could not be traced,
resulted in a 2.5-3.0% loss of households at each successive
quarter.

Table 1 summarizes household nonresponse for the five waves

of the 1978 Panel among households remaining at their sample
addresses or moving within the Panel's sample areas. The table
shows, as expected, an increase in nonresponse through the life
of the Panel, but at a declining rate. Refusals constitute the
main cause of nonresponse at each wave of the Panel.

Table 1. Household Nonresponse in the 1978 Panel

Household	April 1978 %	July 1978 %	October 1978 %	January 1979 %	April 1979 %
Interviewed	93.5	90.5	88.9	85.4	85.0
Refused	5.0	7.2	8.7	11.8	12.3
No one home, temp. absent, other	1.5	2.2	2.5	2.8	2.7
Total	100.0	100.0	100.0	100.0	100.0
(No. of households)	(2048)	(2091)	(2126)	(2135)	(2112)

The wave-to-wave change in nonresponse rates is the net
effect of obtaining responses in a later wave from households that
failed to provide data in an earlier wave together with the loss
of some households cooperating in an earlier wave but not providing data in a later one. Attempts were made in the 1978 Panel to
secure responses in a later wave from households not responding
in an earlier one. The only evidence we have on the effect of
these attempts comes from the first two waves of the Panel: of
140 nonresponding households in the April wave, 28 (20%) provided
data for the July wave (of the 103 households refusing in April,
15 responded in July).

An examination of household nonresponse rates for various
subgroups identified some variations. These may be found in
Table 2 for four waves of the Panel (the rates for the October
1978 wave are unavailable at this time); like Table 1, this table

relates to households remaining at their sample addresses or
moving within the Panel's sample areas. The table indicates that
at every wave nonresponse rates are higher for the area sample
than for the SSI sample, and that they are higher for those living
within the standard metropolitan statistical areas (SMSA's) than
for those living outside these areas. The last part of the table
shows that the nonresponse rates for white households are higher
than those for households of blacks and other races for the first
two waves of the Panel, but this situation is reversed for the
last two waves.

Person level nonresponse occurred when no data were collected
for one or more household members in an otherwise cooperating
household. In the 1978 Panel, interviewers collected data for
all adults by personal interview if they were present at the time
of interview or by proxy interview with a knowledgeable family
member if they were absent. Data on 31.7% of respondents in the
April wave of the Panel were provided by proxy informants. In

Table 2.   Household Nonresponse Rates in the 1978 Panel for
           Selected Subgroups*

Subgroups	April 1978	July 1978	January 1979	April 1979
Area Sample	7.2% (1656)	10.7% (1696)	15.8% (1708)	16.9% (1713)
SSI Sample	5.3% (400)	6.2% (405)	9.4% (406)	9.3% (407)
SMSA	8.0% (1401)	11.4% (1425)	16.6% (1432)	17.4% (1433)
Non-SMSA	4.3% (655)	6.5% (676)	10.4% (682)	11.5% (687)
White	7.7% (1719)	10.2% (1752)	13.8% (1740)	14.7% (1746)
Blacks and Other Races	2.1% (337)	8.0% (349)	18.2% (374)	19.0% (374)

* Figures in brackets are numbers of households on which percentages are based.

consequence, responses were not obtained for only about 1.5% to
2% of adults in cooperating households, either because the person
refused to cooperate or because he was absent and no other member
of his family felt able to respond on his behalf.

3.2. Item Nonresponse

Item nonresponse occurred when some but not all data were
missing for a sample person. Its extent varied considerably
between items. Table 3 exhibits item nonresponse rates for several
items from the first wave of the Panel.

The table focuses on data items for which amounts of payment
or income were to be reported. Social Security or SSI payments
among those receiving such payments had very low nonresponse rates.
In contrast, item nonresponse on the amount of savings interest
received during the period January to March is substantial, with
missing amounts for almost half the persons with savings accounts.
The dominant cause of the missing data was the respondent's
inability to answer the question. Anticipating a high level of
item nonresponse for this question, a supplementary question
concerning the amount held in savings accounts was asked of those
failing to report the amount of interest. The purpose of the
supplementary question was to provide a basis for imputing the
missing interest amounts. Although the table shows that there
was a sizeable degree of nonresponse to the supplementary question
among those who failed to provide information on the amount of
interest received, the supplementary question was answered by
about two out of three of such respondents. The use of the amount
in the account for imputing the interest received is discussed in
Section 4.

Other items investigated included data on salaries and wages.
For simplicity most of the results in Table 3 for these items are
restricted to the subgroup of individuals paid the same amount
each payday. In order to obtain accurate reports of salaries and

Table 3. Item Nonresponse Rates in the April Wave of the 1978 Panel for Some Amount Items among Persons Known to have Non-zero Amounts

Item	Refused %	Don't know or other nonresponse %	Total %	(No. of persons)
For those receiving program support:				
Social Security payment in January (area sample)	1.6	3.1	4.7	(533)
Federal SSI in January (SSI sample)	1.3	0.7	2.0	(300)
For those with savings accounts (area sample):				
Interest on savings accounts, January-March	2.6	45.0	47.6	(1,329)
Amount in savings accounts at end of March*	11.5	27.0	38.5	(636)
For those paid salaries receiving the same amount each payday – first employer only (area sample):				
Amount of paycheck from records	3.2	39.7**	42.9	(472)
Amount of paycheck from records or estimated	5.7	5.1	10.8	(472)
For those paid hourly (area sample):				
Amount of paycheck from records (for those receiving the same amount each payday – first employer only)	0.6	51.5**	52.1	(466)
Regular hourly rate of pay as of March 31 (all jobs)	1.7	8.0	9.7	(1,144)
Number of hours per week, usually worked at this job during January-March (all jobs)	---	0.3	0.3	(1,144)

*   This item was asked only of those with savings accounts who failed to report the amount of interest received for January to March.
** This percentage includes those who did not have records available.

wages, respondents were encouraged to consult records; however, the table demonstrates that a sizeable proportion did not do so. In the case of salaried individuals for whom records were unavailable, respondents were asked for an accurate estimate or a rough guess of the salary. Thus, salaries were obtained for 57.1% of salaried employees from records, for 26.5% as accurate estimates, and for 5.5% as rough guesses, leaving 10.8% for whom no estimate was obtained.

In the case of individuals paid hourly, the unavailability of records was treated by obtaining data on the person's regular hourly rate of pay and usual number of hours worked per week. Table 3 shows nonresponse for these items to be much lower than that for wages as obtained from records. The use of these two variables for imputing wages is discussed in Section 4.

A comparison of item nonresponse rates for various subgroups of the sample points to some differences. In particular, item nonresponse rates were higher when the data were collected from proxy informants than from self reports. This effect is illustrated for several items with the area sample in Table 4, where the item nonresponse rates are uniformly higher for proxy reports, usually deriving from a slightly lower refusal rate but a substantially higher rate of 'don't knows'. It is conceivable that if the individuals whose data were provided by proxy informants had been interviewed in person, they would have generated higher item nonresponse rates than the self reporters. It seems unlikely, however, that this factor would fully account for the differences observed in the table. The results thus suggest that the use of proxy informants contributes to an increased level of item nonresponse. However, firm conclusions on the effects of using proxy informants can be drawn only from a properly designed experiment, in which persons absent at the initial interview are randomly assigned either to be followed up for an interview or to have their answers provided by a proxy informant.

Table 4. Item Nonresponse Rates for Self Reports and Proxy Informants in the April Wave of the 1978 Panel, Among Persons to Whom the Question Applies (Area Sample)

Item	Self reports %	Proxy reports %
Interest on savings accounts January-March:		
Refused	3.0	1.5
Don't know and other nonresponse	42.1	53.5
Total	45.1	55.0
	(995)	(333)
Regular hourly rate of pay, for those paid hourly (all employers):		
Refused	2.1	1.2
Don't know and other nonresponse	2.9	16.6
Total	5.0	17.8
	(716)	(427)
Amount of paycheck from first employer, for those paid hourly and paid the same amount each payday	43.4	64.6
	(274)	(192)
Amount of salary at March 31 for all salaried workers, from records or estimated, for those paid same amount each payday:		
Response from records	65.8	42.4
Accurate estimate	22.6	33.5
Rough guess	2.7	10.6
Refused	6.0	5.3
Don't know and other nonresponse	3.0	8.3
Total	100.0	100.0
	(301)	(170)

In addition to the differences between self and proxy reports, some other subgroup differences were also observed; however, because of the small sample bases, no definitive conclusions can be drawn about these differences without more detailed analyses. The present findings suggest that there may be a greater proportion of missing data for blacks and other races than for whites. For instance, the following results were obtained from the April 1978 wave of the area sample: for those paid the same amount each payday by their first employer, 61.2% of blacks and other races did not provide the amount from records (n = 152) compared with 45.7% of whites (n = 812); and for those paid hourly, the regular hourly rate of pay was missing for 15.8% of blacks and other races (n = 177) compared with 8.7% for whites (n = 976). Two examples of other subgroup differences are: for those paid hourly, the regular hourly rate of pay was missing for 15.4% of those aged 45 or over (n = 306) compared with 7.8% for those under 45 (n = 835); and the quarterly interest on savings accounts was missing for 52.8% of females with such accounts (n = 615) compared with 43.1% of males (n = 707). For each of these examples similar results were obtained from July wave analyses.

4. TWO POTENTIAL IMPUTATION PROCEDURES

The results in the preceding section indicate that a fair amount of missing data can be expected for certain items to be collected in the SIPP. A common procedure for dealing with this problem is to use some form of imputation procedure to assign values for the missing responses. Two such procedures have been investigated with the 1978 Panel.

The first procedure imputes a value from responses to highly related items for the same wave. For certain items where considerable nonresponse was expected, alternative, closely related information was obtained. Two examples stand out: (1) when interest from a savings account was not reported, a question was asked about the amount of savings in the account, and (2) the earnings of those paid hourly were obtained from paycheck records where possible, and information was also collected on their regular hourly rates of pay and usual number of hours worked.

This procedure is useful if reasonable response rates are obtained for the alternative questions among those not responding to the original ones, and if a close relationship exists between the information collected from the original and alternative questions. Evidence on the first point is given in Table 5 for the area sample. The table shows that about three out of five of those failing to provide answers to the savings interest question did answer the question on the amount in the savings accounts. The position on earnings for those paid the same amount each pay-day was even better: about 5 out of 6 of those failing to provide earnings from records (i.e. paycheck amount and frequency of payment) did answer the two questions on hourly rate of pay and usual number of hours worked. There remains a significant proportion of persons for whom information was unavailable from either source both for savings and earnings; however, if the alternative information can be used to provide accurate estimates, the missing data problems for these two items would be substantially reduced.

Little evidence is available to quantify how well responses to the alternative questions can predict those to the original questions. Twenty-five persons in the April wave of the 1978 Panel (both area and SSI samples) were incorrectly asked and then answered both questions on savings account interest and the amount in the accounts. The correlation between the amount of interest and the amount in the accounts was 0.63 for these persons. In the case of the paycheck item, there were 162 persons in the area sample paid hourly, and paid the same amount each payday, for whom earnings could be estimated both from records (Y) and from the hourly rate of pay and usual number of hours worked per week (X). The correlation between these two estimates was 0.83, and

Table 5. Availability of Information on Amount in Savings Account
          for Imputing Savings Interest, and of Regular Hourly Rate
          of Pay and Usual Hours Worked per Week for Imputing Wages
          for Those Paid Hourly the Same Amount Each Payday
          (April 1978 Wave, Area Sample)

Response	Savings Interest %	Amount of Paycheck %
Response obtained to original item*	52.4	46.5
Response not obtained to original item, but information available from alternative item(s)	28.2	45.5
Response not obtained to either original or alternative items	19.4	8.0
Total	100.0	100.0
(Number in receipt of such payment)	(1329)	(402)

* Amount of savings interest for savings item, and amount of
  paycheck from records, together with frequency of pay, for
  paycheck item.

the regression equation was $Y = -55.4 + 1.05X$, with an intercept
not significantly different from zero, and a slope not significantly different from 1. While there remains room for improvement,
it appears from this evidence that the alternative questions may
provide a reasonable basis for imputing for missing data on
these two items.

   The second imputation procedure investigated uses the panel
feature of the survey design to impute for a person's missing
response to an item in one wave from his response to the same item
in another wave. Such cross-wave imputation can in principle be
applied in either direction, using data available for April to

impute missing data for July or vice versa. Table 6 summarizes
the patterns of response for three items across the April and July
waves of the 1978 ISDP Panel among persons in households that
cooperated in both waves. With the first item, January SSI payments for the SSI sample, there were very few missing data cases
in April (1.5%), and almost three-quarters of them had data available in July; of the 4.3% of missing data cases in July, 3.9% had
data available in April. The other two items, savings interest
and earnings, have similar response patterns: each wave had approximately 45% missing data, with about two-thirds of it being common
to both waves. In consequence, for these items only about a third
of the persons with missing data on one wave have data available
on the other wave for possible use in imputation.

   An indication of the strength of the relationships between
the amounts reported for the two waves for these three items is
provided by the correlations between the amounts for those
responding on both occasions: for the Federal SSI payments,
$r = 0.95$ ($n = 268$); for savings account interest, $r = 0.67$ ($n = 368$);
and for earnings, $r = 0.89$ ($n = 223$). The high correlations for
SSI payments and for earnings suggest that if a response is available from an adjacent wave, that response may serve as a firm
basis for imputation. While the correlation for savings interest
is lower, the use of a response from another wave in the imputation scheme may still be valuable.

5. ISSUES FOR NONRESPONSE ADJUSTMENTS

   As experience with the 1978 Panel demonstrates, the SIPP
will incur a fair amount of missing data. After every effort has
been made to minimize nonresponse at both the unit and item levels,
an appreciable amount will remain and it will be concentrated in
certain population subgroups. This situation calls for the
application of nonresponse adjustments to reduce the extent of
nonresponse bias in the survey estimates.

Table 6. Patterns of Response for the April and July Waves of
          the 1978 Panel for Three Items*

April Wave	July Wave	(a) SSI payment %	(b) Savings Interest %	(c) Earnings %
Response	Response	94.7	39.4	39.8
Response	Nonresponse	3.9	16.0	16.5
Nonresponse	Response	1.1	16.8	13.0
Nonresponse	Nonresponse	0.4	27.8	30.8
Total		100.0	100.0	100.0
(No. receiving payment)		(283)	(934)	(601)

* (a) Amount of Federal SSI received in January (April Wave)
       and April (July Wave), among SSI recipients in the
       SSI sample;
  (b) Amount of interest received from savings accounts in
       January-March (April Wave) and April-June (July Wave),
       for area sample persons with savings accounts;
  (c) Earnings from paycheck records for all those in the
       area sample paid the same amount each payday, April
       and July.

Several features of the SIPP complicate the task of determining an appropriate strategy for nonresponse adjustment and
imputation. These features are briefly reviewed below.

(a) Variety of analyses: The data collected in the SIPP will be
used for a wide range of purposes by many different researchers.
They will, for instance, be used to provide numerous descriptive
estimates for the population, including household, family and
individual income levels, the number of participants in various
federal transfer programs and the amount of support received by
federal program participants. The data will also be used for
analytic work, especially microsimulation modelling of the federal
tax and transfer system. These analyses will enable the costs

and impacts of new programs or changes to existing programs to be evaluated.

If the SIPP data were to be used solely to produce basic descriptive measures of average levels, the procedure of imputing for an item nonresponse the mean of the imputation class in which the nonresponse occurred would serve well, and so would the procedure of imputing the predicted value from a regression equation of the responses to the item in question on other related items. However, because these procedures distort distributions, reduce element variances and alter covariances, their application is inappropriate to data which are to be employed for analytic purposes, many of which depend on a variance-covariance matrix. In view of the variety of analyses to which the SIPP data will be subjected, the procedures adopted for dealing with the survey's missing data will need to provide good estimates of distributions, variances and covariances as well as of average levels.

In order to retain the distributional properties of the sample, imputations are often made by some form of hot-deck procedure, that is assigning the value of some respondent in the imputation class - the donor - for the missing response (see, for instance, Bailar, Bailey and Corby, 1978, and Coder, 1978); the equivalent procedure with the regression approach is to add a random residual to the predicted value from the regression. However, while these procedures approximately preserve the respondent element variances in the data, they may still not yield good estimates of covariances (see, for instance, Kalton, 1981).

Another aspect of concern in estimating covariances arises when there are several missing responses in a record. If separate hot-deck imputations are made for each of these items, different donors may well be used for different items. In this case the covariances between the responses to these items will be changed. To avoid this happening, the same donor may be used to provide data for all the missing responses on a record, or at least for those in a set of closely related items.

The use of a common donor for sets of missing items on a
record has the advantage of helping to generate consistent data,
that is mutually compatible responses within the record. Logical
edit checks for consistency are of course needed even when no
imputations are made, but the risks of inconsistent data are considerably increased when imputed values are inserted for missing
responses (see Hill, 1978, for instance for the imputation of
consistent data in the 1976 Canadian Census). The requirement
that an imputation procedure for the SIPP should generate consistent data is an extremely difficult one to satisfy: there are an
extremely large number of consistency checks that could be specified for the many detailed interrelated items of data that will
be collected in any one wave of the survey, and this number is
increased substantially when cross-wave consistency checks are
also considered. Both within- and between-wave checks are needed
to ensure a reasonable and consistent data set.

(b)   Panel imputation: A major concern for nonresponse adjustments
and imputation for the SIPP is the issue of cross-sectional imputation (each wave taken separately) versus cross-wave imputation
(several waves being treated as a single combined data set).
The former procedure has the practical convenience that it can be
carried out as soon as each wave's data are available. However,
it fails to use all the data, and may therefore produce imputed
values inferior to those that would have been obtained from the
combined data set. For instance, as illustrated in the previous
section, a response for an item on one wave may be a good predictor
for that item on another wave. (Ashraf and Macredie, 1978,
describe a cross-wave imputation procedure used in the Canadian
Labour Force Survey.)

Another disadvantage with the cross-sectional procedure is
that records with imputed data may be inconsistent from wave to
wave. This feature is particularly damaging for analyses requiring
more than one wave of data, as for instance in the case of microsimulation modelling of the federal tax and transfer system.

In many surveys, adjustments for unit nonresponse are carried out by some form of reweighting procedure. The procedure may be a relatively simple one of defining subgroups of the sample - usually in terms of some combination of characteristics - and then allocating different weights to the responding elements in the various subgroups to adjust for the subgroups' differential response rates (see, for instance, Thomsen, 1973; Chapman, 1976; Bailar, Bailey and Corby, 1978; and Platek and Gray, 1979); or it may be a more complex procedure, such as raking ratio estimation which adjusts for differential response rates across several characteristics, but in relation to the marginal distributions of these characteristics not to their cross-classification (see, for instance, Oh and Scheuren, 1978a,b). Whichever form of reweighting was used, separate wave adjustments for household nonresponse in the SIPP would cause problems for cross-wave analyses. The varying pattern of household nonresponse across waves would result in respondents being assigned weights that changed from wave to wave. The problem would then be determining what weights to use when several waves were merged for longitudinal analysis. (Since the sample inclusion probabilities of some persons in the 1978 ISDP Panel changed between waves, a factor which also caused a variation in weights across waves, the avoidance of varying nonresponse adjustments would, however, not entirely eliminate this problem.)

The disadvantages of the cross-sectional approach to nonresponse adjustments and imputation make the cross-wave approach an attractive one. Household or person nonresponse on one wave could, for example, then be treated as a collection of item nonresponses in the combined data set. However, the full cross-wave approach cannot be applied until all the waves of the panel have been completed and merged tapes produced, a feature which impairs the production of timely adjusted estimates from individual waves. The demand for timely estimates suggests the use of a two-fold strategy, first making preliminary adjustments for immediate

individual wave analyses, and then later making final adjustments
using data available in other waves. A variant on this scheme
would be to generate merged tapes as each wave's data became
available, making nonresponse adjustments and imputations for the
latest wave using all the data collected to that point; this
variant may also be treated as a preliminary one, with the final
adjustments for all waves being made from the combined data set.
While the two-fold strategy improves the timeliness of the data,
it also has drawbacks. The development and implementation of its
two nonresponse adjustment strategies, cross-sectional and cross-
wave, make heavy demands on professional and systems time; and
considerable efforts are required to explain that the results
obtained from the data set with the cross-sectional adjustments
are provisional ones, subject to later revision when the final,
cross-wave, adjustments are made.

(c) Aggregation: The SIPP approach for obtaining accurate
estimates of income amounts is to obtain the information at the
micro-level and then to add the responses together in a variety
of ways to provide estimates of different income components: for
instance, unearned income can be derived by adding together
incomes from interest on savings accounts, dividends, rents, etc.
There are clearly a number of income estimates of general interest
which will be computed and recorded on user tapes. As a means of
reducing the number of variables for which imputation is needed,
the imputations could be carried out only for the aggregate
measures, thus avoiding dealing with all the components of which
these measures are comprised. This procedure faces, however, a
potential problem that various aggregates, based in part on a
common set of components, may not be consistent with one another.
The imputations may also be less accurate because relationships
that exist between individual components and available characteristics of the nonrespondents may be hidden in the aggregates. In
addition, there is a question of how to deal with missing data for
only some of the components in an aggregate: if the aggregate is

treated as missing in this case, the effect is to ignore the
partial information available. For these reasons, it is probably
better to carry out the imputations at the component level, and
then form aggregates as required. However, with this approach
there is a risk of producing some poor aggregate estimates, and
even perhaps some impossible ones. A way to avoid this risk is
to fully incorporate the information available on known components
of the aggregates in the imputation process - and to use a single
donor for the missing components - but, in view of the varied
patterns of missing data that will occur, this would be a complicated operation.

Another aspect of the aggregation issue is that summations
will sometimes be required across several persons, to form
aggregates for households, families or other groups. In making
component imputations, account will also need to be taken of this
type of aggregation.

(d) Weights: In dealing with missing data it is common practice
to avoid large weights in reweighting procedures (e.g., greater
than 2 in the Current Population Survey - U.S. Department of
Commerce, 1978) and to avoid multiple use of donors for imputations.
The reasoning behind these practices involves a trade-off between
a reduction of nonresponse bias and an increase in variance. If
large weights are applied or if a donor is used several times, a
sizeable increase in variance results; it is generally considered
preferable to avoid this increase by accepting a possibly lesser
reduction in bias. This may be achieved with reweighting by
enlarging the weighting class and hence reducing the weights and
with hot-deck imputation by using an alternative donor, perhaps
one less well matched with the nonrespondent.

The above argument is made in terms of samples in which the
elements are selected with equal probabilities (epsem). The SIPP,
however, will be a non-epsem sample; it will combine an area
probability sample and samples from administrative record files
selected at higher rates. In consequence, the avoidance of high

weights for nonresponse adjustments and multiple use of donors
for imputation needs some reexpression. For imputation purposes,
the allocation of a donor's response to a nonrespondent effectively
resembles adding the sampling weight for the recipient to that of
the donor. If the recipients' weights are large relative to those
of the donors, a sizeable increase in variance would result. One
way to tackle this problem is to split the nonrespondents' weights
into parts, using different donors for each part (see Scheuren's
discussion of Coder, 1978; Kish, 1979; Kalton, 1981). This
procedure, a kind of multiple imputation, may also aid in the
examination of sampling errors of estimates formed from the full
data set padded with imputed values.

(e) Sampling errors: If imputed values are substituted for missing data, and the augmented data set is then treated as if there
had been no missing data, any sampling error estimates computed
will be underestimates. Since a fair amount of imputation is
likely to be needed for the SIPP, the underestimation may be
appreciable. In order to obtain sampling errors that take into
account the imputation of part of the data, multiple imputations
may be used (Rubin, 1978, 1979). The procedure outlined in (d)
above, of partitioning the weights of nonrespondents with separate
imputations for each part, may form the basis of a valid sampling
error estimation procedure.

(f) Choice of donors: The first step in making imputations with
a hot-deck procedure is the construction of imputation classes;
a missing response is then assigned the value from a donor within
that class. To be effective the classes should be formed so that
they account for a good proportion of the variance in the variable
for which the imputations are being made, and they should also
have varying response rates. As has been noted in (a) above, it
is often desirable for one record to have necessary imputations
for a set of related items taken from the same donor in order to
create consistent responses and retain the covariance structure

for the items. To achieve this objective, a single division into imputation classes needs to serve for the set of variables.

The construction of imputation classes is made more complicated by the fact that the variables used to define the classes may themselves be subject to nonresponse. A manageable solution for handling this complication may involve imputing for sets of items sequentially: first, take some basic (e.g. demographic) items with little missing data, filling in the "small holes" (Coder, 1978) with some simple procedure, then use these items to form imputation classes for the next level of variables, make these imputations, etc. (see also Hill, 1978).

With SIPP there may be the possibility of using certain administrative data in making imputations. The Master Beneficiary Record (MBR) of Social Security beneficiaries and claimants, the Supplemental Security Record of SSI benefit amounts, payment history and demographic data, and the Summary Earnings Record (SER) of an individual's lifetime earnings and quarters of Social Security coverage are candidate data systems to be linked to the SIPP. An advantage of using data from such systems in making imputations is that they should have relatively little missing data, and hence largely avoid the complications mentioned above. However, the linking of these data systems to SIPP will be a lengthy operation, and in practice it may not be completed in time for the administrative record data to be available for imputation purposes.

After imputation classes have been formed with a hot-deck procedure, donors have to be chosen for missing responses. As has already been noted in (d) above, one consideration involved in choosing donors is the avoidance of multiple uses of a single donor. Two other considerations for the SIPP are what to do about outliers and about data collected from proxy informants.

As indicated in Table 3, item nonresponse in the SIPP is mostly concentrated in the amount items, that is items which are

numeric in character. An issue with such data is whether a
respondent with an extreme value should be available to serve as
a donor or not. If he is, there is the risk of duplicating the
extreme value. If he is not, the variance of the data's distribution is reduced. The procedure of partitioning the weights of
nonrespondents described above helps to alleviate this problem,
since if respondents with extreme values are used as donors, they
are used with smaller weights.

There is some evidence to suggest that not only do proxy
informants generate a greater amount of item nonresponse but also
that the quality of their responses is lower than that of self-
respondents. Thus, for instance, the results in Table 4 show that
a higher proportion of salaries reported by self-respondents were
obtained from records, the proxy informants relying more heavily
on estimates. The possibly lower quality of proxy reports suggests
that some preference might be given to the use of self-respondents
as donors for imputation. This argument can be extended to other
cases where differential quality of responses can be identified;
thus, for instance, preference might be given to the use of
salaries reported from records, and perhaps to accurate estimates,
over rough guesses in imputing for missing salaries. One way to
give preference to the more accurate reports is to choose the
donor for a nonrespondent as the respondent closest in terms of
a distance function, with the distance function incorporating a
penalty for lower quality reports. (This suggestion follows the
lines of Colledge, Johnson, Pare, and Sande, 1978, who describe
a distance function designed to reduce the multiple use of donors
by incorporating a penalty based on the number of times a respondent has already served as a donor.) Whether this strategy is
effective depends on a possible trade-off between quality of
response and closeness of the matching of donors; more accurate
imputed values may in fact be obtained by taking lower quality
responses from closely matched donors than higher quality responses
from less well matched donors. Taking account of the quality of

responses would also complicate the imputation procedure.

In conclusion, it should be reiterated that this paper reports only some initial thoughts about the problems of non-response that the SIPP will face, and about possible methods of dealing with these problems. The purpose of the paper is to raise these problems for a workshop discussion, not to suggest any firm solutions.

REFERENCES

Ashraf, A. and Macredie, I. (1978). Edit and imputation in the Labour Force Survey. *Imputation and Editing of Faulty or Missing Survey Data,* U.S. Department of Commerce, Washington D.C., 114-119. *Proceedings of the Section on Survey Research Methods,* American Statistical Association, Washington, D.C., 425-430.

Bailar, B.A., Bailey, L. and Corby, C.A. (1978). A comparison of some adjustment and weighting procedures for survey data. In *Survey Sampling and Measurement* (N. Namboodiri, ed.), Academic Press, New York, 175-198.

Coder, J. (1978). Income data collection and processing for the March Income Supplement to the Current Population Survey. In *Survey of Income and Program Participation, Proceedings of the Workshop on Data Processing, February, 1978* (D. Kasprzyk, ed.), Income Survey Development Program, U.S. Department of Health, Education, and Welfare, Washington D.C., Session II, 1-32.

Colledge, M.J., Johnson, J.H., Pare, R. and Sande, I.G. (1978). Large scale imputation of survey data. *Survey Methodology 4,* 203-224.

Hill, C.J. (1978). A report on the application of a systematic method of automatic edit and imputation to the 1976 Canadian Census. *Imputation and Editing of Faulty or Missing Survey Data,* U.S. Department of Commerce, Washington, D.C., 82-87. *Proceedings of the Section on Survey Research Methods,* American Statistical Association, Washington, D.C., 474-479.

Kalton, G. (1981). *Compensating for Missing Survey Data.* Survey Research Center, University of Michigan, Ann Arbor, Michigan.

Kish, L. (1979). Repeated replication imputation procedure. Unpublished memorandum, Survey Research Center, University of Michigan, Ann Arbor, Michigan.

Lininger, C.A. (1980). The goals and objectives of the Survey of Income and Program Participation. *Proceedings of the Section on Survey Research Methods,* American Statistical Association, Washington, D.C., 480-485.

Oh, H.L. and Scheuren, F. (1978a). Multivariate raking ratio estimation in the 1973 Exact Match Study. *Imputation and Editing of Faulty or Missing Survey Data,* U.S. Department of Commerce, Washington, D.C., 120-127. *Proceedings of the Section on Survey Research Methods,* American Statistical Association, Washington, D.C., 716-722.

Oh, H.L. and Scheuren, F. (1978b). Some unresolved application issues in raking ratio estimation. *Imputation and Editing of Faulty or Missing Survey Data,* U.S. Department of Commerce, Washington, D.C., 128-135. *Proceedings of the Section on Survey Research Methods,* American Statistical Association, Washington, D.C., 723-728.

Platek, R. and Gray, G.B. (1979). Methodology and application of adjustments for nonresponse. *Proceedings of the 42nd Session of the International Statistical Institute,* 1979.

Rubin, D.B. (1978). Multiple imputations in sample surveys - a phenomenological Bayesian approach to nonresponse. *Imputation and Editing of Faulty or Missing Survey Data,* U.S. Department of Commerce, Washington, D.C., 1-22. *Proceedings of the Section on Survey Research Methods,* American Statistical Association, Washington, D.C. 20-28.

Rubin, D.B. (1979). Illustrating the use of multiple imputations to handle nonresponse in sample surveys. *Proceedings of the 42nd Session of the International Statistical Institute,*1979.

Thomsen, I. (1973). A note on the efficiency of weighting subclass means to reduce the effects of nonresponse when analyzing survey data. *Sartryck ur Statistisk Tidskrift 4,* 278-283.

U.S. Department of Commerce (1978). *An Error Profile: Employment as Measured by the Current Population Survey.* Statistical Policy Working Paper 3. U.S. Government Printing Office, Washington, D.C.

Ycas, M.A. and Lininger, C.A. (1980). The Income Survey Development Program. A review. *Proceedings of the Section on Survey Research Methods,* American Statistical Association, Washington, D.C., 486-490.

IMPUTATION TECHNIQUES: GENERAL DISCUSSION

G. SANDE, *Statistics Canada:* Thank you for the opportunity
of being a discussant here. One is never quite sure what the role
of the discussant is, or why one has been invited. So let me give
a little bit of the background of some of my biases on this topic.
I might also say that it's not necessarily evident that they in
fact picked the right Sande for doing the discussing in terms of
the amount of experience actually obtained.

My involvement in imputation was that it was presented to me
as more or less a self-contained problem, and I was able to look
at the problem without having to worry about the rather heavy
burden of the overall survey operations.

The very general view that I have come to on imputation is
that imputation is the process of somehow plugging in values.
This usually has the connotation of clearing partial nonresponse,
but it may also be because one doesn't have the variables one
would like and so has a file merge problem. If one was really
trying to do regression of some sort, one would estimate a mean
vector that is based on some sort of subpopulation. The agency
statistician is very leery of using mean vectors because they
destroy one of the vital aspects of his data - namely its variability. Consequently he faces not only one difficult problem,
but two difficult problems - one in estimating the mean and the
other of estimating some sort of variability for the individual.
There is a rather clever scheme which says that, in fact, the local
mean doesn't matter and this is called hot deck imputation where
you plug the data in.

I came into this problem with a rather different set of tools
than the standard survey statistician and so ended up with views

which are different. Some of the tools with which I arrived are
those of the standard applied mathematician, who spends most of his
life doing approximation. Approximation is really the only thing
an applied mathematician ever does in practice. There are some
things which one learns such as the difference between global
approximation properties and local approximation properties. You
learn very quickly that local matters are rather unimportant and
that global matters are rather important. Those are the sorts of
things that have tended to flavour my views in this area - rather
strongly.

One of the points which is very important, is that there is
a big difference between imputation and missing value methods. It
is very easy to look at elementary examples where no amount of
imputation will correct things. In practice imputation is often
the missing value procedure for complex and multi-purpose situations.

That gives you a bit of the background of my biases. Let me
now discuss the various papers.

I was, on balance, pleased with the Ford paper. It is
important that one, in fact, does some sort of simulation studies
to look at these problems. There were some parts I thought were
weak having to do with the actual pragmatics of the data being
observed. Whenever you try to do imputation you have to understand
the particular body of data at hand.

In the Ford case the two variables which were being imputed
were proportions. They were subconstituents of some larger value
which was known. I presume they were some kind of animal counts
or acreage counts. One should not go to the direct variables but
rather to the proportions or - ratios. In terms of the micro-
economics of the farms, that make sense as it is the pattern of
the activity that matters not the detailed value. In terms of
matching, which is a global property, it makes a difference in
terms of the definition and properties of the sub-population from
which you are going to estimate that local mean and the local
error. So one does have to get in and understand the data and do

things which make sense in terms of the data. I would be interested in looking at the same kinds of simulations done on ratio scaled
values rather than on the direct values. In my experience this is
a rather important distinction.

From this discussion you can see that I am quite curious
about the details of the relevance of the comparison of the imputed
values with the suppressed known values. One really wants to
measure those in some scaling that is a measure of the variability
of the item. In terms of the detailed micro-structure of these
observations, large scale values will have much larger variability
than small ones. We have heteroscedasticity and global transforma-
tions do make a difference. This is another kind of problem that
can be used to address.

About Chris Hill's work: One of the first things I should
say is that in order to discuss that kind of work one has to
realize that there are many layers of the problem to go down through.
The kinds of questions I am going to raise are in fact a major
tribute in that there are many layers of problem which have been
cleared away quite well. So let me say up front, that what I'm
really quibbling about is whether or not the dot on the "i" is
correctly centered and not about whether or not the spelling is
correct.

One has to do with the effects of degeneracy in the identification of the fields to impute or, in the terminology which I
prefer, the error localization problem.

Then there is this interesting case where there are alternate
error localizations for a single record. I have done some computations on discrete data and I have an example of a six field record
with five edits, three of which failed. It turns out that it
requires the clearing of three fields to generate clean records.
There are 20 ways of choosing 3 out of 6 fields, and it turns out
that 12 of the 20 ways lead to possible clearings of the record.
So rather than just two in this case there are 12 alternates -
so the situation does arise. In some of my own experiments with

the numerical work I've done, I have found that the different error
localizations resulted in markedly different qualities of match,
or, in the discrete case, they would correspond to a markedly
different size of donor populations. I am curious to know if there
has been any experimenting done to find out whether or not there
are alternate error localization commonly.

   The other thing that comes up is that one often hears of
correlated response errors. If you have courage in your convictions
you should consider joint failure of fields and whether some fields
are in error more often than others. This is a generalization of
the notion of minimum change - with weights and non-linear functions
When I was a graduate student many of my friends were physicists
who seemed to spend much time doing Monte Carlo studies of their
detectors to find detector efficiencies. It's a shame statisticians
don't do Monte Carlo's on their editing programs to find out how
good they are.

   To move on to Kalton's discussion, all I can do is to echo
what I have already said. Imputation is a much broader thing than
just clearing non-response. It can be used in many other ways.
If one has a very complex situation, typically a large administrative frame and some amount of survey data, in trying to blow the
survey data up to the administrative frame one will encounter
complicated problems. Tax yield is an obvious example. It's not
at all evident what the correct estimators should be and by blowing
up the data - imputing it up - one can in fact construct estimators.
I think that is a kind of modelling activity which will be very
important. The same kind of argument can be given in lieu of
using synthetic estimators for small areas.

   One of the questions that comes up is "What is the difference
between this kind of activity and that of microsimulation or microanalytic simulation?" I was able to come up with about 3 side
inches of reports on microanalytic simulation which were really
quite delightful to read. I had not gotten into that field before.
They talked about file merging - Statisticians talk about imputatior

They talked about imputation - but it's what statisticians would talk about as stochastic modelling.

One of the issues that seems to be very serious is the ergodic properties of the modelling; that is to what extent the cross-sectional work coincides with longitudinal work. These are what the modelling people call file aging problems. Another serious issue is distributions. Some of the work by Nancy Ruggles on the properties of merged files, concerning "what are the joint distributions after you've done this bulk imputation?" suggests that things are quite plausible. From friends who do public policy research, which is another name for microsimulation, I understand that they have an internal problem of generating inconsistent records as a result of doing their merging. Since one of the things that good imputation does is to generate clean records, perhaps they should borrow some technology from the statisticians. As well, the statisticians can learn from some of their experience.

One of the detailed responses to Kalton is the repeated use of donors. I have heard the statement that you should not use donors repeatedly and I've always looked at the formulae which are used to justify that as telling me that I have a bad sample because it has very high variances. I do not understand why you should make it worse by spreading the donor usage around and introducing additional bias above and beyond what you have already got.

I think it is very hard to define exactly which properties one wants the microsimulations to have. I am always reminded of some work done by a physicist who was modelling stellar systems with discrete coordinates. In doing the simulation it turned out that the physics was most insightfully represented by two of the conservation integrals (there are about 18 conservation integrals that physicists can prove on continuous systems). He preserved the two most important. As a result of doing this it turned out that at a detailed level his system did not satisfy the equations of motion exactly, only approximately. But he got physical insight from the system. So I think it is a very hard problem to know

which parts to do exactly and which parts to let do their own
thing and come out approximately true when doing microsimulations.

W. MADOW, *Committee on National Statistics* (to B. Ford):
Did you consider using data dependent missingness? I would think
that this would be one of the key things in many data dependent
imputation problems; particularly where the level of response (the
answer that you would have given) led you, not to give it in relation to something, i.e., in relationship between your using ratio
where let's say everyone else has an average ratio of 6%, but in
your case you would have had 20% of your income from this source
and therefore you might not want to reveal it, so you don't have
randomness that depends on your other responses.

B. FORD, *U.S. Dept. of Agriculture:* Well, the only thing
that we did like that was we set up some of the criterion so that
some of the missing values were larger and some were smaller.
You're asking, "Did we look at the ratios to see if they are
actually different for the nonrespondents than for some of the
respondents?" That is something we would try to do in a survey
context, I think, where you would have some administrative data
or some way to check for nonrespondents and get a handle on what's
going on. We really didn't do that.

W. MADOW, *Committee on National Statistics:* One other thing
that I noticed was that you didn't discuss how good a ratio estimator was in your case. Was the correlation between numerator and
denominator let's say 95% or was it 3% in your data. It seems
this would be relevant information for people deciding whether to
use a ratio estimator or not.

B. FORD, *U.S. Dept. of Agriculture:* That's true - the
relationship that you have between the variables and auxiliary
variables you're going to use in an imputation process, are

important. I think the correlations in this data set were all
fairly high between the survey variables. There were only 14
variables and it was oriented to a very tight situation where all
the variables were related together - they were usually in the 16
range, so that they were all good in explaining each other.

D. KREWSKI, *Health and Welfare Canada* (to B. Ford): When
you were calculating your variance estimators for the sample which
tried to take into account imputation I think you just used
independent replication as a means of getting the variance estimates,
based on, I think, two degrees of freedom. Those are going to be
somewhat unstable and I was wondering whether you had given any
consideration to using balanced half sample or jackknife techniques,
with separate imputations carried out within each replicate in
order to get a more stable estimate of variance.

B. FORD, *U.S. Dept. of Agriculture:* I think when you have
gotten to an operational context you would start thinking (about)
that - the reason we use three replicates, even though you have
that instability of variance estimates, is that we were doing
simulations for a lot of simulated data sets over a lot of situations. That adds more to the stability of this research kind of
procedure. If we were going to get into an operational survey and
we were going to say, "Let's make our standard error estimates", I
think we would go to balanced half samples rather than three
replicates.

A. TERJANIAN, *Statistics Canada* (to B. Ford): In PRINCOM,
for how many variables did you use the shortest distance. We are
not clear whether you said one variable or the other 13 variables.

B. FORD, *U.S. Dept. of Agriculture:* In the Principal
Component Procedure, we used the first principal component since
we only had the situations where either $y^{(1)}$ or $y^{(2)}$ missing, or

$y^{(1)}$ and $y^{(2)}$ both were missing. We had a different principal
component for each of these situations. For example, if $y^{(1)}$ was
missing we would have a principal component which used all of the
variables except $y^{(1)}$.

P.S.R.S. RAO, *University of Rochester* (to G. Kalton): If a
person is a nonrespondent, the reason, I believe, is because he is
different from the respondents in the characteristic under study.
Now, if he is not different I could just take the average of the
respondents and weight it if I like (or not weight it) and I know
I would not lose much. But what bothers me is that since this
person is a nonrespondent and he is not the same, he is not like
the respondent. In fact the respondent is really an outlier in
some sense, and how do we bring his value back?

G. KALTON, *University of Michigan:* I fully agree that nonresponse is not distributed randomly across the population: we
show in the paper, for instance, that in the case of the 1978 ISDP
Research Panel, households in SMSA's have higher nonresponse rates
than those in non-SMSA's. The basis of many nonresponse adjustment procedures -- both weighting adjustments for unit nonresponse
and imputations for item nonresponse -- is to divide the sample
into classes which are more homogeneous than the total sample in
terms of the survey variables and within which the assumption that
the missing data are missing at random is more tenable. I acknowledge that within classes the data are not truly missing at random
but, if the classes are chosen well, applying the assumption within
classes should be an improvement on applying it to the total sample.
In consequence, I think that the purpose of nonresponse adjustment
procedures should be viewed as reducing nonresponse bias, not
eliminating it.

A. ZINGER, *Université du Québec à Montréal:* I'm scared
stiff in some ways by what I have heard this afternoon. You take
real data and when you have some information which you don't know,
you replace this by the same data, assuming some hypothesis which
may be true or not. You add a chance element and you mix facts
and fiction to obtain what you call new data.

Now, suppose that you have a system in which the statistician
decides what the data should be by putting forth the model he thinks
is right. The next thing he does (it will naturally be written
down somewhere, usually in fine print) is to make some transformation on the data. The user will often not read the fine print,
and thus you have managed to hide what you really don't know. Now,
let's think what will be the evolution of such a system. In
practice, since there is no more penalty because you don't know,
the amount of "don't know" will increase. Nobody will know it,
and this new reality will be more and more what the statistician
thinks it should be. Now, it can go very far. Suppose that the
statistician uses for imputation some regression, and he is faced
for some reason, with 50% nonresponse. Somebody will take this
new data and will think, "why not run a regression?" and he will
find out that he has a beautiful regression, and he will be con-
vinced that he has found something from the data. What he will
have in fact found is the 'beautiful regression' the statistician
built in. So I think these techniques are terrible, and that there
is a big danger in doing such things.

W. MADOW, *Committee on National Statistics:* I think everybody in the nonresponse and imputation area is well aware of the
danger, and what they are faced with is how to do something without
creating further problems.

B. FORD, *U.S. Dept. of Agriculture:* I think that anybody
who is doing imputation and trying to get a clean dataset is trying
to help the user rather than disguise what kind of data we are

collecting. I don't think that the latter is the objective at all.
People such as the Bureau of the Census, CPS and so forth, when
they put out a clean data set, will, in fact, flag which items have
been imputed so that any user can be aware of it and if he wants to
use his own system for imputing then he is perfectly free to do so
- it's up to him. I think your remark is good in the sense that
survey organizations haven't really come up with a standard. You
feel that there should be some sort of cutoff level. If I'm
imputing 60 or 70% of the items it would be better not to publish
period, than to publish estimates with 70% nonresponse. There
should be some kind of cutoff that we haven't really discussed or
hit upon yet. The problem is often on output tables. Sometimes
it is in fine print and may be the nonresponse rates need to be
publicized more exactly.

        C. HILL, *Statistics Canada:* One very brief point in relation
to Mr. Sande's summary. I don't want to take up one or two of the
"i" dottings he made to my presentation. We are well aware of the
fact that item nonresponse, or nonresponse generally, is not in
fact drawn at random from the population. In fact these systems
that we've put in place in Canada and the other imputation systems
we are going to be utilizing in '81, do as much as they can to
take account of all the matching variables, to take account of
this problem of differential item nonresponse.

        Of course, the one thing we cannot take account of is when
the actual variation in item nonresponse occurs over and above all
the matching variables we have. I, in fact, reported some evidence
on the problem of response rates to our labour force data in the
Census, and it was quite clear, that when the non-participation
rate goes up so the tendency to nonresponse goes up. For example,
the over 65 population has a much higher nonresponse rate to the
labour force data than the under 65. We did, in fact, have some
evidence from a micromatch to show that although we don't adjust
for this perfectly, in fact, by matching very thoroughly on age, sex

and other factors that are related, we do account for a substantial
part at least of this problem. But undoubtedly we don't account
for all of it. The only way we can account for all of it is to
have evaluation studies that will go in detail into the variability
in response rates.

R. GRAVES, *Statistics Canada:* I would like to offer some
comments from the viewpoint of a computer systems professional who
has worked on interdisciplinary project teams with survey statisticians and subject matter experts.

I see the primary definitional driving force for the design
of a survey processing system in the form of a "semantic data model"
or "conceptual scheme". This model is a formal description of
what is to be measured and is defined by the subject matter experts.
This model is then elaborated by the survey statistician to create
a feasible measurement scheme. This composite model determines
both the *what* and the *how* of the survey measurement.

The system designer must then devise a mechanism for testing
individual survey response against the composite model as well as
designing a medium and mechanism for storage and retrieval of the
data.

The semantic data model specifies the things in the real
world to be counted (detection of existence) their intrinsic
properties and relationships between things. The measurement
scheme views these same observations as a set of either quantitative
or qualitative characteristics. A view of data editing and imputation in terms of quantitative and qualitative characteristics can
therefore (and is typically) decoupled from and insensitive to the
semantics of the measurements.

The EDP industry is actively engaged in addressing this
problem in the context of data base management and information
management system design.

It would seem appropriate for survey statisticians to take
part in the development of "semantic data modelling" concepts that

address the requirements of statistical survey processing systems.

W. MADOW, *Committee on National Statistics:* All the statisticians agree with you. They, too, would like this. This is not a
laughable matter. It's a very serious question, however, the
working with people in substantive areas is not too easy to achieve.

M. LAWES, *Statistics Canada* (to G. Kalton): Related to the
Labour Force Survey, which is the kind of survey where we are now
constructing longitudinal data files by which, after the households
rotate out of the sample, we collect the six months worth of information together. We are now developing procedures to impute
for the nonresponse. We have looked at methods for imputing for
a single month of missing data and then we will take care of some
of the item nonresponse. We clean that up on a longitudinal imputation procedure. Basically the Labour Force questionnaire is very
structured document, so that if we impute a path, we follow that
then on the detailed item imputation - that is, impute the missing
elements for that path. We look at various things - the "similar
record" substitution type approach, reweighting, use of transition
probabilities, many of these things. I am saying this because I
sympathize with some of the problems that you are having or will
have in your survey.

My real question is the following. Something we are tackling
now is, at what point do we throw the record out. If we have only
one month of data, and have to impute all five, the question arises
as to when it is better to throw it out than try to fill in all
the missing blanks and live with the possible problems of imputing
spurious or superficial moves in the population? I would appreciate
any comments you might have on that.

G. KALTON, *University of Michigan:* I am extremely interested
to learn of your work on longitudinal imputation, which seems to
follow the type of approach we are considering.

With a panel survey the question of when to throw a record out rather than impute for many missing values is a difficult one. The 1978 ISDP Research Panel had five waves of data collection, with many possible different patterns of response and nonresponse across the waves: some households may respond on only one wave, some on two waves, etc. As yet, we have no information on the frequencies of the various patterns. At some point we will need to consider how to handle the various patterns of missing data, but we have not yet done so.

T.M.F. SMITH, *University of Southhampton*: The one thing that most statisticians agree on in statistical inference is the value of sufficiency - it's the one concept we do all agree on. It seems to me that in terms of imputation the one thing that all surveyors must keep is the record of the sufficient statistic. In this case, the sufficient statistic is the complete data record including the pattern of item and individual nonresponse. So, I urge you, whenever you produce imputed data to maintain the flags that itemize the nonresponse items. At least statisticians who want to use other imputation procedures are able to use them.

G. KALTON, *University of Michigan*: I think we all fully agree on that point. Flagging imputed values is often -- but not always -- done. In addition, with imputation procedures that assign respondents' values to nonrespondents, it is useful to make a count of the number of times each respondent is used as a donor or a value to a nonrespondent.

I. SANDE, *Statistics Canada*: I feel obliged to make a comment following the last comment about the flags. For myself, I am very much opposed to issuing a microdata file with imputed data on it. If we had to issue a microdata file, I would simply issue the data, with flags saying that these items are missing. If the statistician wanted to, he could do the imputation for himself.

Now, as a statistician, the conclusion that I came to a long time ago, when I was involved with imputation projects, was that the reason why you wanted to do imputation was not because you wanted to fake up missing data, but because we were faced with a very interesting problem: We had a large data file and it had missing data on it, and we had several potential clients who all wanted different types of tabulations. They didn't tell us in advance what kind of tabulations they wanted. They are inclined to pop out of the woodwork at odd times and request new and different kinds of tabulations, and you have to satisfy them all. Now, if you were simply to reweight your file every time you would get inconsistent tabulations, then you would have people asking questions about why your tabulations were inconsistent. If you simply leave unknown blocks on your file then this for some reason the public finds annoying - why don't you have this data and why don't you have that data. Also, you have a feeling that the partial information that you have on a record somehow does contribute some information about the missing pieces of the record that you should utilize.

Now because of this, a clean solution to this particular problem is to impute the missing data by some plausible method, although I'm not about to dictate what method you should use. Then you will have a complete data file which you can run through an arbitrary tabulation program and therefore drop out any tables you want, and they will be consistent. This is the principal purpose of imputation. If you want to do an economic analysis, if you want to do regression, if you want to do anything else please take a data file without imputation and do it by yourself. Then you will understand fully what problems you face. As a statistician, I am not going to hand to you my imputed file which was developed for a rather different purpose.

P.S.R.S. RAO, *University of Rochester:* I would just like to make one comment. It is true that one reason to fill up the missing values is to make the records clean. But as Dr. Smith pointed out,

I think basically we are also interested in somehow getting closer
to the true values we are missing. So in that case if there is a
red flag and we know it was imputed, perhaps next time I use it I
may use a different procedure to estimate the true value. That is
also behind the minds of the statisticians who do all these adjustments. Filling of the records is of course a major problem as you
say.

W. MADOW, *Committee on National Statistics*: May I comment
on that? In the first place it is correct that you want a clean
record. In the second place it is also correct that everyone doing
imputation is doing it to reduce the biases that are involved. The
problem is that there are the biases because you do not have random
nonresponse. The people doing this work are well aware of the
problem of bias - and they are all working on the question of what
methods could be used to reduce the bias without increasing the
variance too much because they are equally aware that every imputation method will increase variance. One of the techniques that
has been mentioned here, for example, multiple imputation, is
being done precisely in an attempt to reduce the bias.

But what we are most missing is validation data because
these people are nonrespondents and efforts to get real validation
data have been very troubled. In the few cases where this has been
done, it has turned out that, for the most part, poststratification
has been a disappointment. It has not really helped anywhere near
as much as people would have liked. Now and again, we face the
problem of, "You can't get rid of nonresponse, you must live with
nonresponse; you know it is not random". What do you do in such a
case? It is not that people don't know what's happening when they
just clean up a data record. But they have to do that and more
and that is why an awful lot of research is going on now.

I. FELLEGI, *Statistics Canada*: I would like to comment on
the points raised by Fred Smith and P.S.R.S. Rao regarding the use

of imputed or "unprocessed" data. In the case of both our largest
household survey (the monthly Labour Force Survey) and the 1976
Census, a conscious effort has been made to preserve in machine
readable form both the imputed and unimputed data. I am yet to
hear of a single request regarding access to the unimputed data
base -- nobody seems to want to look even at a sample of those
records to get an impression of how the imputed and unimputed data
compare. This applies not only to substantive analysts but also
to mathematical statisticians. Having become "virtuous" and having
preserved the unimputed data, I invite my more learned colleagues
to "get busy".

P.S.R.S. RAO, *University of Rochester:* May I venture to
answer your question on why they did not approach you to verify
the data? They trust you too much.

G. SANDE, *Statistics Canada:* I guess the one set of things
that sometimes appears is that people are fencing against various
straw men which are put forward in the past. One of the problems
is that there is often a great difference in the technology which
is used and available in various places for doing some of these
things. Internally, in some of the work I've been doing, I have
found that rather sophisticated file search techniques are of the
essence and I end up being able to do things which other people
claim are impracticable because I have software for doing these
things that they don't know exists. The retrieval of the five best
matching records out of a file of 3000 is an easy thing to do, if
you have the right machinery. One of the things which is very
distressing is the fact that this type of machinery, in this case,
developed at Stanford - the initial technical reports are dated
about '75 - is still viewed as rather substantial esoterica. And
it is probably going to be another ten years before these methods
get out into rather widescale use. The formation of the imputation
matrices that Barry Ford was talking about is an example of a

valiant attempt to do things which don't work too well because of lacking technology. It's often very hard in this type of discussion to know exactly which level of technology is being discussed, because good ideas poorly implemented are not very good. It is alarming the extent to which detailing, which is often very hard to discuss in a short time, is often exceedingly crucial on these things. You have to ask a large number of very technical and insistent questions to find out whether or not you should believe something is done well. I'm left with a rather pessimistic feeling about the rapidity with which high grade technology is going to spread throughout the community. I think that is unfortunate because some of the clever and good things are now accessible but they are not being used as much as they might.

C. HILL, *Statistics Canada:* The only comment I would like to make is, in fact, to reinforce what Ron Graves and Gordon Sande have been saying. In these very complex survey operations - the sort of things that Graham Kalton was talking about - some of these computer techniques that are being developed are doing a great deal toward aiding us in solving some of our problems.

W. MADOW, *Committee on National Statistics:* I'm sure that I speak for all of us here when I thank the organizers and planners of these sessions for what has been, I'm sure for all of us, a very lovely several days. Thank you, the meeting is closed.

ABSTRACTS OF CONTRIBUTED PAPERS

A BAYESIAN MODEL FOR SAMPLING AN ATTRIBUTE AT MULTIPLE SITES

Richard W. Andrews and James T. Godfrey

University of Michigan

Consider the situation in which there are multiple sites
or locations. At each site there are a finite number of units on
which a zero-one attribute is defined. It is the purpose of this
paper to develop a one-sided Bayesian tolerance interval for the
maximum occurance rate over all the sites. This will be based on
sampling less than all sites. The model uses a prior distribution
to relate the occurance rates at different sites. The methods of
Bayesian analysis on finite populations will be employed. The
results given by the posterior distribution are compared with
conventional finite sampling. The methodology is demonstrated
through the use of an auditing example which investigates compliance with an internal control procedure.

AREA ESTIMATION BY POINT COUNTING TECHNIQUES

D.R. Bellhouse

University of Western Ontario

The classical approach to estimating areas by point counting techniques, systematic sampling, is compared to stratified
random sampling and to systematic sampling with multiple random
starts. Under an assumed model, we find that stratified sampling
is usually more efficient than systematic sampling which in turn
is more efficient than the multiple random starts method.

A DIFFERENT VIEW OF FINITE POPULATION ESTIMATION

Cathy Campbell

Aurther D. Little, Inc.

Hinkley, in 1978 <u>Biometrika</u>, and others have contributed considerably to the understanding of jackknife procedures for iid random variables by expressing parameter estimates as functionals of empirical distribution functions. This procedure can be extended to unequal probability sampling from finite populations. Some commonly used sample designs and corresponding estimators of the population mean are examined in terms of the implied estimate of the population distribution function. Implications of these findings for jackknifing sample survey data are discussed.

SAMPLING OF LANDSAT DATA FOR CROP SURVEYS

Raj S. Chhikara

Lockheed Electronics Co.

With the development of remote sensing technology, it is now feasible to attempt agricultural surveys by sampling of satellite data. A sample survey methodology is being developed at NASA/JSC, Houston, for crop estimation using multispectral scanner (MSS) data acquired by the Land Observatory Satellite; Landsat.

Development of both sampling frame and efficient sampling strategy for multicrop estimation will be presented. Discussion will include methods of stratification, optimum sample allocation, and large area crop acreage estimation.

The problem of non-response results from non-acquisition of MSS data by satellite due to clouds or haze. Another problem that arises is measurement error caused by the difficulty in separating various crops in a Landsat image of a sampled area segment. These and other problems likely to be encountered in satellite agricultural surveys will be discussed along with their potential solutions.

SAMPLING WITH REPLACEMENT

Keith R. Eberhardt

National Bureau of Standards

Suppose a sample of size n is drawn with replacement from a finite population. For the case of equal probability sampling, it is argued that the formula σ^2/n does not represent the appropriate variance for the ordinary sample mean. The alternative proposed is the variance of the conditional distribution given an ancillary statistic which describes the pattern of repeated units in the sample. When we look at the usual variance estimator, s^2/n, in the light of this conditional distribution, a bias is revealed which does not appear in the unconditional theory. A similar contrast is seen between the conditional and unconditional theory for the sample mean of distinct units and estimators of its variance. For the case of unequal probability sampling with replacement, analogous results may be easily derived using a simple superpopulation model.

FINITE SAMPLE PROPERTIES OF BEALE'S ESTIMATOR

V.K. Srivastava, T.D. Dwivedi, Y.P. Chaubey and S. Bhatnagar

Concordia University

In this paper we derive exact expressions for the relative bias and relative mean square error of Beale's ratio estimator when the characteristic under study and auxiliary characteristic follow a bivariate normal probability law. Numerical investigation is carried out for making comparison with the large sample approximations to the appropriate quantities.

STUDY OF TELEPHONE SAMPLING TECHNIQUES

Victor Tremblay

Université de Montréal

The increasing use of the telephone in sample surveys is
certainly due, at least partly, to the developing concern of
optimizing the total survey design. This commands the need to
implement better methods for selecting valuable telephone samples.
This paper attempts to present an overall view of sampling
techniques based on random generation of telephone numbers
and describes the approach that was recently put to
test at the Centre de sondage. An assessment of it is now
possible after a one-year experience.

PARTICIPANTS

M. Abouel-Ata
A. Al Saleh
M. Ali
R. W. Andrews
B. Armstrong
A. Ashraf
Barbara A. Bailar
Mike Bankier
D. Bates
David Bayless
A. Beauchamp
D. G. A. Beckstead
David Bellhouse
P. Y. Blain
J. Booth
B. A. Boyes
David F. Bray
Judith M. Brennan
R. Burgess
Cathy Campbell
S. Cantin
M. Cardenas
G. Catlin
Y. P. Chaubey
A Chaudhuri
T. Cheney
S. Cheung
R. Chhikara
N. P. Chinappa
G. H. Choudhry
James Chromy
M. Collidge
C. L. Craig
Douglas K. Dale
Tore Dalenius
G. Davidson
J. B. Davis
Y. Deslauriers

J.-D. Desrosiers
T. DiCiccio
G. Dinsdale
L. Dionne
D. Dodds
D. Dolson
D. Domoney
Doug Drummond
M. R. Dunn
Stephen Earwaker
Frank Eaton
Keith R. Eberhardt
S. T. El-Helaly
A. Farrell
M. Feingold
Ivan P. Fellegi
D. Ferguson
Barry L. Ford
Wayne A. Fuller
J. Garrett
P. D. Ghangurde
P. Giles
P. A. Gimotty
J. Godfrey
Maria E.Gonzalez
J. F. Gosselin
Jack Graham
Gerry B. Gray
R. P. Gupta
Joan J. Hahn
F. Hardy
H. O. Hartley
L. Heckmann
Irene C. Hess
Michael A. Hidiroglou
J. Higginson
Chris J. Hill
Geoff Hole

Elaine Hoskins
Edward Hughes
Gek Choon-Lee
Hui Tak-Lee
G. Hunter
Y. Hwang
S. Ismail
S. Job
G. Jones
Elizabeth Junkins
Graham Kalton
Rose S. Katapa
Daniel R. Krewski
Karol P. Krotki
S. Kumar
S. Kumar
Andre Lamontagne
Murray Lawes
Naomi Lee
Liu Tzen-Peng
Lu Wen-Fong
L. Lui
R. Lussier
Urst Maag
L. H. Madow
William G. Madow
F. Maranda
M. March
R. Martin
F. Mayda
M. McCloskey
A. McGrath
Y. J. Messeri
P. Mette
K. Miller
O. Molake
Nash J. Monsour
H. Morin
W. Murdyk
M. S. Nargundkar
S. Nesbitt
Rita Nesich
A. Ogunsola
R. Okafor
R. Oni
Charles Patrick
P. Peskun
Richard Platek

H. D. Potter
M. Rahman
S. Raman
J. N. K. Rao
P. S. R. S. Rao
Georgia Roberts
C. A. Rocha
D. Boyce
R. Rustagi
R. Ryan
G. Saaed
Innes Sande
Gordon Sande
Robert L. Santos
Carl Erik Särndal
W. Saveland
W. L. Sahible
A. R. Sen
Wilma Shastry
S. N. Sharma
A. C. Singh
M. P. Singh
T. M. F. Smith
K. P. Srinath
T. Stuken
L. Swain
J. L. Tambay
F. Tarte
Teo Kar-Seng
R. Tessier
J. Thompson
M. E. Thompson
J. Tomberlin
J. Tourigny
Victor Tremblay
H. Tully
Rodger Turner
J. Valdivieso
J. Vallejo
N. Varma
G. Villeneuve
H. D. Wightman
L. J. Wilson
L. Woelfle
Kirk M. Wolter
T. Wright
A. Zinger

INDEX